Einstein's Relativity
and the Quantum Revolution:
Modern Physics for Non-Scientists,
2nd Edition

Richard Wolfson, Ph.D.

THE
GREAT
COURSES®

PUBLISHED BY:

THE GREAT COURSES
Corporate Headquarters
4840 Westfields Boulevard, Suite 500
Chantilly, Virginia 20151-2299
Phone: 1-800-832-2412
Fax: 703-378-3819
www.thegreatcourses.com

Richard Wolfson, Ph.D.

Professor of Physics
Middlebury College

Professor Richard Wolfson is Professor of Physics at Middlebury College, where he has also held the George Adams Ellis Chair in the Liberal Arts. He did undergraduate work at MIT and Swarthmore College, graduating from Swarthmore with a double major in physics and philosophy. He holds a master's degree in Environmental Studies from the University of Michigan and a Ph.D. in Physics from Dartmouth College. His published research includes such diverse fields as medical physics, plasma physics, solar energy engineering, electronic circuit design, observational astronomy, and theoretical astrophysics. Professor Wolfson's current research involves the sometimes violently eruptive behavior of the Sun's outer atmosphere, or corona. He also continues an interest in environmental science, especially global climate change. As a college professor, Wolfson is particularly interested in making physics relevant to students from all walks of academic life. His textbook, *Physics for Scientists and Engineers* (Addison Wesley, 1999), is now in its third edition and has been translated into several foreign languages. Wolfson is also an interpreter of science for nonscientists; he has published in *Scientific American* and wrote *Nuclear Choices: A Citizen's Guide to Nuclear Technology* (MIT Press, 1993). The original version of this course on physics was produced in 1995, and in 1996, Wolfson produced another Teaching Company course, *Energy and Climate: Science for Citizens in the Age of Global Warming*. Although he has been at Middlebury for his entire post-Ph.D. career, Professor Wolfson has spent sabbaticals at the National Center for Atmospheric Research in Boulder, Colorado; at St. Andrews University in Scotland; and at Stanford University. ∎

Table of Contents

INTRODUCTION

Professor Biography i
Course Scope 1

LECTURE GUIDES

LECTURE 1
Time Travel, Tunneling, Tennis, and Tea 3

LECTURE 2
Heaven and Earth, Place and Motion 21

LECTURE 3
The Clockwork Universe 38

LECTURE 4
Let There Be Light! 57

LECTURE 5
Speed *c* Relative to What? 76

LECTURE 6
Earth and the Ether—A Crisis in Physics 94

LECTURE 7
Einstein to the Rescue 115

LECTURE 8
Uncommon Sense—Stretching Time 133

LECTURE 9
Muons and Time-Traveling Twins 153

LECTURE 10
Escaping Contradiction—Simultaneity Is Relative 175

Table of Contents

LECTURE 11
Faster than Light? Past, Future, and Elsewhere 197

LECTURE 12
What about $E=mc^2$, and Is Everything Relative? 217

LECTURE 13
A Problem of Gravity 238

LECTURE 14
Curved Spacetime 259

LECTURE 15
Black Holes 283

LECTURE 16
Into the Heart of Matter 303

LECTURE 17
Enter the Quantum 325

LECTURE 18
Wave or Particle? 345

LECTURE 19
Quantum Uncertainty—Farewell to Determinism 365

LECTURE 20
Particle or Wave? 385

LECTURE 21
Quantum Weirdness and Schrödinger's Cat 407

LECTURE 22
The Particle Zoo 427

LECTURE 23
Cosmic Connections 447

Table of Contents

LECTURE 24

Toward a Theory of Everything..467

SUPPLEMENTAL MATERIAL

Timeline 488
Glossary 491
Bibliography..498

Einstein's Relativity and the Quantum Revolution: Modern Physics for Non-Scientists, 2nd Edition

Scope:

The 20th century brought two revolutionary changes in humankind's understanding of the physical universe in which we live. These revolutions—relativity and the quantum theory—touch the very basis of physical reality, altering our commonsense notions of space and time, cause and effect. The revolutionary nature of these ideas endows them with implications well beyond physics; indeed, philosophical debate continues to this day, especially over the meaning of quantum physics.

Is time travel to the future possible? You bet—but if you don't like what you find, you can't come back! Are there really such bizarre objects as black holes that warp space and time so much that not even light can escape? Almost certainly! And do the even weirder cousins of black holes, wormholes, exist—perhaps affording us shortcuts to remote reaches of space and time? Quite possibly! Is the universe governed by laws that strictly predict exactly what will happen in the future or is it governed by chance? In part, by chance! All these and other equally strange consequences flow from relativity and quantum physics.

Many people think that relativity and quantum physics must be far beyond their comprehension. Indeed, how many times have you heard it said of something difficult that "it would take an Einstein to understand that"? To grasp these new descriptions of physical reality in all their mathematical detail is indeed daunting. But the basic ideas behind relativity and quantum physics are, in fact, simple and comprehensible by anyone; for example, a single, concise English sentence suffices to state Einstein's theory of relativity.

This course presents the fundamental ideas of relativity and quantum physics in 24 lectures intended for interested people who need have no background whatsoever in science or mathematics. Following a brief history of humankind's thinking about physical reality, the lectures outline rigorously the logic that led inexorably to Einstein's special theory of relativity. After an exploration of the implications of special relativity, we move on to Einstein's general theory of relativity and its interpretation of gravity in terms of the curving of space and time. We see how the Hubble Space Telescope provides some of the most striking confirmations of general relativity, including near-certain confirmation of the existence of black holes. Then we explore quandaries that arose as physicists began probing the heart of matter at the atomic and subatomic scales, quandaries that led even the great physicist Werner Heisenberg to wonder "can nature possibly be as absurd as it seems to us in these atomic experiments?" The resolution of those quandaries is the quantum theory, a vision of physical reality so at odds with our experience that even our language fails to describe the quantum world. After a brief exposition of quantum theory, we explore the "zoo" of particles and forces that, at the most fundamental level, comprise everything, including ourselves. We then bring together our understanding of physical reality at the smallest and largest scales to provide a picture of the origin, evolution, and possible futures of the entire universe and our place in it. Finally, we consider the possibility that physics may produce a "theory of everything," explaining all aspects of the physical universe.

A first version of this course was first produced in 1995. In this new 1999 version, I have chosen to spend more time on the philosophical interpretation of quantum physics and on recent experiments relevant to that interpretation. I have also added a final lecture on the "theory of everything" and its possible implementation through string theory. The graphic presentations for the video version have also been extensively revised and enhanced. But the goal remains the same: to present the key ideas of modern physics in a way that makes them clear to the interested layperson. ■

Time Travel, Tunneling, Tennis, and Tea
Lecture 1

This is a course in the great ideas that shaped the physics of the 20th and 21st centuries. It's not a course in all of physics and it's not a course in the really latest, most contemporary ideas, although we will touch on those. Rather, it's a course in the true, fundamental ideas that set the tone for all the physics that's occurred since the year 1900.

The two big ideas of modern physics are relativity and quantum physics. Relativity radically alters our notions of space and time, while quantum physics reveals a universe governed ultimately not by strict determinism but, in part, by pure chance. Modern physics stands in contrast to classical physics, developed before 1900 but still applicable to everyday phenomena. Although modern physics forces radical changes in our philosophical thinking about the physical world, its basic ideas are nevertheless accessible to non-scientists.

I'll start with two impossible tales—or are they? The first tale involves time travel, where I'll use a set of twins as an example. The twins start out at the same age. One makes a high-speed round-trip journey to a distant star while the other stays home on Earth. When the traveling twin returns, they're different ages! The second tale involves tunneling; I'll

Ultraviolet image of Venus's clouds.

NASA/NSSDC/Pioneer Venus Orbiter.

use a prison as an example. You're trapped in a concrete-walled prison cell. You pace back and forth all day, confined by the walls. Suddenly, you find yourself on the outside! It seems impossible, at least according to anything your common sense tells you. But it definitely happens with subatomic particles.

Physics is the subject that describes our physical environment, from the smallest subatomic particles to the entire universe. Physics is important to everyone, not just physicists! Classical physics is the realm of physics developed before 1900, which is still applicable to most everyday phenomena (such as driving a car, engineering a skyscraper, designing a telescope, launching a satellite, generating electric power, predicting weather and planetary motion, and so on). Modern physics has been developed in the 20th century and describes phenomena at very small (i.e., atomic) scales or when relative speeds approach that of light. This course is about the two big ideas at the heart of modern physics: relativity and quantum physics.

Physics is the subject that describes our physical environment, from the smallest subatomic particles to the entire universe.

Relativity, developed largely by Einstein beginning in 1905, is a great equalizer. It asserts that everyone experiences the same laws of physics, regardless of their location or state of motion. In this way, relativity builds on the Copernican notion that Earth does not occupy a privileged position in the universe. Despite its simple content, relativity radically bends and blends our notions of time and space.

Quantum physics reveals a noncontinuously dividable, or "grainy," universe at the smallest scales and with it, bids farewell to strict determinism. In other words, at the fundamental levels of matter, causation is a matter of statistical probabilities, not certainties.

Physics is a human activity, which most scientists nevertheless believe is a quest toward understanding an underlying objective reality. Developing new ideas of physics, especially with fundamentals such as relativity and quantum physics, is often more like a creative artistic process than the stereotypical "scientific method" emphasized in school science courses.

It doesn't take an Einstein to understand modern physics! The goal of this course is to make the key ideas of modern physics accessible, comprehensible, and even simple for the interested non-scientist. For example, you can play tennis, or brew a cup of tea, equally at home, on a cruise ship, on Venus, or on a planet in a distant galaxy. Trivial as this example is, understanding it means grasping and accepting the essential idea of Einstein's relativity, namely, that the laws that govern physical reality are the same everywhere in the universe. ■

Suggested Reading

Hoffmann, *Relativity and Its Roots*, Chapter 1.

Questions to Consider

1. Articulate why it is, in terms of your own understanding of time, that you find the example of the time-traveling twins disturbing.

2. Why do you suppose it is that you don't have to take into account the ship's motion in the example of playing tennis on a cruise ship?.

Time Travel, Tunneling, Tennis, and Tea
Lecture 1—Transcript

Welcome to *Einstein's Relativity and the Quantum Revolution: A Course in Modern Physics for Non-Scientists*. I'm Professor Rich Wolfson from Middlebury College. And before I begin, I'd just like to say a little about my title, to emphasize two terms in that title. First of all, the term modern physics. This is a course in the great ideas that shaped the physics of the twentieth and twenty-first centuries. It's not a course in all of physics. And it's not a course in the really latest, most contemporary ideas, although we will touch on those. Rather, it's a course in the true, fundamental ideas that set the tone for all the physics that's occurred since the year 1900.

Before I describe the course and give you a brief introduction, I'd like to tell you two stories that give you a sense of what modern physics is about and how radically it's reshaped our notions of the physical universe, in particular our notions of space and time and causality.

The first involves a couple of twins. Imagine you are a twin, and you're 20 years old—in your youth—and you have a twin sister. And you and your twin sister go out in the backyard and you build a space ship. And on your 20th birthday, your twin sister hops in the space ship and travels off to a distant star. And she goes out there, and she spends some time exploring that star—and you wait, and you wait, and you wait. And eventually your twin sister returns, and you find the time of her return is your 80th birthday. So she has been away for 60 years. And here comes the space ship, and it lands back in your backyard, and out hops your twin sister, and she's a youthful 25. Five years have elapsed for her. Sixty years have elapsed for you. In some sense she's time traveled into your future. And that is the first phrase in the title of this lecture, "Time Travel, Tunneling, Tennis, and Tea," because one of the things that modern physics seems to make possible is time travel, at least into the future in the sense I've just described in the story of the two twins.

Is this really possible? Could it really happen? Is time so strange that two twins who start out at age 20, together, can find themselves later 55 years apart? The answer is, yes. We've confirmed this with subatomic particles.

We've even confirmed it with real-sized clocks, although there the time difference was billionths of a second instead of decades. But this happens. It can happen, and it tells us something remarkable about the nature of time.

Let me tell you another seemingly impossible story that relates to the second phrase in my title. Again, the title of this lecture, "Time Travel, Tunneling, Tennis, and Tea." This is the phrase, "Tunneling." Imagine you are stuck in a prison. You have concrete walls on all sides. You're trapped there. You're in for a long sentence. And so, what do you do? Well, you have nothing to do, but you pace back and forth. And occasionally you lean up against those hard concrete walls of your prison. Of course, you can't get out. And then one day, when you're leaning against the walls of the prison, you suddenly find yourself on the outside. You've somehow inexorably penetrated a seemingly impenetrable barrier. Is this possible? It seems not. Not according to anything your common sense tells you.

In fact, it's the phenomenon of quantum tunneling. Again, we've verified it. Again, with subatomic particles. We believe, in principle, it could happen in other circumstances, but we have not seen it happen, and it's very unlikely that it would happen with a macroscopic sized object like a prisoner. But it definitely happens with subatomic particles. In fact, we're alive today because protons in the core of the sun tunnel through an insurmountable barrier—analogous to that concrete wall of the prison—and are able to get together and fuse to make helium, and in the process releasing the energy, which tens of thousands of years later emerges from the surface of the sun as sunlight, and eight minutes after that reaches Earth, and is the energy that sustains life on our planet. So, this is a real phenomenon, and it's one on which our lives directly depend.

Those are two stories of seemingly impossible events. And yet, they're events that occur. They could occur as I've described them—although are unlikely or difficult to make happen—they do occur regularly in the subatomic realm. They're ideas that suggest to us that our basic common sense notions of time, of space, and of what it means to be rigid and solid and insurmountable are all in some way compromised, or modified, or extended, or enriched by the ideas of modern physics. So, there are some stories. And we'll come to understand how both those things are possible in the course of this lecture series.

Let me pause now and say a little bit about what physics is. People have a lot of misconceptions about physics. I often tell people I'm a physics professor. And they say, "Oh, physics," and they kind of think that's something for a rarefied breed of people who occupy specialized laboratories and think about things that have nothing to do with what's important to the common person. Actually, nothing could be further from the truth. Physics is the essence of the way the physical universe works. We live in the physical universe, like it or not. We walk. We drive cars. We fly airplanes. We have weather happen to us. We swim. We turn on water. We cook. All these basic, everyday processes involve the fundamental laws of physics. Physics is the fundamental science. The laws of physics underline everything that goes on in the physical universe.

Physics is not just for physicists. It's not just for people who occupy laboratories and invent bizarre new devices or think about esoteric thoughts like time travel, for example. Physics is for everyone, because physics is the essence of what most of us do—in fact, what all of us do all of the time. In fact, I'm going to give you another example at the end of this lecture, which will convince you that you already know a lot more physics than you think. In fact, you already know the theory of relativity. And I'll get to that in just a few moments.

So physics is something that describes the entire physical universe—from the smallest subatomic particles (the quarks and the electrons) to the grand scale of the entire universe, and how it seems to be evolving, and whether it's going to have an ending, and when it had a beginning. All these questions are in the realm of physics. So, it's a great big subject, and it's important to every one, not just physicists.

Physics, as a subject, divides into two very, very broad classes. And these are historically based, and they have names that can be slightly misleading. So I want to make a clarification here about the names. One subject is classical physics; the other is modern physics.

Classical physics is, basically, the physics that was developed before the year 1900. Now, you might say, well that's old-fashioned physics. That's ancient physics. We don't really want to know about that. Well, you're going to need to know a little bit about classical physics to understand how modern physics differs. Furthermore, classical physics—which is what we tend to teach, primarily, in introductory physics courses at the college level, and certainly in high school physics, and any earlier physics we teach—classical physics is very much relevant physics, and it's even very much contemporary physics. You can't do something like design a skyscraper that will stand up, or develop an anti-lock braking system for cars and understand how it works, or build a heart monitor, or an electrocardiograph machine, or engineer an electric power distribution system, or build a compact disc player—many, many, many modern technologies depend very much on classical physics.

So that old-fashioned physics that was developed before the 1900s, so called classical physics, is really very contemporary physics, it's important physics, and it describes many of the things we do today—even launching a spacecraft to Jupiter; or putting satellites in Earth orbit; or sending people to the moon; or any high technology things like that that don't involve subatomic physics, semiconductor electronics, or things like that—they're still in the realm of classical physics. So classical physics is an important branch of physics.

Next comes modern physics. Modern physics was developed in the twentieth century. It arose out of what seemed to be very slight discrepancies between classical physics and the reality of the physical world that were noticed in the second half of the nineteenth century. Those ideas grew into crises in physics—and that's what we'll be exploring in this course—and those crises eventually led to the development of two key ideas in what are called modern physics.

Now, before I go on with that, let me again distinguish: Modern physics does not mean the physics of the last decade or the last few decades. Modern physics has a distinct meaning. It means the key physical ideas that were developed to explain physical reality in the twentieth century. The main ideas were developed early in this century between 1905 and about 1930. We will explore those ideas in great depth in this course, and we will also look a little bit at some of the newer more contemporary ideas, or some of the observations in, say, astrophysics that have been used to verify those early ideas. But this course is primarily about the key ideas of twentieth century physics. The big ideas. The two great big ideas. And what are those big ideas?

Well, the two big ideas are *relativity*—developed primarily by Albert Einstein, although with help from others—and *quantum physics*, which was developed by a whole slew of physicists. Einstein first told us about relativity in 1905. He expanded on relativity in 1914. And the rest of the twentieth century constituted of an exploration by physicists trying to verify that Einstein was correct. And in the last 10 years, and even in the last five years, we've had some remarkably dramatic confirmations of Einstein's theories, although the essence of those theories was confirmed very early in the century.

What does relativity do? Well, in a nutshell, relativity is a great kind of equalizer. A democratizer. Most people think of relativity and they think of strange things like the time travel twins story I told you earlier. Or they think about $E=mc^2$. And maybe they think about nuclear weapons, which really have nothing to do with relativity. They think about bizarre, esoteric things happening to space and time. That's true. Relativity tells us about bizarre, esoteric things that happen to space and time.

But the essence of relativity is very different. The essence of relativity is that all of us are equal. And we're going to explore and understand how that developed. But in a nutshell, let me just tell you what relativity says. It says your point of view doesn't matter. It says everybody has equal access to the physical laws that govern the way the universe works. And by everybody, I mean it doesn't matter where you are—what planet you're on, whether you're in interstellar space, whether you're in some distant galaxy—physics looks the same to you as it does to me here on Earth. It doesn't matter how you're moving. You could be in an airplane moving at 600 miles an hour and physics acts just the same there. You read a physics textbook. You learn the laws of physics. You do experiments. And they work just the same in that airplane as they do here on Earth. You could be in a spacecraft whizzing by Earth at 90 percent of the speed of light and the laws of physics would still work.

Relativity is in essence a great equalizer. It makes us all equal. It puts all places and all states of motion in the universe on an equal footing. In the process, it leads to these bizarre notions like time travel, like stretching of space, squeezing of time, and so on. It leads to $E=mc^2$ and to the equivalence of matter and energy. But the basic idea of relativity is a simple one. It says we're all equal. It says we all see physics the same. There's no special place. There's no special state of motion. And, in telling you that, I've told you the essence of relativity. You know relativity, once you know that.

In a way, relativity has elaborated on the Copernican revolution—the idea that Earth is no longer the center of the universe. And relativity says there's no special place. Earth is not special as a place, it's not special as a state of motion. And, furthermore, there's no place or state of motion that is special.

The second of the enormous ideas of twentieth century physics—the modern physics ideas—is quantum physics. And quantum physics is a little bit harder to pin down in one basic idea. But if I can try to do it, it's the notion that at the fundamental scales the universe is grainy in a sense. It's built of little discrete parts. It isn't continuously subdividable.

Think about a glass of water. Pour off half of it. You're still left with half a glass of water. Pour off half of that. You're left with a quarter of a glass of water. Pour off half of that. You're left with an eighth. And keep doing that. You've always got liquid water. Will you ever reach a point where you don't have liquid water? Well, at some point in human history that was an open question. Today we know the answer. The answer is that eventually you get down to the point where there's one water molecule in there. One H_2O molecule, consisting of an oxygen atom and two hydrogen atoms. And once you've reached that point, you can't subdivide the water any further. You've got one water molecule. You can take it apart. If you take it apart, you have different entities—oxygen and hydrogen. So the universe is not continuously subdividable. That was known before the twentieth century.

What's new about quantum physics is it quantizes or makes discrete—not only matter, which is broken up into molecules and atoms and subatomic particles—but it also quantizes energy—the other essential stuff of the universe. And as we'll see later, relativity makes matter and energy somewhat interchangeable. So, it's not surprising that this happens.

Quantum physics tells us that the world at the most fundamental level consists of discrete, little bundles of stuff. Bundles of matter. Bundles of energy. It's like buying eggs in the store. You can buy a dozen eggs. You could probably buy a half a dozen eggs. You might even convince the grocer to sell you one egg. But there is no way in the world you're going to buy half an egg or a quarter of an egg. An egg is simply not further subdividable and still have it be an egg. And that's the essence of quantum physics. The universe, including the energy that makes it go, is not infinitely subdividable but comes in discrete little hunks. And although that idea—like the idea of relativity, that no place or motion is special—the idea of quantum physics that the universe is fundamentally discrete, that sounds like a very easy idea to grasp. That egg analogy is a good one. But that leads to very bizarre consequences. For example, the tunneling situation I described. For example, the fact that the, at the fundamental level, the laws that govern the universe are not even deterministic. They don't say, "This will happen if you do this." They say this may happen with a 70 percent chance of this outcome, and a 30 percent chance of this outcome, or whatever. The universe at the fundamental level—this discreteness of matter and energy tells us—is fundamentally statistical. And we have a whole new way of looking at the universe in causality, as a result of that.

So those are the two big ideas of twentieth-century physics. And what this course is about, again, is to expose you to those ideas in a way that makes them understandable to people who have no particular scientific background. Let me give you just the briefest hint of how we're going to do this.

The course is going to be divided into basically two big halves. The first half is going to take us through the *special theory of relativity*. This was the first of the two relativity theories; the one that Einstein developed and published in the year 1905. And it has been very, very highly confirmed by all experiments that have been done since then.

We're going to spend a good deal of time on special relativity. We'll take about four lectures of history. Understanding classical physics. Understanding the contradictions that rose from classical physics. Contradictions that we'll worry about both with relativity and quantum physics.

But after about four lectures, we'll begin to plunge into the essence of special relativity. And we're going to spend through Lecture Twelve on special relativity. And the reason for so much emphasis on this particular subject is special relativity is the most easily understandable of the modern physics ideas. It requires almost no mathematics. And this is a course that's basically devoid of mathematics. So I'm going to present all this material in a non-mathematical way. But if you wanted to understand special relativity thoroughly and mathematically, high school algebra would suffice.

So relativity, special relativity, is an easy to understand subject. I can get you to understand thoroughly and logically why it has to be true. I can get you to believe in that time travel. I'm not just going to tell you that, I'm going to make you believe that that has to be the case. And you'll understand the full logical progression.

After we do special relativity, we'll spend a few lectures on *general relativity*, which is Einstein's extension of special relativity to cover all states of motion. And it turns out to be a theory of gravity. And it's the theory that predicts the existence of bizarre things like black holes, and wormholes, and white holes and the kind of time travel that occurred in the movie *Contact*. If you've seen that where you go through a wormhole tunnel through space and time. General relativity talks about very strange things like that. It's harder to grasp mathematically. It's harder for me to show you the full, logical flow. We aren't going to spend as much time on it. But, you will get the essence of what general relativity is about.

And then the remaining lectures will be about quantum physics—most of them dealing with, again, the fundamental ideas that were developed by about the year 1930 when people realized just how strange the world, at the subatomic level, was.

And then we'll move on from there and look briefly at some very modern ideas that are outgrowths of quantum mechanics. We'll look at some experiments that have been done in the last few decades that are still causing controversy about the philosophical interpretation of quantum physics—because quantum physics is subject to interpretation, which is still a matter of great philosophical debate. And then, in the very end, we'll look at very contemporary searches for a so-called *theory of everything*—ideas in physics that would be able to explain everything from gravity to the structure of an electron and everything in between with one theory of everything. We aren't there yet, but there are some people who think we're close. And we'll get a glimpse of what those theories are.

Now, I want to say a few more things about physics before I actually plunge into the material. First of all, I want to say that physics is a human activity. I want to remind you that although we're trying to talk here about the content of physics primarily, there's the human touch behind it. And I don't have a lot of time to bring that in, but I will mention it where possible. It's very definitely a human activity. There's been a great deal of controversy in the last few years in fact about just whether physics is another human activity and the laws of physics are like texts we human beings have made up, or whether they reflect some underlying physical reality. I happen to believe, and most of my fellow physicists believe, that the laws of physics, as we describe them, are either fairly accurate reflections or very good approximations of an underlying objective physical reality. We happen to believe that. I think it would be difficult to motivate doing physics if we didn't believe that.

That doesn't mean physics isn't influenced by human characteristics. Relativity probably would have been discovered soon after 1905 by somebody other than Einstein if he hadn't done it. But Einstein's unique genius, perhaps, made it come out sooner. It's also probably true that those subjects we choose to study in physics, those problems we choose to investigate, are influenced by the kinds of people we are. They're influenced by gender. They're influenced by cultural backgrounds. They're influenced by things like that. So, this is not to say that science, although it studies an objective reality, isn't affected by human issues. I think it is. But I think what it ultimately tries to study is something that is underlying, objectively real.

I also want to dispel another notion about science, and physics in particular. There's this notion about the scientific method. And it's actually, I think, enhanced by the way science is taught—not necessarily always as creatively as it might be—in the public schools.

We're given this notion that the scientist is somebody who sort of thinks up a hypothesis, and writes it down very carefully, and then puts on the white lab coat and marches into the lab, and does all kinds of experiments, and either verifies or throws out the hypothesis. It's a very rigid, rigorous process; very carefully spelled out what you do.

Now, some science is done that way. And probably 99 percent of scientists are working in established areas of science where that is an appropriate way to do things. Does this drug kill cancer? I think it does. So I invent an experiment. I say, I'm going to test this drug. I'm going to put it on cancer cells. I'm going to see if they die. I go in the laboratory and I do that. Can I make a smaller computer chip by using this quantum mechanical property of the electron to build this particular device? I think I can. So I make a hypothesis. I put on my lab coat. I go in and I try to build the device and I see if it works. Those are manifestations of the, sort of, basic scientific method as it's taught in schools.

But I think the great science, the big science, the science that takes us leaps ahead in our understanding of physical reality is very different. I think that's a creative activity much more akin to art, or composition of music, or other creative human endeavors than it is to the sort of dry sense people have of science as this dull scientific method kind of thing. Science is a very creative activity, and it's most creative, in particular, when it's taking these big leaps forward.

Einstein didn't do experiments. Einstein thought up relativity in his head. He said this is the way the world has to be. And he wrote it down. And, later, the world proved to be that way when people did experiments. Science is a very creative activity, and I hope in this course you'll get a sense of that. Because we'll see how some of the great minds of modern physics struggled with the bizarre consequences of what they were studying.

So remember that this is a human activity, and I'll try to emphasize that occasionally. But I won't give it as much emphasis as it deserves in a course of this relatively short length for this big a topic.

Now, to end this lecture I want to say one thing, again, to all of you who are sitting there saying, "Well, gee, this is not something I'm going to understand." It is something you're going to understand. My goal and my whole professional goal as a science teacher is to make science, particularly physics, comprehensible to the average person. You need to be intelligent. You need to be open-minded. But you do not need to know anything about science to understand what I am about to describe. It doesn't take an Einstein to understand modern physics. It may have taken an Einstein to come up with modern physics. But once the genius of Einstein had done that, all of us can understand what modern physics has to say. And that's my goal. So, if you're an interested, non-scientist—you gotta be interested—but if you're interested, stay with me and you will understand this material.

And I'm going to end by giving you an example that I think will convince you that you really know a lot more about modern physics than you think. You even know relativity. Let me imagine that you're playing tennis. You're playing tennis on a cruise ship. Now, if you're not into tennis, make it ping-pong, make it soccer, make it just tossing a ball back and forth with someone, tossing a bone to your dog. I don't care what you're doing, but you are doing some kind of activity that involves maybe tossing a ball around. And you're doing it on a cruise ship. And we're going to put you below decks, so there isn't a problem of wind from the motion of the ship. And the ship is moving through calm water, and it's moving at a steady rate of speed. It's not turning. The water is not rocking it. There aren't big waves. It's just plowing steadily along. Now, since you know how to play tennis, I assert that you already know a lot about the laws of physics in your muscles. You know what a ball is going to do when it's hit with the racquet—where it's going to move. Your muscles can judge the laws of physics that describe the motion of that ball. And that is what makes you a decent tennis player. So you know the laws of physics in your muscles even if you think you don't know them in your brain.

So now you go below decks into this enclosed tennis court. The ship is plowing along through the water at 40 miles an hour or something, and you proceed to play tennis. Let me ask you some questions. Does it matter to you whether you're the tennis player that's facing the forward direction of the ship or the backward direction of the ship? Does it matter that the ship is moving? Do you have to say to yourself, "Gee, I'm facing the forward direction. The ship's moving 50 miles an hour, I gotta take that in to account when I swing?" Of course not. Does the ship's motion make any difference to you? Of course not. You play tennis on that steadily moving cruise ship exactly like you would back home on solid Earth. Makes no difference at all. If it did, you'd have a real problem. Because when you sat on an airplane, for example, and tried to eat the little peanuts they bring you, you'd have to worry about the fact that those peanuts are going 600 miles an hour with the airplane. How am I going to get them to my mouth? That doesn't matter. And it doesn't matter playing tennis on the cruise ship.

Now, you're done playing tennis, and you want a nice hot cup of tea afterwards. If you're not a tea drinker, have coffee. But it's got to be instant because I want you to put it in the microwave to heat the water. Do you think that when you push the start button on the microwave you've got to worry, "Oh, wait a minute, the microwaves in this oven are going to be upset by the fact that the ship is moving this way at 50 miles an hour. I gotta worry about that. I gotta set the time differently. Maybe I gotta turn the microwave oven. Maybe I gotta do something strange." Do you worry about that? Of course not. You stick the teacup in the microwave oven. You press three minutes. And your tea water gets hot. Doesn't make any difference about the motion of the ship.

Now, I really want you to believe that. And it's very obvious. If that isn't obvious, let me give you two other examples that, I think, will make it even more obvious. The last of which is really only possible to understand in the contemporary era.

For the second one, let's imagine we now, have a base on Venus. Not hard to imagine—30, 50 years from now. And you're in the recreational room of that base on Venus. You're stationed there as one of the scientists working away or something, or maybe you're one of the maintenance people, or maybe you're one of the cooks. You're doing anything there. But you're going to go play some tennis in the recreation hall. And the recreation hall consists of a big inflated dome, like you've seen some tennis courts in. And inside it is an atmosphere exactly like the Earth's, because Venus's atmosphere—with a super runaway greenhouse effect like we hope we don't get on Earth—is way too hot and thick and so on to nurture us or to play tennis in. So inside this dome is an Earth-type atmosphere. And Venus's gravity is essentially the same as Earth's. So you go in this thing, and you play tennis.

Does it matter to you that you're on Venus? Nope. The tennis works just as it would on Earth. Gravity is the same. The air is the same. The tennis ball obeys the same laws of physics. Works just the same. However, you're on Venus, tens of millions of miles from Earth. And, depending on where and when Earth and Venus are in their orbits, they may be moving, relative to each other, at speeds of 20, 30, 40, 50 miles per second. Do you have to sit in that tennis court in Venus and say to yourself, "Gee, I'm moving at 45 miles a second relative to Earth. It's a lot faster than I can hit a tennis ball." Do I have to worry about that fact, in order to play this game? Well, of course you don't. The tennis ball, and Venus, doesn't care what Earth is doing. There's nothing special about Earth. Physics works the same on Venus as it does on Earth.

And then you get done with your tennis game on Venus, and you step over to the microwave oven to heat your tea. You put it in the microwave oven. You say to yourself, "Gee, Venus. Forty-five miles a second relative to Earth. It's going to do something strange to the microwave oven. Maybe it won't even work. Maybe the microwaves will get left behind and they won't heat my tea." Nonsense. You press the start button, you go for three minutes, and you have boiling water. Works just the same as on Earth.

Now, the last example is not one I could have given you in the earlier part of the century when people were first discovering modern physics. But now I can. Because now we know through astronomical observations, particularly recent ones with the Hubble space telescope, that there are objects so distant from us, and I'll explain later in the course why distance matters here, so distant from us that they are moving away from us at speeds of 80 percent or more of the speed of light. That's fast. The speed of light is about 200,000 miles a second. So these things are moving away from us at maybe 160,000 miles a second.

These are galaxies. We don't see them as galaxies when we look at them in space, because we're seeing light from a long, long time ago. But now they've presumably evolved into galaxies. Maybe some of them have stars around them like the sun. And maybe some of those have Earth-like planets around them.

So imagine there's yourself or some other creature on an Earth-like planet around a star in a distant galaxy that's moving away from Earth at 80 percent of the speed of light. And you're going to play tennis there. And you get out your tennis balls. And do you say to yourself, "Gee, Earth, 12 billion light years away, matters a lot, and I've got to figure out the fact that I'm going at 80 percent of the speed of light relative to Earth in order to play this tennis, right?" Nonsense. The laws of physics are the same for everybody, and you sit there on that distant planet and you play tennis just like you would normally. And then you pop over to the microwave oven. And you put your teacup in. And you press the button. And you don't worry about the fact that the microwaves, which move at the speed of light, are in an oven, which itself is being whisked along, relative to Earth, at 80 percent of the speed of light.

Why don't you worry about that? Because it all makes sense to us that the physics that happens, the laws that govern the way the physical world works, should be the same for everybody.

Does that idea make sense to you? It should. Ask yourself, if you think it doesn't make sense, are you some kind of closet Aristotelian who believes the Earth is the center of everything. Well, if you don't believe the Earth is the center of everything and the only special place in the universe, it follows pretty logically that the laws of physics ought to be the same for everybody, and that your motion, relative to Earth, ought not to matter.

If you believe that—and I assert that you ought to believe that if you accept these examples I've just given you—if you believe that, then you already believe in your gut and in your brain the essence of Einstein's relativity. Because the essence of Einstein's relativity is that Earth is not a special place and Earth's state of motion is not a special state of motion. And anyone out there anywhere in the universe who chooses to do physics experiments— playing tennis is a physics experiment because it's an experiment with the laws of nature and how they act on the tennis ball; and heating tea in a microwave oven is a physics experiment because it's an experiment in that case about the laws of electromagnetism, which are behind the operation of a microwave oven and how they behave. And if you choose to do those physics experiments, no matter where you choose to do them, they will come out the same.

Now at this point you could shut off the tape and walk away and tell all your friends, "I know all there is to know about relativity," because in essence you do.

Relativity is simply the statement that the laws of physics are the same for everybody. Why is it so hard to understand? Because it has implications that are not obvious and not easy to grasp and seem to run counter to common sense. And so it's going to take a while to explore those notions and understand them. It's also going to take a while to understand how humankind came to the idea that the laws of physical reality are the same for everyone. That's what our task is in the lectures to come. But as we end this one, let me remind you that you do, in fact, already know and believe the essence of the theory of relativity—that everybody has the same laws of physics, that the universe behaves the same way for everyone.

Heaven and Earth, Place and Motion
Lecture 2

In this lecture I've got to take you back ... further to what I'll call pre-classical physics; physics before the time of Isaac Newton because we want to understand how ideas evolve from that physics into classical physics and then set the stage for the several quandaries I mentioned that then led to the introduction of modern physics.

Understanding motion is the key to understanding space and time, because to move is to move through space and time. Is there a "natural" state of motion? The ancients thought so; they thought that there were different "natural states" for terrestrial and celestial motion. Aristotle (c. 349 B.C.) described a geocentric universe with planets and the Sun orbiting Earth in perfect circles. Ptolemy (c. 140 A.D.) added circles-on-circles ("epicycles") to represent planetary motion more accurately. Terrestrial objects naturally assume a state of rest close to the center of the universe (Earth); force is required to maintain motion.

Copernicus (1543) posited a Sun-centered universe, but maintained the celestial/terrestrial distinction and the idea of perfect circular motion in the celestial realm.

Galileo discovered that the Sun was blemished with sunspots, found moons orbiting Jupiter, and observed the phases of Venus.

Through careful study of planetary observations collected by Tycho Brahe (1546–1601), Kepler (c. 1610) showed that planetary orbits are ellipses, not circles. He developed mathematical laws describing the orbits, but gave no explanation for why the planets moved as they did.

Galileo (1564–1642) discovered that the Sun was blemished with sunspots, found moons orbiting Jupiter, and observed the phases of Venus. These discoveries helped dispel the notion of celestial perfection and lent support to Copernicus's heliocentric theory.

Experimenting with motion on Earth, Galileo concluded that all objects fall with the same acceleration. Through a "thought experiment," he developed the law of inertia—that an object continues in straight-line motion at constant speed unless disturbed by an outside influence (force). Thus, he redefined the "natural state" of motion as straight-line motion at constant speed. According to Galileo, force is needed for *change* in motion, not for motion itself. ■

Aristotle.

Essential Reading

Hoffmann, *Relativity and Its Roots*, Chapter 2.

Mook and Vargish, *Inside Relativity*, Chapter 1, Sections 1–5.

Suggested Reading

Einstein and Infeld, *The Evolution of Physics*, Chapter 1 through p. 33.

Questions to Consider

1. For the ancients, objects in the heavens moved naturally in circular motion. Would Galileo consider circular motion a "natural state" of motion? Why or why not?

2. Why isn't Galileo's conclusion that objects naturally move in straight-line motion at constant speed obvious from our everyday experience? Can you describe an environment in which it would be obvious?

Heaven and Earth, Place and Motion
Lecture 2—Transcript

Welcome to Lecture Two: "Heaven and Earth, Place and Motion." I advertise this is a course about modern physics, but in the first lecture I also talked at some length about classical physics as a way of contrasting the physics before the year 1900 with the modern physics that this course is primarily about.

In this lecture I've got to take you back even further to what I'll call ancient physics. Pre-classical physics. Physics before the time of Isaac Newton. Because we want to understand how ideas evolve from that physics into classical physics and then set the stage for the several quandaries I mentioned that then led to the introduction of modern physics.

I'm going to concentrate in this lecture and in subsequent lectures for a while on the question of motion. Now, this sounds like kind of a dull, dry topic. And many students going through introductory physics courses, unfortunately, come away thinking, "Gee. Motion. Why do I really care about that?" Well, if you think for a minute about motion, you'll realize it's the key to everything that happens. If there were no motion everything would stop. There would be no change. There would be no time. You wouldn't get from one place to another. Electrons wouldn't jump across the synapses in your nerves, in your brain, and you wouldn't think. There'd be nothing happening. Motion is what makes everything happen.

Why are we going to concentrate on motion? Now, one reason physicists concentrate on motion is because they'd like to predict how things are going to move. If I launch this rocket with this velocity, will I get it into orbit around Venus? for example, is a question. If I slam on the brakes with this much force, will my motion stop in time to avoid hitting the car in front of me? How does the electron move in this complicated new molecule I've synthesized that may be a drug that cures cancer or whatever. These are questions about motion.

We need to understand motion to understand how the world works. But, that's not the reason in this course for our being interested in motion. The reason we are interested in motion here is because what does it mean to move? It means to move from place to place. That is, through space. And it takes time to move. There is no such thing, we don't think, as instantaneous motion. Being here and then, instantaneously being here. So, motion is inextricably tied with the key ideas of space and time.

And space and time, if you think about them, are the fabric of physical reality—the underlying stage in which all physical events take place. So, if we understand motion, we have a clue to the nature of space and time. And it's no secret, as you understand from the first lecture with the time traveling twins, it's no secret that relativity does strange things to our common sense notions of space and time. So we've got to understand how those common sense notions were developed. And then we've got to understand what a true understanding of motion tells us about the real nature of space and time. So we're going to be emphasizing questions of motion.

And we're going to ask a useful question. And I ask you to keep this question in mind throughout at least the first two-thirds of the course, because it's going to pop up again, and again, and again. And the question is this: Is there a natural state of motion? Is there some way things sort of left to their own devices like to move—or maybe not move? Is there a natural state of motion? If there is a natural state of motion, what is it? If there isn't a natural state of motion, does everything just want to stop? You might have thought that was the natural state. But, think again. Things don't just stop. Electrons are whizzing around. Cars are moving. The air is moving. The planets are going around the sun. Maybe being at rest is not such a natural state. We'll come to understand that.

To answer the question, is there a natural state of motion, as we now understand the answer, we've got to go back though and look at what the answers were in ancient times and then in classical physics times. And so that's what I intend to do. And in this lecture I want to look particularly at the ancient's answer. And the ancient's answer is interesting because it's a blend of science and philosophy and theology.

In ancient times, the universe was divided into two realms. There was the earthly realm. That was the realm of the base, the imperfect, the crude, the solid, the material stuff—like us, and rocks, and water, and oceans—the Earth. And there was one set of rules that governed how things happened on Earth.

And then there were the heavens. Hence my title, "Heaven and Earth, Place and Motion." There were two places—the heavens and the Earth—and they had different laws of physics. If you want to think about these ancient ideas as laws of physics.

And the answer to the question, what's the natural state of motion, had a different answer on Earth and a different answer in the heavens. And the answer given for Earth was very simple. The natural state of motion for earthly entities is to be as close as possible to the center of the Earth—which was the center of the universe. And the Earth was a realm, again, of the material, and the base, and the solid, and so on.

And so that explains a lot of things. It explains the fact that if I take this ball, for example, and drop it, after a while it ends up at rest as close as it can possibly get to the center of the Earth—which happens to be the floor in this instance. So, the philosophical sense that things on Earth tended toward the center of the Earth and to come to a state of rest as close to the center of the Earth as possible seems to be born out by experiments. Again, dropping a ball is a physics experiment. I'm challenging physical reality to tell me how it behaves, and I'm learning something about it. That experiment seems to verify the ancient notion that motion on Earth ultimately has a natural state, which is to be at rest as close as possible to the center of the Earth.

Notice that that is also, in a sense, a philosophical explanation for the phenomenon that we now call gravity, because it sort of gives a teleological, a purposeful, a philosophical explanation for why that tennis ball ended up on the floor. It wants to be as close as possible to the center of the Earth and to be at rest—and it got into that state.

So the answer, what's the natural state of motion, to the ancients for objects on Earth was they want to be as close as possible to the center of the Earth and they want to be at rest. And that is easy and done with and explains a lot of physical phenomena that we observe.

In the heavenly realm, on the other hand, things were very different. The heavens are the realm of the ethereal, the spiritual, the stars, the planets. People really didn't know what they were. They were things up there in the heavens. And there there's a very different, natural state of motion. The natural state of motion of the heavens, according to the ancients, is circular motion. Move in a perfect circle. A circle is about the most perfect geometrical form you can think of. It has no corners. It has no beginning or end. It's everywhere the same. It's a beautiful, perfect thing. And the idea was motion in the heavens was circular motion. And objects in the heavens simply continued in circular motion. That was the natural thing for them to do.

There were various theories about why that happened. There were also various theories to explain funny phenomena of earthly motion. For example, if I throw a ball, it keeps moving for a while before it ends up in that natural state. Well, there were theories about air rushing from the front to behind to keep it moving and things like that. But the basic idea was it wanted to end up in that state of rest, whereas objects in the heavens wanted to be moving in those perfect circles just continually.

I want to emphasize one other aspect of this. Why do things move? That's another question we can ask. Well, we've just given a kind of philosophical explanation. Things on Earth move toward the center of the Earth because they'd like to be at rest as close as possible to the center of the Earth. If they're not in that state, according to the ancients, it takes some kind of push to get them going or to keep them going. If you see an object that's moving, for example, an ox cart is moving down the road, the ox is pulling it. That's what makes it move. If the ox stops, it stops moving. If you take your foot off the gas pedal of your car, the car comes to a stop. Albeit, not instantaneously. But sooner or later it ends up in a stopped state. So the idea was you needed to have some kind of push, in that case the push of the tires against the road, ultimately driven by the engine of the car.

You needed to have some kind of push to keep something moving. Absent that push, the thing stops moving. So a fundamental idea of the ancient view of motion is that it takes a push to keep something moving. At least in the terrestrial realm. In the heavenly realm, you have this different situation with the perfect circles and they just continue.

Let's look for a minute more at the heavenly realm. Out of that idea came a sort of cosmology. A picture of what the universe looked like. And let's take a look at the first visual here. This is the picture of the universe as the philosopher Aristotle envisioned it. Aristotle is working in the 300s B.C. and he has a vision of a universe in which the Earth is sitting at the center.

The Earth is surrounded by concentric rings that hold the different planets that are thought to go around the Earth and that also hold the sun, which is also thought to go around the Earth. And farthest out are the so-called fixed stars that stay in the same relative position in the sky, and they are thought to move as one giant sphere around the Earth.

By the way, for those of you who are "watching" this course on audio, I do have a number of visuals. And I've selected the most important ones and put them in the booklet. So I urge you to refer to those. If you're driving your car, you might want to wait until you come to a stoplight or something. But this course is definitely aimed at you audio people as well. And when I describe a visual that isn't in the book, and this is an example, I'll just tell you what we're looking at. So we're looking at this picture of the Earth at the center of the universe surrounded by concentric rings holding planets, and in one case the sun.

Well, there is a problem with this picture. One of the things we'd really like our laws of physics and our cosmologies and descriptions of the physical universe to do is correspond to reality. And, unfortunately, the picture Aristotle described of objects going around the Earth doesn't quite make it. And the reason is if you look at the motion of the planets in the sky, they do not seem to describe perfect circles around the Earth. The sun, essentially, does; the distant stars essentially do; but the planets don't. They do strange things.

If you watch a planet, Mars say, or Jupiter, night after night, you'll see it changing its position in the sky relative to the fixed stars. For a while, looking like it's going in a nice circular path around Earth. But then it will pause and it will go backwards for a while—in a little loop. And then it will pick up and go forward again. It isn't describing perfect circular motion around the Earth.

But Aristotle and the other ancients were hung up on the notion that celestial motion must be perfect circles. So it was up to Ptolemy, in about 140 A.D., to come along and modify the Aristotelian picture. And here is what Ptolemy did. He simply took the planets and instead of having the planets moving in perfect circles around the Earth, they moved in little tiny circles—they were called epicycles—around other perfect circles that themselves went around the Earth.

So Ptolemy made a cosmology, a picture of the universe that fit physical reality better. If you study how this system works in which planets are going around the Earth, but then they're also moving in little circles around your path, you'll find that they do in that situation pretty much what they appear to do in the sky. That is, they exhibit this occasional backwards or retrograde motion. And this could be embellished with additional circles upon circles to make it more accurate, and so on.

But the point here is we are emphasizing the notion of perfect circles in the celestial realm. So the universe, according to Ptolemy, is basically like that according to Aristotle. But the circular motions consist not of a single circle around the Earth but circles upon circles. We don't need that for the sun, which appears still to describe a perfect circle.

So that's the picture, according to Ptolemy. And that picture reigned for a couple thousand years. Basically, until Copernicus, in 1543, published the heretical idea that no, the Earth is not the center of the universe. Instead, thought Copernicus, the sun is the center of the universe.

And this is the great Copernican revolution. And it's the first step in what is ultimately going to lead us to Einstein's general theory of relativity. It's the first step in saying Earth is not a special place, and Earth's motion is not a special state of motion.

Copernicus emphasized the business of place. I'm going emphasize also the business of motion, because that's what's tied up with relativity. So what does the universe, according to Copernicus, look like? Well, Copernicus was loath to give up the perfect circles upon circles. So his cosmology also had a sun at the center, it was surrounded by orbits of planets, and a circular orbit of the Earth, and the planetary orbits needed to have circles upon circles, because they didn't seem to be describing even quite perfect circles around the sun. And Copernicus was really insistent on maintaining this view that the celestial motions are perfect. So he came up with a cosmology—again, circles upon circles. But here the sun is at the center.

Well, the next bold step was taken by Kepler, who was working in 1610 roughly, and he was working with data that had been collected painstakingly by the astronomer Tycho Brahe. And Tycho had amassed years and years and years of data on the positions of the planets. And, on the basis of that data, Kepler concluded something quite different. He concluded that motion in the heavens, the motion of the planets around the sun, at least, were not perfect circles, but they were ellipses. And I'm showing here a picture of an ellipse, for you audio people it's simply an egg shaped curve with the sun not at the center, but off at a place called the focus, which is at one side.

This is a highly exaggerated picture. The orbits of the planets are not very elliptical at all. They are almost perfectly circular. The Earth's, for instance, differs from perfect circularity by about three percent. If I were to draw it, you'd have a hard time distinguishing it from a perfect circle. So this is highly exaggerated. Orbits of comets do look exaggerated like this. So do orbits of the satellites that are used to provide television service to Siberia, for example. They describe great looping orbits that take them hanging for a long time over the extremes of the Northern Hemisphere. And then they whiz down around in these elliptical orbits.

So elliptical orbits do happen, but they aren't the orbits that the planets follow. The orbits the planets follow are circular, almost. But they are measurably different. And Kepler said they are ellipses not circles, and not only did he say that, he made up laws that describe how fast the planets were moving at different parts of the ellipse in relation to the period—the time it takes to go around—and the size of the elliptical orbit.

So he described mathematically what these orbits looked like and how the planets moved in the orbits. But he gave us no physical explanation. He didn't say, "Why do they move in these orbits?" He just said, "I think they do."

So, where have we moved? We've moved away from the notion of an Earth-centered universe. That was Copernicus's big step. We now have a sun-centered universe, but in Copernicus's idea still with perfect circles.

Now we've moved away from that further to a universe in which the orbits of the planets appear to be ellipses not circles. And Kepler is telling us how to describe mathematically what those elliptical orbits are. But we have no explanation for why the ellipses are there. And we still have this dichotomy between motion on Earth and motion in the heavens.

And now we move to someone who is probably the most important figure in all of science. And that's Galileo Galilee. Galileo, who lived from 1564 to 1642, was a real experimentalist. People like Kepler were looking at observational data. Galileo also got his hands on real things. He purportedly dropped objects off the Leaning Tower of Pisa. Whether that's true or not, he certainly did experiment with terrestrial motion on Earth and tried to come up with an explanation and an understanding of how that motion behaved. He tried to answer the question—and succeeded in answering the question—what's the natural state of motion on Earth.

Galileo did several things. We'll talk a little more later about his experiments with earthly motion. But he also did some observational astronomy. And he developed the first astronomical telescope. He was alive at the era when the whole idea of lenses for eyeglasses, on vision correction, had just been invented. And he made a telescope, and he pointed it heavenward. And he saw some things that had never been seen before.

I just want to describe for you three things Galileo saw, because they shook the foundations of this philosophical idea that the heavens up there were somehow perfect and different from the base and coarse and crude Earth down here. The first thing Galileo saw, I don't know if it was the first thing temporally, but one thing he saw when he looked at Jupiter—and you can do this observation with a pair of binoculars—if you look at Jupiter, you'll see what Galileo saw. He saw four little dots lined up in a line on either side of Jupiter. You look on different nights—they're in different places. Sometimes you can't see one or two of them because they are behind or in front of Jupiter. But if you watch Jupiter, when it's up in the sky at night, on different nights, you'll see these little dots that are in line right across Jupiter's equator, on either side of Jupiter. Those are the so-called Galilean satellites of Jupiter. They are clearly, and were clearly to Galileo, a kind of miniature solar system. They represented moons of Jupiter, going around Jupiter in the same way the Earth's moon goes around the Earth.

And so, Galileo had, for the first time, an example of another solar system–like thing, another miniature cosmology. And it clearly had Jupiter at the center and the moons going around it. And that helped confirm the notion that there was nothing so special or unusual about the Earth—being sort of like a moon of the sun going around the sun along with all the other planets.

By discovering another miniature solar system, if you will, Galileo had confirmed the Copernican idea that the sun was a solar system with the sun at the center and the planets going around it. Galileo's discovery of Jupiter's moons, I think, is a little bit like what will happen when eventually perhaps we discover life somewhere else in the universe. We'll have a second example. All of a sudden what happened here on Earth, that many scientists don't think is unique, will be proved not to be unique. Similarly, Galileo discovered the moons going around Jupiter, and for the first time the solar system idea was not unique. It was a commonplace or, not commonplace, but at least it was a physical occurrence that had occurred more than once. So that was an important observation.

The second observation Galileo made was of the phases of Venus. You're all aware of the phases of the moon. Sometimes it's full. Sometimes it's a new moon, and you can't see it. Sometimes it's a crescent. Sometimes it's a half moon. Sometimes it's gibbous—not quite full—and so on.

Why does the moon have phases? Because the moon goes around the Earth, and at different times in its orbit it's in a different place in relation to the sun. For instance, when it's a full moon, you'll notice the full moon rises at sunset. Why? Because the sun is on one side of the Earth. The moon is on the other side. And it is getting full sunlight. That's why the full moon always rises at sunset. And always sets at sunrise, for example. At other phases, the moon is being lit—only part of it is being illuminated. Only the part of it you can see from Earth is being illuminated. And we get the phases. In the moon's case, although the moon goes through different phases, the phases are always the same size.

Galileo observed the phases of Venus. Venus goes through full and crescent, and so on. But there's something else that happens with the phases of Venus. When Venus is full, it's a very small circle. When Venus is a crescent, it's much larger. Why would that be?

Well, there's an easy explanation for that in terms of the Copernican picture. In the Copernican picture, Venus is full when it's on the opposite side of the sun from the Earth, and therefore, it's farthest away, and therefore, it looks smaller. It's a crescent when it's relatively near the same side of the sun as the Earth is. The Earth is only seeing a small sliver of the lit up part of Venus. And, therefore, we see a crescent. And it looks much bigger because Venus is much closer.

It would be very difficult to explain that observation in terms of the Aristotelian, Ptolemaic idea that the Earth is at the center of the universe. Because there Venus is going around the Earth. And, except for that little circle on a circle, there's very little change in its distance from Earth. So, the observation of the phases of Venus was a very important one in helping to verify the Copernican notion, again, that the Earth is not special. The Earth is not the center of everything. We're moving away from Earth-centeredness with that observation.

The final observation that Galileo made is that he looked at the sun and he observed sunspots. Timing here, by the way, is remarkably lucky because the sun, very soon after Galileo's time, went into a phase called the Maunder Minimum that we solar physicists, and I happen to be a solar physicist in my research field, we don't understand why that occurred, but the sun went through a period when it had almost no sunspots for a long time.

What are sunspots? Well, to Galileo they were dark blemishes on the surface of the sun. He hadn't actually discovered sunspots. The ancient Chinese had discovered them much earlier and called them crows and didn't know what they were. And Galileo didn't know either. But the important philosophical point was they were blemishes on the surface of the sun. Here's this celestial body, and it's no longer perfect. And, by the way, you look at the moon and you see it has craters and mountains. It's not perfect either.

Here's a break in the notion that the celestial realm is somehow perfect and somehow distinct from the terrestrial realm. All of a sudden, there are mountains and craters and there are spots on the sun. And it's not perfect up there either. So why should it obey some kind of different, perfect laws?

Galileo didn't get any further than that. But he established the imperfection of the heavens, if you will. And that was a break in this notion of a separate terrestrial and celestial realm.

Then Galileo turned to motion on Earth. And he did several important experiments. The experiment we are most famous with, the experiment in which he supposedly dropped objects off the Leaning Tower of Pisa. That was an experiment, which, if it occurred, verified the notion that all objects fall toward the Earth with the same, not speed, but acceleration. They gain speed at exactly the same rate. It's a difficult experiment to do accurately. Air resistance gets in the way. But Galileo, if he did it from the Tower of Pisa, was able, by dropping objects of different weight, to confirm that. They didn't fall at grossly different rates. That was an important understanding of gravity.

Then Galileo carried out an important "thought experiment", which I will describe here and show you in a visual. What Galileo did was to imagine a ball rolling in some kind of trough. The ball starts at some height in the trough. It rolls down to the bottom of the trough. And then it rolls up to the other side. And how far up would it get? Galileo asked. Well, he reasoned that if there were no friction and no air resistance and so on—everything were perfect—the ball would roll to the same height it started out at. And, absent any friction or air resistance, it would simply roll back and forth in that trough continuously. Back and forth. Always rising to the same height.

Of course, it doesn't. It eventually slows down because of friction and stops at the bottom.

But, that was Galileo's thought. So, if he rolled a ball from one side of the trough, it will rise to the same height on the other side. What if we make the other side of the trough somewhat longer? What if we take this trough, which is now symmetrical, and we stretch out one side of it? Well, the ball will still rise to the same height, reasons Galileo, in this so-called "thought experiment." He didn't actually do this. But he said what will happen if we did this. It will rise to the same height, vertically. But it has to roll further horizontally to get there because the trough that was originally symmetric is now longer on one side—the side it's going to rise up to—after it rolls down.

And then, said Galileo, "What if we go one step further?" What if we take that trough and lower it completely—make it horizontal at the bottom. So the ball rolls down the trough and then it rolls continually horizontally along the bottom of the track. What happens? Galileo said absent friction and air resistance, it never stops.

Now, the historians of science will tell me I'm being a little bit simplistic here. And I am. Galileo was worried about the curvature of the Earth and other things like that. But, in principle, this is the essence of his idea. Galileo said if this trough never goes back up again, the ball will just continue rolling horizontally without its speed changing. And he reasoned that from this idea of taking the trough and gradually lowering the one side, the ball rolls farther and farther horizontally while rising to the same height. Finally, if it can't ever rise to the same height it started at, it will just continue to go horizontally forever.

And Galileo, therefore, concluded something very, very different about terrestrial motion. He reasoned this way. He said, "Look, if an object is moving at a constant speed in a straight line, what's it going to do? It's going to keep doing that. It's going to move forever in a straight line at constant speed. It's never going to stop."

That idea is crucial to classical physics, and to modern physics as well. That idea is one of the key ideas in understanding motion. What Galileo is saying is: The natural state of motion is not to be at rest on the surface of the Earth. That's a state of motion you get to because of things like friction and air resistance and things that you bump into, like the floor when the ball falls. The natural state of motion is to move in a straight line at constant speed.

It doesn't take a push to have something be moving. This is a crucial point. This is one we teachers of introductory physics spend a lot of time trying to get our students really to believe. It's really true folks. It doesn't take any push to make something move. If you see something moving, the question is not why does it keep moving. The question might be, how did it get moving in the first place? Or, if you see something that's stopped, how did it get stopped if it was earlier moving? The question is not, what keeps it moving? Because it takes no push. No pull. What physicists call a force. It takes no force to keep something moving. That's what Galileo says. And that's what appears to be a correct description of the way motion works.

The natural state of motion is to be moving uniformly in a straight line. And there is a name for that statement. It's a name attributed to both Galileo and Newton. It's called the law of inertia. It says that an object has an inertia. A natural tendency to keep moving. If I take this big ball over here, and I set it into motion, it keeps moving. And it would fall on the floor and crush itself if I didn't stop it. That's because of its inertia. It takes a push to get it started moving. And it takes a push or a pull to stop it. But what it naturally wants to do is move in a straight line at constant speed.

That's the natural state of motion. That's the law of inertia. By virtue of having mass, by being heavy, by having, if you will, weight, although those terms are not quite synonymous, this object, once moving, will keep moving. It takes a push to get it moving. It takes a push to stop it moving. It also takes a push to change the direction of its motion. The motion that is natural, according to Galileo, is uniform motion. That means motion in a straight line at constant speed. If you see any deviation from that, if you see an object speed up or slow down or, and this is the point our introductory physics students sometimes have trouble with, or change its direction, in all those cases you can assume that a force—a push or pull—has acted on it. But if you see it moving in a straight line at constant speed, there is no reason to explain that. That is what it naturally wants to do. Nobody is pushing it or pulling it.

Why is that such a hard idea for us to grasp? Because in most of our world, the experience—push this tennis ball, you see it slowing down. Eventually it stops. Take your foot off the gas of the car again. It ought to just keep rolling forever if what I've just said is true. Well, it is true except there are pushes and pulls working on the car. There's the push of the air against it. There's the friction pushing against the tires. In the case of the ball, there's friction with the table. And again, there's the air pushing back on it. We don't see those forces. They don't manifest themselves the way my hand does as a forcible push or pull. But they're there, and they cause the motion to change.

If we could abstract those forces away, and that's what Galileo's thoughts-on-motion thought experiment did. Or, if we could go to a physical situation like, say, an ice hockey rink. Look at the hockey puck. It whizzes across the ice. There's almost no friction. It would go much further than the size of the ice rink before it slowed down appreciably. Look at the skaters themselves. Unless they dig their skates into the ice to try to force themselves to turn or to start or stop, they simply coast at essentially constant speed. The forces of friction and air resistance, in that case, are small. The pushes and pulls and the natural state of motion reveals itself.

But the reason we have trouble teaching this to introductory physics students and getting them to believe it—it actually makes the world much simpler—and the reason we have such trouble accepting it in our everyday lives is we're all sort of closet Aristotelians, in a sense. I use that term with my students. I tell them they are closet Aristotelians. They sort of believe really in their gut that Aristotle was right. That it takes a push or a pull to make motion, that the natural state of motion is to be at rest on the surface of the Earth. It's not true. The natural state of motion is to move in a straight line at constant speed.

Perhaps the very best example of this I can give you is the Pioneer and Voyager spacecraft, which are now beyond the outer most reaches of the solar system. And they've been going for something close now to 30 years—20 to 30 years. And they are just moving in almost straight lines.

There are some slight pushes and pulls from the gravity of the sun and the planets. But they are essentially moving in straight lines at constant speeds. They have no rocket motors going. There's no fuel involved. They're still broadcasting back to us with the radio transmitters, though. And they're way out there. And they're just going to move forever until something pushes or pulls on them—like the pull of a distant star or an alien hand reaching out and grabbing them and trying to figure out where they came from. Otherwise, straight line, constant speed. It's the natural state of motion. It takes no effort. It takes no force. It takes no push or pull.

This is the essential change that Galileo has made. He's developed the law of inertia. He said the natural state of motion is straight-line motion at constant speed. Pushes and pulls are required only for changes in motion. We are going to make a great deal of those two points. If you see a change in motion, you know a push or a pull has occurred. If there is not change in motion, that's perfectly natural. There is no push or pull involved. The law of inertia holds. An object moves in a straight line at constant speed unless a push or a pull acts on it. And out of that is going to come a whole host of classical physics. And that's ultimately going to lead us to the questions that will bring us to relativity.

The Clockwork Universe
Lecture 3

Before we get into the mechanism of the clockwork universe, let's look a little bit about the individual whose name is most associated with this idea, Sir Isaac Newton.

Isaac Newton was born in 1642, the year of Galileo's death. Newton developed his famous three laws of motion, quantifying Galileo's earlier idea that uniform, straight-line motion is natural and changes only if outside influences (forces) act. Newton also considered gravity and had the brilliant insight that the same force that pulls an apple to Earth is also what holds the Moon in its orbit—thus, putting to rest the false dichotomy between celestial and terrestrial motion.

Newton developed the concept of *universal gravitation*, suggesting that every object in the universe attracts every other object, with a force that depends on their masses and the distance between them. Together, Newton's laws of motion and gravity showed that the planets must move in Kepler's elliptical orbits and hinted at the possibility of artificial satellites. The predictability inherent in Newton's laws suggests a "clockwork universe" in which all that happens in the universe is completely determined by the initial motions of its constituents. (On a historical note, Newton did some of his most productive work while away from Cambridge to escape the plague. Today, Stephen Hawking occupies Newton's chair at Cambridge.)

Such statements as "I am moving" or "I am at rest" have no absolute meaning; they are only meaningful when they are about motion or rest relative to something else.

Inherent in the ideas of Galileo and Newton is the *Principle of Galilean Relativity*: The laws of motion work exactly the same way for anyone as long as he or she is moving uniformly. In other words, the laws of physics known to Galileo and Newton preclude such statements as "I am moving" or "I am at rest" from having any absolute meaning.

Newton's three laws:

- Newton's first law restates Galileo's discovery that objects move uniformly unless acted on by outside forces.

- Newton's second law, $F = ma$, tells quantitatively how a given force (F) produces changes in motion (acceleration, a) in an object of mass m.

- Newton's third law, "for every action there is an equal and opposite reaction" says that forces always come in pairs; if object A exerts a force on object B, then B exerts a force of equal strength back on A.

Newton's law of gravity:

- The famous story of Newton and the apple may be a myth. But if it is true, its significance lies in Newton's realization that apple and moon are attracted toward Earth by the same force, which Newton named *gravity*.

- Thus, Newton subsumed celestial and terrestrial motion under the same laws. He generalized to the idea of *universal gravitation*: that every object in the universe attracts every other, with a force that depends on their masses and the distance between them.

- Using his law of gravity and his newly invented calculus, Newton proved that the planets must move in elliptical orbits, just as Kepler had observed.

Nicolas Copernicus.

- Newton anticipated artificial satellites, showing that an object, given enough speed, will "fall" around Earth, pulled by gravity out of the straight-line path it would otherwise follow. Today, we are highly dependent on satellites for communications, weather prediction, navigation, science, and other applications.

To review, ideas about motion evolved until, by Newton's time, all motion in the universe was assumed to be governed by the same deterministic laws.

The Principle of Galilean (Newtonian) Relativity (we might call it the "original principle of relativity"):

- The laws of motion are the same for anyone, provided that he or she is in uniform motion.

- Such statements as "I am moving" or "I am at rest" have no absolute meaning; they are only meaningful when they are about motion or rest *relative* to something else.

- There are many ways to say this! Remember the cruise ship, Venus, and the distant galaxy. There is simply no experiment you can do in, say, a uniformly moving cruise ship, train, plane, or even planet that will answer the question of whether or not you are moving. (You can answer the question "Am I moving relative to Earth, or to my star, or to whatever?"—but that's a question about *relative* motion, not *absolute* motion.) ∎

Essential Reading

Hoffmann, *Relativity and Its Roots*, Chapter 3.

Mook and Vargish, *Inside Relativity*, Chapter 1, Sections 6–8; Chapter 2, Sections 1–5.

Suggested Reading

Einstein and Infeld, *The Evolution of Physics*, Chapter 1, pp. 34–67.

1. Many people think astronauts in an orbiting spacecraft are "weightless" because "there's no gravity in space." How is this view inconsistent with Newton's ideas of gravity and motion?

2. You're on a plane flying through calm air. You eat, read, and relax just as you would on the ground—you can't tell that you're "moving." Yet when you look out the window, you see the ground slipping backwards. Why can't you conclude definitely that you and the plane are "moving"? What can you conclude?

Note: Figures explaining key concepts presented in each lecture are placed immediately following each lecture. The graphical part of each figure is the same as the corresponding graphic used in the video version of the course, thus giving audio customers a chance to visualize what Dr. Wolfson is discussing. In addition, each figure contains explanatory text to further reinforce the concept under discussion.

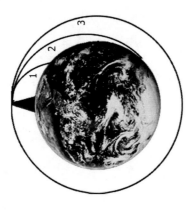

Orbits

Newton's thought experiment describes the motion of an object when thrown off a very tall mountain.
Object 1, thrown with modest force, falls in a curved path toward the earth.
Object 2, thrown with greater force, falls farther than object one, but still eventually falls to earth.
Object 3, thrown with such force that it falls towards the earth at the same rate that the earth is
   curving away beneath it, describes a circular orbit. Absent air resistance or other mitigating
   factors, it will continue in this circular motion forever.

The Clockwork Universe
Lecture 3—Transcript

Welcome to Lecture Three: "The Clockwork Universe." Why that title? Because part way through this lecture, I hope to convince you that if you believe the Newtonian paradigm of physics, the essence of classical physics, you'll believe that the universe is completely deterministic, that whatever conditions were established at the start, the universe then would evolve like a giant clockwork with every move of every particle completely and inexorably determined. The clockwork universe.

Before we get into the mechanism of the clockwork universe, let's look a little bit about the individual whose name is most associated with this idea, and that's Sir Isaac Newton. I advertised that I would get in the personal touch, occasionally. And I'll do that a little bit in Newton's case.

Newton was born in 1642, which, coincidentally, is the year Galileo died. But he very much carried on where Galileo left off and extended Galileo's ideas. And the physics of Newton is really the physics of Newton and Galileo in many ways. Newton had some interesting history. In his early twenties he had to leave Cambridge University because of the plague. And he actually moved back to his family place in Woolsthorpe, England. And he spent several years there, which were among the most productive years of his life. So the plague drove him away from Cambridge, and he was very productive.

He returned to Cambridge and the Chair that he held is now held by Professor Stephen Hawking, who many of you know about. You may have read his book, *A Brief History Of Time.* So Hawking holds Newton's Chair at Cambridge University. What Newton is most famous for are his laws of motion and his law of gravity. I want to spend most of this lecture looking at those laws, but then at the end, getting us to a sense of how they're going to lead us on to Einstein's relativity.

Newton's three laws of motion are something everybody studies who deals with introductory physics. There are three laws that describe how objects move in response to forces, that is, to pushes and pulls. And together, if you know the pushes and pulls and you know Newton's laws of motion, then you can determine exactly what an object is going to do. Hence, the clockwork universe idea.

So, what are Newton's three laws of motion? Well, the first is something we've already seen. The first law of motion is essentially the law of inertia that I described in the last lecture and attributed to Galileo. And it says, basically, in somewhat formal terms that an object moves uniformly. And again, remember what uniform motion means. Uniform motion means it's a straight-line path and it's at constant speed. Any deviation in direction or speed is not uniform motion. An object moves uniformly, that is, in a straight line at constant speed, unless it's acted on by a force. That is, by a push or a pull. And we call this the law of inertia. And it's the statement, basically, that straight-line motion at constant speed, uniform motion, is the natural state of motion.

So, here Newton is simply recapitulating what Galileo said—that objects have inertia, and that inertia keeps them moving, makes uniform motion a natural state of motion. And that you have to do something to change that state of motion. You have to exert a force, a push or a pull, to start an object moving, to stop an object moving, to change its speed, or to change its direction. That's the first law of motion.

The second law is the really meaty one. The second law describes quantitatively how forces, that is, pushes and pulls, cause changes in motion. The first law is really a subset of the second law. Because the first law says unless there is a force there won't be a change in motion. Motion will continue in the absence of a force.

Newton's second law quantifies that with the famous formula $F = ma$. And I'm not big on math here, but I do want to describe for you what that formula means. F is the force, the push or the pull; a is the acceleration, the rate at which the object's motion is changing. That's what acceleration means. When you step on the accelerator of your car, what that does is change your speed.

When you turn the steering wheel—this is less obvious—but the steering wheel is also an accelerator. When you turn the steering wheel, you're changing the direction of your motion, and therefore, you're accelerating your car. You may not be going faster or slower. But you are accelerating because your motion is changing; a is the acceleration—the change in motion.

And m is the mass. If you will, the weight of the object. And so what the second law says is simply this: Well, first of all, let me tell you what it says in the context of the first law. It says if, after zero, if there is no force, then a will be zero. There will be no change in motion. And we get back to the first law that says, absent a force, an object will continue in straight-line motion. There'll be no change in the motion.

But Newton's second law goes beyond that. It tells us, for example, that if m is big, if an object is very massive, it would take a big force to give it an acceleration. If an object is not very massive, it doesn't take as much force to give it the same acceleration. As an example, over here I have two objects of different mass. I've got a tennis ball and a basketball. The basketball is not only bigger; it's also more massive. That's the important thing here. And I'm going to give them each the same push. The same acceleration. Here goes the basketball. I'm going to give it a little push. I'm not going to keep pushing it. I don't need to keep pushing it. I'm going to give it a push to change its state of motion. Right now it's at rest with respect to this table. I'm going to change its state of motion. I'm going to make it move with respect to the table. Here I go. I put it into motion. And it then moved in essentially constant motion until another force acted on it—namely me, grabbing it at the other end. So, that was the basketball. Massive. I gave it a force. I didn't change its motion that much. That is, it didn't get moving that fast.

I'm going to give the same sized tap to the tennis ball. Much less mass. Same F. Newton's equation says I'm going to get a bigger a. It got up to a much greater speed.

Force, mass, and acceleration are related. The bigger the mass, the bigger the force it takes to change an object's motion. And that's true whether you're changing the object's motion, as I did here by changing its speed, by speeding it up, by getting it moving, or changing its motion by stopping it. It takes a bigger force to stop the basketball than it does to stop a rolling tennis ball. It takes a bigger force to stop a truck than it does to stop a car.

The acceleration and the force and the mass are all related. And that's what Newton's second law says.

And if we know how an object is accelerating, if we know how its motion is changing and we know how it was initially moving, we can predict everything it's going to do. So, in principle, if you know the force that's acting on an object, the push or pull that's acting on an object, and you know it exactly, and you know what the object is doing at some point, then, in principle, you can predict what it will be doing at every other point in the future. Provided you continue to know the force. That's the essence of the clockwork universe idea. The clockwork universe says, if we know Newton's laws of motion, which we do, and we know the forces, the pushes and pulls, that are acting on objects, and we knew the positions of all the objects that make up the universe, every planet, every electron, every proton, every atom, and we knew their motion, in principle, even though we don't have any computer powerful enough to do the calculation, in principle that calculation is doable. And in principle, everything that happens from then on is predictable. The universe is completely deterministic.

It's kind of a bleak picture, in a sense, philosophically, because it leaves no room for our free will. Unless you want to postulate that our brains operate outside the laws of physics. But that was the picture that emerged from classical physics. A clockwork universe, a universe that once God or some initial, random event, or whatever it is, puts it into motion, the rest of it is completely determined, and it simply runs on its own like a clockwork obeying Newton's laws of motion. Well, that's two of the laws of motion.

The third law of motion you've also heard of. I'm going to phrase it in slightly different terms than you may have heard. Forces come in pairs is what the third law of motion says. It says if I push on something, for example, I'm going to push on this table now, the table pushes back on me. In fact, it pushes back on me with a force that is equal to the force I used in pushing on it. If I push on this lectern, it pushes back on me. And I can feel that push in my hands. When I hit that basketball to get it rolling, I felt that I was hitting it with my hand. That's because the basketball was pushing back on me. That's also an essential part of the set of laws on motion.

For those of you who have studied any physics, the third law of motion is very closely tied with the idea of conservation of momentum. But we don't need to worry about that. The point is, these three laws provide a coherent whole that describe everything there is to know about motion according to Newton, according to classical physics.

So, we have three laws of motion. But we don't yet understand much about the forces, the pushes and pulls, that might act to cause objects' motions to change. Well, there Newton made another great step forward. Newton studied one of the most important forces known.

There is a famous myth about Newton, and we are going to take a look at a picture that describes that myth. Here's a picture of Newton sitting under an apple tree. And the myth goes, apple falls on Newton, hits him on the head, and he discovers gravity. Well, I think that myth stated that way has to be nonsense. It didn't take Newton to discover gravity. It took some cave person sitting under a cliff outside their cave. And rocks come falling down or something. Or a wild animal jumps over the cliff or something. They realize there's gravity. When our ancestors swung from trees, they knew inherently about gravity. You don't discover gravity by having an apple hit you on the head.

The story of Newton, if it's true, is much more subtle, and much more sophisticated, and much more profound in its implications. So, here is Newton sitting under the apple tree. And if this story were true, the moon is also up in the sky at the same time. And Newton sees the moon. And the realization that's associated with the apple tree myth or story, if it's true, is that Newton recognized that the motion of the apple and the motion of the moon are one in the same motion.

That's the essence of Newton's genius. He said, "The moon, what's it doing? It's moving around the Earth. The apple, what's it doing? It's falling toward Earth. What do both those motions have in common?" What they have in common is that in both cases the objects are being accelerated—deviated from straight line, constant speed motion in a direction toward the Earth. It's as though, in both cases, there's a pull toward the Earth acting on both those objects. In the case of the apple, the pull is causing the apple to move, actually physically, toward the Earth. In the case of the moon, the motion is not actually making the moon come any closer to the Earth. But it is pulling it away from its straight line path it would follow according to the first law of motion if there were no force acting on it. That's the key point to realize. It's that simple.

Also, physics is simpler than most people realize. This is a case in point. You ask, what keeps the moon in orbit? What keeps an artificial satellite? And why don't they fall down? They are falling. They're falling toward Earth in the sense that their motion is deviating from a straight line because the Earth is pulling on them. It just happens to be the case with the moon that that pull and the motion of the moon are such that the pull keeps the moon in a circular path.

I have another example of that here. I've got just a tennis ball attached to a string that I'm going to whirl around over my head here, and I'm going to just ask a couple of questions about this. So, here goes the tennis ball. Is it moving in a straight line at a constant speed? No way. It's not moving in a straight line. It may be moving at constant speed, but it's still accelerating, its motion is still changing. It's not going in a straight line at constant speed. It's not undergoing the "natural state of motion." Why isn't it? Because the string is pulling on it, through me, at the center of the string here, pulling. I'm exerting a force, which is communicated through the string to the ball. And the ball is going on a path that deviates, always, from the straight line it would go if I were to let it go flying off, which I won't do. But if the string were cut or something, or I let go, it would go flying off in a straight line at constant speed. It's falling, if you will, toward my hand. It's being pulled, always, toward my hand out of the straight-line path it would otherwise follow.

It's that simple. Objects move in a straight line at constant speed unless they are acted on by a pull. In this case, the pull has just the right strength to keep the object moving in a circular path. And so it does. It deviates from the straight-line path it would otherwise follow. And the moon, in its orbit around the Earth, is doing exactly the same thing. And that is exactly what Newton recognized if, in fact, it's true that the apple hit him on the head and he discovered gravity.

And he didn't only recognize that qualitatively, he recognized it quantitatively as well. He formulated, on the basis of that observation, the so-called law of universal gravitation. Newton said, "I believe every object in the universe attracts every other object with a force, a pull that depends on how massive the two objects are and also on how far apart they are. The further apart they are, the weaker the force." And it falls off very rapidly. It goes as one divided by the square of the distance, if you've heard the inverse square law. We don't need to get in to that. The point is it falls off with distance.

That means the acceleration that the Earth gives to the moon, the great big, massive moon, very far away, is much less than the acceleration it gives to the apple. Why? Because the apple is closer to the Earth. Well, is that true? Newton went and calculated. He said, "What kind of rate of change of motion would it take to cause the moon, roughly 250,000 miles away, to go around the Earth once every, roughly, 27 days?" Which is roughly the period of the moon's orbital motion. And he found that the force was roughly one thirty-six hundredth of the force that acts on the apple. And he calculated that the moon was about 60 times as far from the center of the Earth as the apple is, and that all works with the mathematics of universal gravitation.

So, Newton came up with the law of universal gravitation and demonstrated that it indeed seemed to be true with regard to the apple and the moon. The apple is close to the Earth and falls with a large acceleration. The moon is much further away and "falls"—again, in the sense that it is falling out of the straight line path it would otherwise take in describing a circle—with an acceleration that is weaker in just the exact relation that Newton's law predicts, given how far away it is.

By the way, universal gravitation really has been verified in the laboratory with very delicate apparatus. We can now do experiments where objects like this tennis ball and this basketball—we typically use lead objects or steel objects or something because they'd be more dense and more massive—over scales like this we can measure the strength of the gravitational attraction. In fact, that was first done in the 1700s by Cavendish. So this is an experiment we could actually do. We can measure the universal gravitation between two ordinary sized objects. It's very, very small and very difficult to measure.

When we accumulate a lot of matter in one place as we do with the planet Earth, star, then gravity becomes a big significant force. It's actually rather weak, but it's one of the fundamental forces that makes the universe work—one of the fundamental pushes and pulls.

Newton realized something else with his universal gravitation. He realized the possibility that we could have objects in orbit. And Newton had a thought experiment. He imagined the Earth with a giant mountain on it. Again, one of these thought experiments where you think about what would happen if you could do something, which maybe it's not so easy to do. So we have a picture here of the Earth with a giant mountain on it. We're going to throw a ball, horizontally, off that mountain, and it's going to fall to Earth. Very familiar with that. If I threw that tennis ball, it would fall to the ground pretty soon. If I threw it horizontally, it would describe a curved path. Well, this ball we throw off this high mountain, also describes a curved path. The only difference is, the Earth is curving away beneath it. So it perhaps goes in a slightly different route and lands somewhere else than it would if the Earth had been completely flat. But it goes around and eventually hits the Earth. If we throw it faster, it goes farther before it hits the ground. Just like if I threw a tennis ball horizontally in this studio, it would go further before it hit the ground. But if I throw it fast enough—and here's what Newton's genius again realized—it could be falling toward the Earth at exactly the same rate that the Earth is, if you will, falling away underneath it, because the Earth is curved, and it would go in a circular path and come right back to its starting point. And, absent air resistance or other factors that would tend to slow it down, that ball that we threw would continue in a circular orbit around the Earth forever. It would never stop. In fact, that was Newton's realization. That we could, perhaps, put artificial satellites in orbit. And, today, of course, we do that all the time. And we rely on those satellites for weather prediction and communication and television and navigation—global positioning system receivers in our car these days. All these are examples of satellite technologies. And it was Newton's genius to realize that such satellites were possible. He didn't have the technology to put them up. And it was not until the middle of the twentieth century that we were able to do that. But we routinely put satellites up. And we not only put them in circular orbits, we also put them in elliptical orbits. And Newton also anticipated that.

Newton took his newly discovered universal law of gravitation and he said to himself, what kind of paths will objects follow. What kind of paths, particularly, will a planet follow as it moves in the vicinity of the sun subject to this force, which gets stronger as it gets closer to the sun and weaker as it gets further away. And Newton said, "How do I figure that out?" Well, he discovered he didn't know how to figure that out so he had to invent the branch of mathematics that it takes to figure that out. And that branch of mathematics is called calculus. So Newton had a real incentive to invent calculus, which was also independently invented by Leibnitz.

But Newton invented calculus because he wanted to understand how his universal gravitation would describe the motions of planets. And lo and behold, out of that mathematics came the prediction that the motions of planets should be elliptical orbits. The sun should be at one focus of the ellipse. And the laws that Kepler had described about how fast the planets move around the sun or how much area they sweep out and, the period in relation to the radius of the orbit, size of the elliptical orbit, all those laws that Kepler had described with no physical basis, fell out of Newton's universal gravitation when he applied the calculus and asked the question, how will a planet move in the vicinity of the sun.

So Newton had brilliantly verified Kepler's observations that the planets moved in elliptical orbits. Newton had explained it. He had said the law of universal gravitation—that all objects attract with this force that depends on distance—explains that. So Newton had made a brilliant explanation.

Well, Newton had done more than that. He made a philosophical leap. Because in recognizing that the motion of the moon and the motion of the apple are the same thing, Newton had forever unified terrestrial and celestial motion. That distinction that I emphasized with the ancients, that distinction is gone. No longer do we have a separate law of physics for the crude, base Earth and a different law for the perfect, ethereal heavens up here. They're all described by the same law. It happens to be Newton's universal law of gravitation, coupled with his three laws of motion. They describe the motion of the apple on Earth. They describe the motion of a ball I throw here in this room. They describe the motion of the moon around the Earth, the Earth around the sun, the sun around the center of the galaxy, and an artificial satellite around the Earth. They described motion in the heavens, and they described motion on Earth. And those realms are no longer distinct.

Before we go on, let's just take a moment and review the evolution to this point because now we're quite advanced, beyond the stage of Aristotle. We now have a universal physics. We have laws that seem to govern the behavior of motion everywhere in the universe. Something very different from what Aristotle posited. And those laws are thought out mathematically. They're written in rigorous mathematical language. So we can make predictions, and those predictions agree with reality, as in the case of Newton comparing the acceleration of the moon and the acceleration of the apple, or Newton predicting the motion of a planet and discovering it was exactly what Kepler had described with no basis for understanding why that description was true. Newton has now explained it. So we have a very powerful statement about how the world works, and it seems to apply everywhere in the universe.

So, let's take a look at the history that got us there. We had a natural state of motion, in the case of Aristotle in around 350 B.C., that was rest on Earth. Terrestrial motion, in a natural state, was being at rest on Earth. In the celestial motion, the natural state of motion was circular. And Earth was at the center. Ptolemy embellished that a little bit with the circles on circles, but it was the same basic idea. Natural state of motion, terrestrial, is at rest. Natural state of motion, circular, is the Earth at the center.

We next went to Copernicus, in around 1540, he didn't change the ideas of motion on Earth, but told us that the natural state of motion in the heavens was still circular, but the sun was at the center. So, a step away from the Earth being special, but a conservatism also in maintaining the notion of circular motion.

Galileo comes along, working around the year 1600 and made a radical change in the natural state of motion on Earth. For Galileo, for the first time, the natural state of motion was uniform motion in a straight line. Galileo stated the law of inertia, which became Newton's first law of motion, which was then embellished in Newton's second law to give it a quantitative substance and talk about how forces, how pushes and pulls, change one's state of motion. Galileo, also, in the celestial realm, set the groundwork for

Newton by saying the celestial realm is imperfect. I see a miniature planetary system in Jupiter and its moons. I see blemishes on the sun. I see sunspots. I see Venus's phases. And I see the size of Venus changing with phase. And I interpret that to be evidence that, in fact, the Copernican system with the sun at the center seems to be correct.

Then along comes Newton, working around the year 1700, a little earlier, a little later. He says terrestrial and celestial motion, for the first time, obey the same physical laws. Same laws. One universal physics that applies everywhere in the universe. So Newton has done the first of the several great syntheses we're going to talk about in this course. Taking seemingly disparate regions, disparate areas, and disparate fields—in this case astronomy and terrestrial physics—and uniting them. They're one. They're one because they obey the same laws. They can be understood as aspects of the same thing. What Venus is doing in its motion around the sun can be understood as the same thing as what the apple is doing when it falls on Newton's head, or what the tennis ball is doing when you hit it on your game of tennis on the cruise ship. Major philosophical shift. One universal set of laws of physics. And there is something very special about those laws—the Newtonian laws of motion. The Newtonian/Galilean scheme. And I want to take the last few minutes of this lecture and emphasize that very important special point.

Let's ask the question, for whom do these laws of motion work? For whom are Newton's laws true? Just for Newton? No. They're true for anyone. Are they true just on Earth? No. Look at the creature on Jupiter, if there is one, who sees these moons going around. That creature, if a Jovian fruit falls on their head, if there's such a thing as a Jovian, they're going to make the same conclusion. The same laws work on Jupiter as they work on Earth. The same laws work on Venus. We already know this. We were doing this tennis thing on Venus bouncing a tennis ball back and forth. We understand that the laws of physics that work on Earth work elsewhere. And, equally importantly, they work regardless of your state of motion. That's the important point. Not so much your place, but your state of motion.

And I now want to get a little more specific about that and be a little bit more careful when I make that statement. The statement I want to make is this: The statement is that the laws of motion are the same for everyone. At this point I have to qualify that statement somewhat further. And until the year 1914, we're going to keep that qualification on. That qualification applies to Newtonian physics. And it also applies to Einstein's special relativity. It's an important qualification. And it says this: The laws of physics are the same for everybody in uniform motion. Everybody who is in uniform motion. If you're moving in straight line motion at constant speed and you do experiments to determine what are the laws that govern physics, you'll get the same, govern motion anyway, you'll get the same answers that I do if I'm sitting at rest on Earth, or if I'm in a spaceship in interstellar space—without my engine going so I'm not accelerating, so my motion isn't changing—or I'm on that planet, in that distant galaxy, moving away from Earth at 80 percent of the speed of light. As long as we're moving uniformly, we'll get the same results for the laws of motion. We'll all agree that Newton's laws work, that universal gravitation works, and that they have exactly the same form. We could read the same textbooks, and do the same physics, and we'll get it all right.

You may be wondering to yourself, "Wait a minute. Earth isn't in uniform motion. It's spinning. And it's going around the sun." And you're right. Earth isn't exactly in uniform motion. But the spinning of the Earth is slow enough, and the motion around the sun, well, that's a little more subtle. In the context of Newton's theory, I'm going to say that's also slow enough. That Earth is approximately in uniform motion. It's not in a perfect state of uniform motion. And later in the course, in Lecture Eleven, no, Lecture Thirteen rather, we are going to be very, very concerned about how we find a perfect state of uniform motion. And I'll give you an answer to that at that time. But for now, accept that Earth, and the other planets, and the sun, which is going in a very big, slow orbit every 250 million years around the center of the galaxy, these are essentially in uniform motion. Close enough. Newton's laws hold almost perfectly. Go on a rotating merry-go-round, or in a car rounding a corner at high speed or something, or in a car that's accelerating, or a jet plane when it's heading down the runway, and Newton's laws distinctly don't apply in those environments because those are not situations where you're in uniform motion. But here at the surface of the Earth, we're close enough. Not perfect. But close enough. And so, we have this principle that the laws of motion are the same for anyone in uniform motion. And I spent a good part of the end of the first lecture convincing you that you, in your gut, know that. And it makes a lot of good sense to you. It would be a ridiculous universe in which we had these nice, simple laws of motion that worked only here on Earth. And if you go to Venus, there's something more complicated, and if you go to the distant planet, on that distant galaxy, it's even more complicated. No way. It makes good physical sense. The laws of motion, and I'm going to emphasize motion, I'm going to use the word motion, instead of physics, for the next few lectures, because at the time of Newton, motion was the only branch of physics really known to any great extent. We knew a little bit about optics. A little bit about heat. But not a whole lot. Motion. It's the laws of motion that are the same for anyone in uniform motion. And that statement has a name. It's called the *Principle of Galilean Relativity*. Sometimes called *Newtonian Relativity*, because it flows out of the ideas of Newton and Galileo.

This is the first time we've seen the word relativity in a real physics context. I've talked about relativity before, how this is going to be a course, largely, about Einstein's relativity. Here's that word "relativity" cropping up. Einstein's relativity is actually not such a new idea. It's actually a very old idea recycled and embellished a little bit. We're going to come to understand that in a few lectures. The original relativity principle was Galileo's and Newton's. It's the Principle of Galilean Relativity, and it states what I just said. It says the laws of motion are the same for anyone provided they're in uniform motion. And that's a kind of formal way of saying it.

There are lots of other ways of saying it, of which I'll just give you a few examples. You can play tennis on a cruise ship, or on Venus, or on this distant planet orbiting a star in a galaxy that's so far away that it's moving away from us at 80 percent of the speed of light. These all make good, gut sense. As long as the motion is uniform. If the cruise ship gets hit by a tidal wave, things get real different. If Venus crashes into another planet and breaks in pieces, or something, or if you're in an airplane and it encounters turbulence, then it doesn't apply because those aren't uniform motion. But you could play tennis on a cruise ship, or on Venus, or on that distant planet. It's all the same. Because the laws of motion work for you just as well there and in that state of motion as they do here on Earth. Other ways to say it: Only relative motion matters. If you say to me, if you're driving down the road at constant speed in a straight line and I'm standing by the roadside, I'm prone to say you're moving. But you have every right to say, no. I'm just sitting here in my car, you're moving the other way. We tend to be very prejudiced to thinking the Earth is this special place. And so whoa, it's really the person who knows Earth that is at rest and the other person moving. But what Galilean relativity, not anything fancy, and Einsteinian, just old-fashioned Galilean and Newtonian motion theory tells us is that you really can't distinguish. We can't say one person is moving and one is not. Anyone in uniform motion has equal access to the laws of physics. They are equally good for anyone in uniform motion, and so no one can say, absolutely, I am moving and you are at rest, or I'm at rest and you're moving. Those statements make no sense.

What we can say is I'm moving down the highway at 50 miles per hour relative to the ground. Or, you're moving—you standing at rest relative to the Earth—are moving past me at 50 miles an hour the other way. We can talk about relative motion. That's why it's a relativity principle. Because it says relative motion is a meaningful concept, but absolute motion is not. Only relative motion matters. That's what that statement means. You can eat dinner on an airplane because you don't worry about the fact your food is moving at 600 miles an hour. As long as the plane is moving uniformly, no problem. You put your fork in the food, just like you would at home, and you eat the food. Now, you hit turbulence and the food goes flying all over the place. That's different. You're no longer in uniform motion and the laws don't apply there. I am moving is a meaningless statement. You can say, I'm moving relative to you, but you can't say, I am moving. That's a meaningless statement. And to put it more succinctly in the final terms, and we're going to come back to the statement again and again in the next few lectures, there is simply no experiment you can do involving the laws of motion that will answer the question, am I moving? Put you in that airplane. Pull down all the shades so you can't see the Earth going by, and there is no way you can tell, for sure, whether or not the airplane is moving or not.

There's no experiment you can do with tennis balls and billiard balls, and throwing things around and measuring forces and pushes and pulls that will give you any sense of physics other than Newton's laws of motion. There's no experiment you can do that will answer the question, am I moving? And that's where we sit at Newton's classical physics with a law of relativity, a principle of relativity that says, uniform motion is the natural state, and motion doesn't matter. As long as you're in uniform motion, we all see the same laws of physics.

Let There Be Light!
Lecture 4

This lecture will be an attempt to give you, in half an hour, what I normally teach in an entire semester, namely, the subject of electricity and magnetism.

The study of motion is not all there is to physics. The ancient Greeks and Chinese knew, respectively, about electricity and magnetism. By the 18th century, scientists engaged in serious experimental studies of electricity and magnetism. The two phenomena turned out to be related; magnetism can produce electrical effects and vice versa.

Early experiments with static electricity showed that electric charge is a fundamental property of matter. Like charges repel; opposites attract. Early experiments with magnets showed that all magnets have two poles. Like poles repel; opposite poles attract.

The *field concept* describes electric and magnetic interactions in terms of invisible *fields* that exert forces on charges and magnets. This view contrasts with the earlier *action-at-a-distance* concept, in which electric charges and magnets somehow "reach out" across empty space to influence other charges or magnets. Experiments led to two laws showing how electric and magnetic fields arise from electric charges and magnets, respectively. The field concept can also be applied to gravitational attraction.

Maxwell showed that his equations implied the existence of *electromagnetic waves*, structures of electric and magnetic fields that travel through empty space.

Electricity and magnetism are intimately related; hence, *electromagnetism*. A moving electric charge produces magnetism—a phenomenon at the basis of many technologies, including electric motors. Such a moving electric charge is constituted by the electrons moving in atoms. Changing magnetic fields produce electric fields. This process is an integral part of how computer disks, audio and videotapes, and electric power generators work.

In the 1860s, Scottish physicist James Clerk Maxwell suggested that if magnetic fields produce electric fields, why not the opposite as well? Maxwell incorporated this suggestion into the laws of electromagnetism, completing four equations that describe all electromagnetic phenomena. He showed that his equations implied the existence of *electromagnetic waves*, structures of electric and magnetic fields that travel through empty space.

Marchese Guglielmo Marconi.

He calculated the speed of such waves from quantities appearing in his equations and found it was equal to the known speed of light! Then he concluded that light must be an electromagnetic wave, making optical science a branch of electromagnetism. Other electromagnetic waves now known include radio, infrared, ultraviolet, x-rays, and gamma rays, which differ in their frequencies (and therefore wavelengths).

In 1887, Heinrich Hertz generated and received electromagnetic waves in a laboratory. In 1901, Guglielmo Marconi transmitted radio waves across the Atlantic Ocean. ■

Essential Reading

Hey and Walters, *Einstein's Mirror*, Chapter 2, through p. 29.

Mook and Vargish, *Inside Relativity*, Chapter 2, Section 7.

Suggested Reading

Casper and Noer, *Revolutions in Physics*, Chapter 12.

Einstein and Infeld, *The Evolution of Physics*, Chapter 2, through p. 33.

Questions to Consider

1. Why was Maxwell's assertion that a changing electric field should produce a magnetic field crucial to the existence of electromagnetic waves?

2. Maxwell's realization that the phenomena of optics can be explained by electromagnetism is an example of scientific synthesis, in which hitherto unrelated phenomena are found to be related. What are some other examples of such syntheses?

Maxwell's Equations

And God said...

Equation	What it says...
$\nabla \cdot \mathbf{E} = \dfrac{\rho}{\varepsilon_0}$	How charges attract/repel
$\nabla \cdot \mathbf{B} = 0$	No isolated magnetic poles
$\nabla \times \mathbf{E} = -\dfrac{\partial \mathbf{B}}{\partial t}$	Changing magnetism produces electricity
$\nabla \times \mathbf{B} = \mu_0 \mathbf{J} + \mu_0 \varepsilon_0 \dfrac{\partial \mathbf{E}}{\partial t}$	Changing electricity produces magnetism

...and there was light.

Let There Be Light!
Lecture 4—Transcript

Welcome to lecture number four: "Let There Be Light!" This lecture will be an attempt to give you, in half an hour, what I normally teach in an entire semester, namely, the subject of electricity and magnetism.

If motion were all there were to the study of physics, our story of the history of physics before relativity would be essentially over. But there's a whole other branch of physics—electricity and magnetism—that actually was known about for a long, long time. The ancient Greeks coined the word electron for the substance amber, which, because it's a very good insulator of electricity, somewhat like Styrofoam, tends to exhibit a lot of static electricity and things cling to it the way socks cling to each other today when they come out of the dryer, for example.

The ancient Chinese also knew about magnetism. And magnetic compasses were already in use by the twelfth century. Nevertheless, in terms of a rigorous base of knowledge on electromagnetism, electricity and magnetism came after mechanics, after the study of motion. And we need to take a look at electricity and magnetism because relativity, really, more than most people realize, is a direct outgrowth of the study of electricity and magnetism, and the questions that were asked about it in relation to the kinds of questions we've already asked about the study of motion. So, we want to spend most of this lecture looking briefly at the subject of electricity and magnetism, and, in the end, coming to a very important conclusion from that subject, and a conclusion that you'll need to realize is true. You'll need to appreciate it. And you'll need to grasp its significance if we're going to go forward and understand where relativity comes from.

So, I want to begin with just a brief description of some early experiments with electricity. People studied things like they would rub a piece of glass with the fur of a cat. And something would stick to it, little particles of dust or paper would stick to it. You've seen things like this happen. You rub a balloon and you can stick it on the wall. This was the phenomenon of static electricity.

And through a number of experiments like this people gradually realized that a property called electric charge is a fundamental aspect of matter. And that electric charge comes in two varieties. And it was Ben Franklin, the American statesperson, who actually studied this in some detail and coined the terms positive and negative for those two kinds of electric charge. There's nothing positive about positive charge, and nothing missing about negative charge. They're just two words for two basic, fundamental properties of matter.

If you ask me as a physicist what charge is, I'm going to have to say I really don't know. It's a fundamental property of matter. The most basic particles we know about—the electrons and the quarks, more about them later in the course—they have electric charge. Some have positive charge. Some have negative charge. We don't know what charge is. It's a fundamental property.

If you ask me what mass is or weight, I don't really understand what that is either. But I know what it is from feeling heavier objects and lighter objects. And, similarly, from doing experiments with electricity and magnetism, and from building electromagnetic devices, we have a feel for what electric charge is. It's a fundamental property of matter. And one important aspect of electric charge is that like charges—imagine these two tennis balls are two identical electric charges—like charges repel. So it's hard to push like charges together. They want to fly apart. There's a repulsive force between them. It's actually somewhat similar to the gravitational force that Newton came up with centuries ago. It falls off in strength with the inverse square of the distance between the objects. So, when they're far apart they don't feel much force. When they're close together they feel a very big, strong force. The force is repulsive if the two charges are the same. The force is attractive if the two charges are different. A positive and a negative charge will be attracted together. And, absent any other forces, they will simply come together. So, like charges repel. Opposite charges attract.

That's not something we can have with gravity, because there's only one kind of matter. One kind of mass. It's all positive, basically, and masses all attract each other.

But electricity works a little bit differently. There are two kinds of charge. Opposite charges attract. Like charges repel. So, that was basic knowledge of electricity.

And in the 1700s and into the 1800s, people experimented with other aspects of electricity. They developed batteries. They could make electric current flow through wires. Frogs legs twitched when touched with pieces of metal. And that led to the discovery of the battery and to electric currents and everything we know today about electric circuits. So, there is a lot of work being done in the 1700s, early 1800s, on the study of electricity.

Now, I'm using the word electricity here. The title of my lecture, "Let There Be Light!" I then introduced as a lecture about electromagnetism, or electricity and magnetism. But, so far, I've said only electricity.

People also knew about magnets and magnetism. And I have here a couple of magnets, and I just want to review what you probably already know about magnets. The typical magnet has two ends called poles—north and south. And magnets have the properties that north and south poles are attracted. So, if I bring the north pole of this magnet and the south pole of this magnet together, they come together. And it takes work to pull them apart. And then that force falls off rapidly as I move them further and further apart. Here, I feel no force between them. Here, I just begin to feel an attractive force. Here, they come together, and they come very closely together, and are attracted.

If I turn one of the magnets around, so I'm putting two south poles together, now I get a repulsive interaction. Again, opposites attracting, likes repelling. So here we go—two likes. I try to push them together, and it gets very hard to do. They want to fly apart. So we have these magnets with their poles— like poles repel, opposite poles attract—similar to what happens with electric charge.

But magnetism is not electricity. Magnets are not electric charges. The ends of the magnets are not charged. We seem to have two different phenomena. We have the phenomenon of electricity that involves this fundamental property of matter—electric charge—and we have the phenomenon of magnetism that seems to involve things like magnets. And, by the way, there are naturally occurring materials that are magnetic.

Before I go on and introduce the idea of electromagnetism, let me say a little bit more about how we conceptualize this interaction between either electric charges or electricity or magnetic pulls. Here are two electric charges. One of them is positive and one of them is negative, say. As I described their interaction, I said, okay these two charges are attracted to each other. The green one exerts a force on the yellow one. The yellow one exerts a force on the green one. Newton's third law, by the way, tells us that whatever force the green one exerts on the yellow one, the yellow one exerts an equal and opposite force on the green one. That was Newton's third law. And they come together.

There's a problem with that, a philosophical problem, and also a mathematical problem, with trying to describe the interaction that way. When I say the green charge exerts a force on the yellow charge, I'm saying that somehow this green charge reaches out across empty space and grabs onto this yellow one. Now, how does that happen? Newton called that "action at a distance." How does "action at a distance" happen? How does an object that's way over here exert an influence on one that's over here? And a particularly disturbing question about that issue is this: What if this one suddenly disappeared? Suddenly disappeared. Ceased to exist. Would the yellow one know that immediately? Would it know immediately to stop feeling an attractive force in this direction? Or wouldn't it?

And if you want to go to a bigger picture, imagine we are talking about gravity. And this is the Earth. And this is the moon. And if the Earth suddenly disappeared, would the moon know immediately to stop moving around in its orbit? How did that information get to it? It's a problem. It's going to be a big problem when we get to relativity and we find we don't want there to be ways where information can get from one place to another instantaneously. So this description of the electric interaction or the magnetic interaction in terms of a force that somehow reaches out across a distance and pulls on the other object—there is something a little bothersome about it, philosophically.

There's also a bit of a problem mathematically, which really isn't as deep a problem, but still bothered people. And that is this: Here's this green charge—positive or negative, I haven't decided which, but it's one of those two. And it's attracting this yellow charge, which is the opposite. What if I took the yellow charge away and put another green charge there? Well, then it would be repelled. And I'd have to describe the interaction all over again. What I'd really like to know is, what electrical influence does this charge exert on objects I might put anywhere in its vicinity? And there's a way to describe that with a new concept. And it's a somewhat abstract concept, but it's important to understand it for purposes of this lecture and then what follows. And the idea is the field, the *field of force*. And I'm going to introduce the field of force, first in terms of gravity and then go back and describe it in terms of electricity and magnetism.

Let me use gravity as an example. This is now not going to be a charge but what it is—a tennis ball. Again, there are two ways of describing the interaction between this tennis ball and the Earth. One is to say the Earth, way down there, all the mass that makes up the Earth—the stuff right below the floor, Mount Everest, on the other side of the planet, all the water in the Pacific Ocean, all the water in the Atlantic Ocean, the continent of North America, the continent of Africa, all the ice in Antarctica—all those aspects of the Earth reach out and pull on this tennis ball, and somehow result in a force, which beautifully—and Newton proved this to be the case—is directed toward the center of the Earth. And as a result, the tennis ball falls. The Earth reaches out, pulls on it, and down it goes. Not just the Earth right here, but all parts of the Earth. And similarly, by Newton's third law and by the universal law of gravitation, the tennis ball reaches out on all parts of the Earth and pulls on them. And the net result is a force upward on the Earth.

It's just the Earth is so massive. Force equals mass times acceleration. Its motion hardly changes. When I let go of the tennis ball—because of the force of the tennis ball and the Earth, the tennis ball is very light—its motion changes fairly dramatically when I let it go, and it falls to Earth.

But they're actually both pulling on each other. And the action at a distance view would say the Earth reaches out and grabs on the tennis ball. Earth reaches out and grabs on the moon. Earth reaches out and grabs on a satellite and exerts a force that swings in its orbit.

The field concept says, "No, there's a more sophisticated way of looking at that." And the more sophisticated way is to say the Earth creates in its vicinity a thing called a field of force. In this case, a gravitational field. At every point in the Earth's vicinity there exists this invisible entity. The gravitational field. And it has the property that if I put matter there, that matter will experience a force. The stronger the gravitational field, the stronger the force it will experience. And the bigger the mass of the object, also the bigger the force it will experience. And the direction of the gravitational field—it has a direction—is directed toward the center of the Earth. So there's a gravitational field right at this point, and it's directed toward the center of the Earth. And there's a gravitational field right at this point, and it's directed toward the center of the Earth. And there's a gravitational field above Antarctica, which is directed in a different way—toward the center of the Earth, but that's a different direction, at Antarctica. And you could imagine the entire Earth surrounded by these invisible arrows, almost, that describe the strength of the gravitational field and tell us what would happen if we put an object there.

What do we gain by this picture? Well, one thing we gain is a description of the Earth's gravity independent of particular objects like tennis balls, or the moon, or satellites that I might choose to put there. The other thing we gain is a way around that philosophical problem I described earlier. Because if the Earth suddenly disappeared, this description says that doesn't mean the gravitational field that the Earth created in space around it disappears immediately. It takes time for that information about the Earth disappearing to get out to the moon. It takes time for the moon suddenly to stop moving in its circular orbit and go in the straight line it goes in if there's no force acting on it.

So, the field concept describes an interaction between two objects—a gravitational interaction in this case, an interaction involving gravity—in a very different way. It says a gravitating object, like the Earth, creates a gravitational field all through space, surrounding it. And objects respond to the gravitational field, not somewhere far away, but in their immediate vicinity. This tennis ball doesn't need to know about the Earth and Mount Everest. It needs to know about the gravitational field they've created at this point in space. The field idea is a local idea. It says objects behave the way they do because of their local environment. Now, that local environment is shaped by the distant parts of the Earth. But it's nevertheless the local environment to which the tennis ball responds.

It's the same thing with electricity and magnetism. Let's go back again to these being two charges. The old fashioned action at a distance description would say the green one reaches out, exerts a force on the yellow one, and pulls it. The yellow one reaches out, exerts a force on the green one, pulls it. They come together. The field concept says nope. The green one creates a field at all points in space. In this case, an electric field, because this is electrically charged. It's a force field. And if I put some other charged object in its vicinity, it will experience a force that depends on the electric field right where I put it. Now that electric field arose, came about, because of the presence of the green one. But, again, there's that philosophical distinction. The force on the yellow one arises because of what's happening in its immediate vicinity. Namely, there's an electric field of force there. I put any charge there, it will experience a force that depends on how charged it is and on how strong that field is.

Similarly, if I want to talk about magnetism, I can say that this magnet creates a field of force. And if you've ever seen the demonstration where a magnet is plunked down on a pile of iron filings, they line up to create a visual picture of that field of force. Which way it points. There's a field of force surrounding this magnet. If I put another magnet somewhere in that vicinity, that other magnet experiences that field of force and gets a force exerted on it. And that's a description of the magnetic interaction.

So, it's very abstract sounding. But we've gone from the point where we said two objects interact—magnetically, gravitationally, or electrically—because each one pulls or pushes on the other, to saying each object creates this mysterious field in space around it, and the other object responds to that field. Very abstract sounding. But it has a tremendous reality. Because as soon as you acknowledge the existence of these fields, and you study electricity and magnetism a little more, you find these fields, for instance, have energy.

They have substance to them, in a sense. We can't see them. We can't touch them. But they're there. And they actually are repositories of energy.

And we find other things about them. We find, in particular, that if we think about the fields, we come to understand a relationship between electricity and magnetism. A really deep and profound relationship. And I want to take a few minutes to describe briefly what some aspects of that relationship are.

It turns out that if I move an electric charge—again, here is my electric charge—if I simply move it along, that creates magnetism. In fact, what's going on inside this magnet is not that there are lots of funny little fundamental magnetic things. But, rather, that there are moving electric charges. There are the electric charges in the atoms that make up this material. And this happens to be a rod of iron, which I've painted one end silver to mark the South Pole and mark things on it. The atoms in iron, in particular, have a property that they all line themselves up so all their little electron motions are going around, basically, together. And that creates a big composite effect of electric charge—moving, which in turn, creates magnetism. So that's where a magnet comes from. It's really arising from moving electric charges.

Similarly, in an electric motor, which is a device you use every day—it runs your refrigerator, it runs your CD player, it runs the fans that keep you cool, it runs your air conditioners, it runs the subway trains you travel in, it runs electric trains, it runs the starter motor of your car, it runs your watch. I could go on, and on, and on, and on. Electric motors simply consist of coils of wire with electric currents flowing through them. They become magnets, and the magnetic forces that are so generated cause them to spin. There's another example where electricity and magnetism are related.

More fundamentally, there's a relationship directly between the electric fields and the magnetic fields. And the relation is something like this: If I have magnetism, if I have a magnetic field arising from, say, a magnet. For example, I'm going to grab this magnet again. This magnet creates a magnetic field in the space around it. If I change that magnetic field in any way, for example by moving the magnet around or something, the fact is that

that creates electricity. It creates an electric field. The fields that I've talked about, these abstract entities, have existence of their own, basically.

Changing a magnetic field by moving a magnet creates an electric field. I've got a little demonstration of that here with a device I borrowed from my daughter. It's a flashlight. But it's a very special flashlight because it has no batteries. And I'm going to run this flashlight by cranking the little crank at the bottom. As long as I do that, the flashlight is lit. Now, what I have here is a miniature electric generator. The electricity that runs your televisions, your computers, your refrigerators, all your household lighting, most of our industrial machinery, is generated by large-scale versions of this. And it consists of nothing more than a coil of wire in the presence of a magnet. The coil of wire spins. It senses changing magnetic field, because it's moving in the vicinity of this magnet. And, consequently, changing magnetism creates an electric field. And there are electrons in that wire and they respond to the electric field. They feel an electric force. And they're driven through the wire and make the electric current that ultimately lights that light bulb. This little device could be an entire course in electromagnetism all itself.

But the point I want to leave you with from this is there is an intimate connection between electricity and magnetism. And the connection is that changing magnetism makes electricity. To be a little more precise about that, a changing magnetic field makes an electric field.

There are many ways to change the magnetic field. I can move a coil of wire around near a magnet. I can wave a magnet in the air like I was doing before. I can turn an electromagnet on or off and that makes a changing magnetic field. In all those cases, I produce electricity.

There are a lot of practical consequences of that. I already described electric generators. But you're all either watching, or listening to this course with video or audio tape. And what's going on there is the information of this course has been impressed as a pattern of magnetization on a thin strip of tape containing metallic oxides that are themselves magnetic. And what's happening in your tape player, whether it's video or audio, is that tape is being moved rapidly past a little coil of wire. And the changing patterns associated with the changing magnetic information on that tape is producing changing electric fields. And those changing electric fields are producing forces on electrons in the coil of wire, and those electrons are flowing in patterns that correspond to the visual appearance of what I'm doing and to the sound of my voice. And that's ultimately being amplified and displayed on your television screen and/or coming out your loud speakers. So audio and video tapes, computer disks, work the same way, your credit card stripe when you swipe your credit card and that magnetic stripe gives out its information, works the same way. The pattern of magnetization on that credit card is swiped past a little coil of wire and the changing magnetization makes electricity.

So a key thing you need to know, without knowing a whole lot more about electricity and magnetism, is that changing magnetism makes electricity. So where are we at this point? Well, we've really seen, sort of, three things. We've seen a brief discussion, not quantitative, no equations or anything, that describes how like charges repel and opposite charges attract. The force between electric charges. We've talked about electricity and described electricity briefly. We've talked about magnetism. We've described how there's an attractive force on opposites and a repulsive force on like poles. We've described magnetism, independent of electricity. So there are two things: I've described electricity, and I've described magnetism. And then I've talked about a relationship between the two, whereby changing magnetism produces electricity, and my little generator-powered flashlight is an example of that.

Something's missing there. Something's missing. I talked about electricity. I talked about magnetism. I talked about a relationship between the two whereby magnetism makes electricity. To be precise, changing magnetism, a changing magnetic field, makes an electric field. Something's missing.

In the 1860s, the Scottish physicist James Clerk Maxwell thought about this a lot. He thought there is something missing here. Now, remember I talked about in the first lecture about how science is a creative activity. It isn't just a bland, go in the laboratory and test your hypothesis kind of activity. Well, Maxwell was displaying the creative, almost aesthetic, side. He said there's something missing. There's not a symmetry there that I kind of expect to see in nature. We know about electricity. We got laws that cover that. We know about magnetism. We got laws that cover that. We know that changing magnetism produces electricity. Why doesn't changing electricity produce magnetism? And Maxwell says, "I think it should."

You know, if I was creating the universe, I would make it that way. It would have more symmetry. It would be more beautiful. There's nothing practical about this. There's nothing technological. This is aesthetics pure and simple. Maxwell said, "I think this should be the case." Now, he had some technical reasons, also, for proposing that. But there was no experimental evidence directly for it.

Maxwell proposed, "I think a changing electric field should also produce a magnetic field." And he formulated then out of what had been known before and this new idea, four laws of electricity and magnetism. And in his honor they are called Maxwell's equations. They represent the second of the great syntheses I've been mentioning in this course. In that synthesis, electricity and magnetism become joined in one subject—electromagnetism. And I'm going to show you these equations. Not because I want to do lots of mathematics. But because to a physicist they're so beautiful.

So here's a picture showing you Maxwell's equations. On the left, I've got the equations. There are four of them. Don't worry about the funny looking symbols. That's not what I'm trying to impress you with. Although I am trying to impress you with the fact that those four equations tell us everything we physicists need to know about electricity and magnetism. At least, at the level before quantum physics becomes important. In the whole macroscopic realm, from things the size of atoms and on up practically, this describes, basically, everything there is to say about electricity and magnetism.

And there are four of these equations. The first one talks about electricity alone. You see the "E" in the equation, that's for the electricity. You see the "B," sorry about the terminology, but that's for magnetism. So what do these equations say? The first one talks about how charges attract and repel. The second one talks about magnetic poles. They always come in north/south pairs. We won't get into that. It's an interesting fact, has some profound implications. But that's what that one says. It's the last two that are important to us.

The third one says that a changing magnetic field produces an electric field. That funny symbol on the right, with the "B," magnetic field, and the "*t*," time, talks about changing magnetic fields. And it says that makes electric field. You see that on the left.

The bottom one says what we knew for a long time. Namely, that electric charges in motion create magnetism. That's what the "J" stands for. But it also tells us that changing electric field makes magnetic field. There's the "E" and the "*t*" on the right. Electricity changing with time. There's the "B," magnetism, on the left.

What's the point of all of this? If you don't know anything else from either this lecture or from a course you might take in electricity and magnetism, come away understanding that changing magnetism makes electricity. And changing electricity makes magnetism. Because Maxwell then went ahead and said, "What are the consequences of this fact?" And he came up with a brilliant prediction. He said, "Look, suppose I had some changing magnetic field. I might make it by taking a magnet and wiggling it a little bit or something. Well, now I have changing magnetic fields here. What do they produce? They produce electric fields. And the likelihood is, depending on how I wiggle this stuff, unless I wiggle it in a very smooth, regular, continuous, steady, fashion, those electric fields will also be changing. And, therefore, what do they produce? They produce changing magnetic field. And what do those changing magnetic fields produce? They produce changing electric fields." And so on.

And Maxwell envisioned, then, there could be a structure of—started out with either electric or magnetic fields, I don't care which—changing, creates changing the other field back and forth. Changing electric, changing magnetic, changing electric, changing magnetic, changing electric. And the thing leap frogs itself through space. It's acquired an independent existence. And it carries with it electromagnetic energy. And it propagates through space as a wave.

Maxwell predicted the existence of electromagnetic waves. And he did something else. He predicted the speed that those waves should travel at. If you look at my picture of Maxwell's equations again, you'll see some funny little Greek letters. There's a mu (μ) and an epsilon (ε). Those are numbers that describe how electricity, the strength of the electric and magnetic fields, behave. Basically, how strong electricity is. How strong magnetism is. They were numbers that were arrived at by laboratory experiments. Doing things like taking a couple of charged objects and measuring how strong the force is. Or taking a current flowing through a wire, an electric current flowing through a wire, and seeing how strong a magnetic field it produces. And so on.

Nothing to do with optics, among other things. Maxwell calculated that the speed of his electromagnetic waves depended on those numbers and nothing else. And he calculated that speed. And it came out to be the known speed of light. Now until this point, nobody knew what light was. People had speculated since about the early 1800s that it was a wave. Newton thought it was a particle. Much more on that later in the course.

But it was known by Maxwell's time that light was a wave. Nobody knew what it was a wave of. All of a sudden, Maxwell, putting together this theory of electricity and magnetism, predicts that there should be waves of electromagnetism. And they should propagate through empty space. They should travel through empty space carrying energy with them—electromagnetic energy with them—and they should do so at the known speed of light. And it immediately became obvious that light was a form of electromagnetic wave.

And Maxwell asked other questions about electromagnetic waves. Are there other forms other than light? And, yes, there are other forms. There are forms like radio waves, and infrared waves, and ultra-violet waves, and x-rays, and gamma rays. These are all electromagnetic waves. They are all basically identical. All that differs in them is how rapidly they vibrate. Light waves vibrate about 10^{14} or 10^{15} times a second. That's a one with 14 or 15 zeros after it, times a second. Radio waves vibrate maybe a million or a hundred million times a second. FM radio is about a hundred million times a second. That's why you tune your dial to a hundred megahertz or 101.5 or whatever. AM radio, tune your dial to 850. That's 850,000 vibrations a second. All those waves are the same. They're basically electromagnetic waves. And they consist of changing electric and magnetic fields. And the important point for our course here is that Maxwell's equations predict the existence of those waves. And equally importantly, predict the speed that those waves should go at.

Today, we know a whole spectrum, as I say, of electromagnetic waves. A whole spectrum from radio waves—the slowest vibrating ones—but they all go at the same speed. It's just a question of how fast the vibrations of the electric and magnetic fields are occurring. The waves themselves are all moving at exactly the same speed. It's the known speed of light. About 186,000 miles a second, 300,000 kilometers a second, or to put it in terms that are better for people who design electric circuits and so on, it's about one foot per nanosecond. So in a nanosecond—a billionth of a second—light goes about one foot. That, by the way, sets a limit to the speed of our computers. Because computers need to move information at speeds that are close to, but can't be faster than, the speed of light for reasons you'll soon see. And, therefore, they can't be very big. If they were too big, they'd be too slow in terms of the time it takes them to get information.

So Maxwell has proven the existence of electromagnetic waves. People set out to prove his hypothesis. And Heinrich Hertz, in 1887, succeeded in generating electromagnetic waves on one side of a laboratory and, with another apparatus on the other side, receiving those waves. So he demonstrated that Maxwell was correct by making, if you will, artificial electromagnetic waves as opposed to the natural ones that are all around us.

That was in 1887. By 1901, the Italian, Marconi, transmitted electromagnetic waves across the Atlantic Ocean for the first time. And so, the electromagnetic wave discovery became a very, very practical thing. And today, electromagnetic waves are used in our cell phones. The mouse I've been clicking to operate my computer visuals over here is transmitting its clicks by electromagnetic waves, by radio waves in that case. All of radio, television, microwaves, infrared devices, all these things, lasers, all these things are generating and using electromagnetic waves.

Electromagnetic waves are ubiquitous. And the important point for us is they all arise in exactly the same way. You somehow make an electric or magnetic field change—maybe by electrons going back and forth in a wire antenna, maybe by doing something to the electrons in atoms, as in a laser—whatever you do, once you've started that process, the electromagnetic wave propagates away, and at least if it's in vacuum, it does so at one unique speed. No matter what kind of wave it is, radio, television, infrared, visible, whatever, x-rays, gamma rays, it goes at the speed which we designate by the letter "c"—186,000 miles a second, a foot per nanosecond, 300,000 kilometers a second.

In fact, if you're religious, and I'm not trying to be religious here, but if you were religious, I think you would embellish Maxwell's equations in the following way. And this way involves the title of my lecture here. Maxwell's equations, if you were creating the universe, and you wanted a universe in which there was going to be light, from what we know of light, you would need to create a universe with electricity and magnetism. A universe that obeyed Maxwell's equations. So if you want to think of it in sort of terms of Genesis, and God said this—all those equations—and there was light. Because that's where light came from. Light comes from electromagnetic waves. Light is electromagnetic waves. It's a result of electromagnetism.

Where are we going with this? What I want you to take away from this lecture is the idea that electromagnetic waves are a necessary consequence of the branch of physics called electromagnetism, and that electromagnetic waves, as predicted by that branch of physics, travel with a known speed. It's the speed of light. It's a known definite number. There's no exceptions to that. Electromagnetic waves, in vacuum, travel with this known speed, which from now on we'll simply call c.

Speed *c* Relative to What?
Lecture 5

I want to introduce a new concept. It's a concept called a frame of reference. What's a frame of reference? Well, you'd probably say it's something like your point of view. And in physics it is something like your point of view. It's basically sort of your state of motion; the things that share your state of motion.

I'm going to use that concept of frame of reference a lot in this and subsequent lectures. It's a crucial concept in relativity, because the relativity principle—Galilean or Einsteinian—is fundamentally about frames of reference. But before we delve into the concept of relativity, I want to summarize what we've covered so far.

Between about 1600 to 1750, Galileo, Newton, and others developed a mechanical understanding of physical reality. The mechanical universe is *deterministic*, and a *relativity principle* holds. Between about 1750 to 1900, Maxwell and others developed an understanding of electromagnetic phenomena, including light.

Now we're ready to discuss frames of reference and the validity of physical laws. A frame of reference is the place that shares your motion (if you're in a car, it's the car; if you're in a plane, it's the plane). A simple question: In what frame of reference are the laws of motion (mechanics) valid? Answer: In any frame of reference in uniform motion (the Principle of Galilean Relativity, from Lecture 3).

Another simple question: In what frame of reference are the laws of electromagnetism (Maxwell's equations) valid? An equivalent question is "In what frame of reference does light go at speed *c*?" or, put another way, "Relative to *what* does light go at speed *c*?" One possibility is that light goes at speed *c* relative to its source. This is ruled out by astronomical observations, especially of double-star systems.

The 19th-century answer was that light goes at speed c relative to the *ether*, a hypothetical medium, or substance, believed to permeate the entire universe. The ether was thought to be the medium through which electromagnetic waves propagate, just as sound waves propagate through air and water waves through water. Ether must have some unusual properties, being at once very stiff to account for the high speed of light, yet letting planets and other moving objects slip through without resistance and being able to permeate the tiniest of spaces. But to 19th-century physicists, ether's existence seemed essential.

One branch of physics, mechanics (or motion), obeys the relativity principle—meaning that the laws of mechanics are the same in all uniformly moving reference frames. As far as mechanics is concerned, the statement "I am moving" is meaningless; only relative motion matters.

The other branch of physics, electromagnetism, does not seem to obey the relativity principle. This means that the laws of electromagnetism, including the prediction that electromagnetic waves move with speed c, are valid only in one frame of reference: the ether's frame. As far as electromagnetism is concerned, the statement "I am moving" is meaningful and means "I am not at rest in the ether's frame of reference."

The other branch of physics, electromagnetism, does not seem to obey the relativity principle.

An obvious question is "Is Earth moving relative to the ether?" One possible answer is that it isn't, which leaves two possibilities. First, earth alone among all the universe is at rest with respect to the ether. This follows because all the other planets, stars, galaxies, and so on are moving relative to Earth. This possibility flies in the face of the Copernican notion that Earth isn't special; this idea is also ruled out by observational evidence.

Secondly, earth "drags" the ether in its local vicinity with it. The observed *aberration of starlight* rules this out. A good analogy is the example of using an umbrella to keep dry. If you run through the rain, then you need to hold the umbrella at an angle to keep dry. But if you "drag" the air in your vicinity with you, then the rain will fall vertically even if you run, and you won't need to tilt the umbrella. Similarly, because of Earth's motion around the Sun, a telescope will need to be pointed at different angles at different times of year to see the same star—provided Earth does not drag the ether with it. But if there is "ether drag," then the telescope angle will not need to be changed. In fact, the telescope angle must be changed—showing that Earth does not drag ether with it.

Where does this leave us? By Copernican principles and sufficient observational evidence, we have rejected the possibility that air is moving relative to the ether. We have also determined by this point that the Earth is not dragging the ether with it. Therefore, Earth must be moving relative to the ether, and we should be able to detect this motion. ∎

Essential Reading

Hey and Walters, *Einstein's Mirror*, Chapter 2 through pp. 29–36.

Hoffmann, *Relativity and Its Roots*, Chapter 4.

Suggested Reading

Einstein and Infeld, *The Evolution of Physics*, Chapter 3, through p. 160.

Questions to Consider

1. You might be thinking, "This ether sounds like a farfetched idea. Maybe it just doesn't exist." How, then, would you answer the question "With respect to what does light go at speed *c*?"

2. Speculate on how the success of Newton's mechanical view of the universe led physicists to embrace the ether concept.

Aberration of Starlight: Umbrella Analogy

A) Standing Still. The man and the umbrella are both at rest with respect to the rain.

B) Running. Tilting the umbrella gives the best protection from the rain. Viewed from the frame of reference of the rain.

B) Running. Similar to B in all respects except viewed from the frame of reference of the running man.

D) If the runner drags a large volume of air with him as he runs, then the rain falls vertically within this volume and the umbrella is best held horizontally.

Speed *c* Relative to What?
Lecture 5—Transcript

Welcome to Lecture Five: "Speed *c* Relative to What?" with a big question mark after it. This is going to be a lecture full of questions, mostly questions about light and its behavior and, particularly, its speed *c*.

Before we get on to that, let me begin with a brief summary of where we've been before now—a brief history of physics through the year 1900. Because at this point we're at the jumping off place for the theory of relativity. We're at about the year 1900. Not quite. And we're ready to jump into relativity, and we have to put together where we've been. So here's a very brief history of physics through the year 1900.

In the years 1600 through 1750 we had Galileo, Newton, a lot of others developing an understanding of motion. Mechanics—the study of motion. They developed a mechanical understanding of physical reality. They could explain the way the world worked in terms of Newton's laws, gravity, and other forces and predict the motion of objects. They had a thorough understanding.

And I want to emphasize two aspects of the Galilean-Newtonian vision of reality that classical physics gave us. One is that the universe was deterministic. We're going to come back to that. We saw that in the clockwork universe. We'll come back to that when we get to quantum physics. The other thing was that a relativity principle holds in Newtonian and Galilean physics. The laws of Newtonian physics or Galilean physics work for any one in uniform motion. That's the principle of Galilean relativity. So, a relativity principle holds in this branch of physics.

Then, from roughly 1750 to about 1900, Maxwell and a whole host of others developed the theory of electromagnetism that I covered very briefly in the previous lecture. And that theory led to an understanding of all electromagnetic phenomena—electricity, magnetism, the interactions of the two, and especially light. Light was understood to be an electromagnetic wave. And Maxwell's equations made very definite predictions about those electromagnetic waves, specifically that they traveled at a known, calculatable speed—the speed of light, 186,000 miles a second, one foot per nanosecond, etc. So that's a brief history of physics to where we are about now.

And now we want to ask some questions about those two different branches of physics that are listed in that brief history. Mechanics—the study of motion—and the study of electricity and magnetism, including light.

Before we do I want to introduce a new concept. It's a concept called a *frame of reference*. What's a frame of reference? Well, you'd probably say it's something like your point of view. And in physics it is something like your point of view. It's basically sort of your state of motion; the things that share your state of motion.

If you're sitting here on Earth and you're doing physics experiments, Earth is your frame of reference. When you say, "I see a car moving at 50 miles an hour," you mean the car is moving at 50 miles an hour relative to the Earth and all the things that are attached to the Earth. If you're in an airplane flying at 600 miles an hour relative to Earth, the airplane is your frame of reference. And if you say, "I think I'll stroll down the aisle at 10 miles an hour," or something, that's relative to your frame of reference. Relative to the plane. If you're in a car moving down the highway and you want to do physics experiments in the car, the car is your frame of reference. Your frame of reference is basically the things that share your state of motion.

So that's the idea of a frame of reference. And I'm going to use that concept a lot in this and subsequent lectures. It's a crucial concept in relativity, because the relativity principle—Galilean or Einsteinian—is fundamentally about frames of reference.

So let's ask a very simple question. Let's ask the question in what frame of reference are the laws of mechanics, that is, the laws of motion, valid? In what frame of reference is this physics of the first half of our brief history of physics valid? We already know the answer to that question. We answered that question in lecture three with the Galilean relativity principle. The Galilean relativity principle said the laws of motion are the same for all observers in uniform motion. Put that in terms of frames of reference, it means the laws of physics, the laws of motion rather, the laws of mechanics, work equally well in all frames of reference in uniform motion.

So if I'm here on Earth, which, as I explained earlier, is a good enough approximation in uniform motion, or if I'm in that airplane moving through level air, no turbulence, at a steady speed, it's a valid frame of reference for doing mechanics of that frame of reference in which Newton's laws, the laws of mechanics, are valid. If I'm on Venus playing that tennis game, or even on that planet in that distant galaxy moving away from Earth at $0.8c$, 80 percent of the speed of light—as long as that motion is uniform, that's still a valid frame of reference for doing mechanics. Newton's laws will be equally valid there. I'll do laboratory experiments. I'll verify Newton's laws. And those experiments will not be able to tell me I am or am not moving. That's a meaningless question as far as mechanics is concerned. That's what the mechanics principle of relativity, the Galilean relativity principle, means. It means that the answer to the question, in what frame of reference are the laws of mechanics valid? is in any uniformly moving frame of reference. Known for hundreds of years. Nothing special about that. But it's such a crucial question I want to emphasize it again. In what frame of reference are the laws of motion valid? Answer: in any uniformly moving frame of reference. It's an easy question to ask. It's an easy question to answer. And the essence of the answer is the Galilean principle of relativity that says the laws of physics are the same for everybody in uniform motion. Only it says the laws of motion, which was all the physics that was known in Galileo's time. And I want to emphasize that distinction. It's the laws of motion that are valid in all frames of reference. All uniformly moving frames of reference.

We now want to ask another question. Equally simple. In what frame of reference are Maxwell's equations valid? In what frame of reference are the laws of electromagnetism valid? In what frame of reference is that second part of physics valid? Is it valid in all frames of reference in uniform motion, like the laws of motion are? Or is there some special frame of reference in which the laws of electromagnetism are valid, and they're only valid in that frame? That's the question we want to concentrate on in this lecture. And that's a crucial question in the evolution of Einstein's thinking and relativity.

So, the question is, in what frame of reference are the laws of electromagnetism, Maxwell's equations that I displayed for you last time, in what frame of reference are they valid? Now that's actually not the question I'm going to ask. I'm going to ask an equivalent question. And I want you to understand why it's an equivalent question. I'm going to ask instead the question, in what frame of reference does light go at speed c? That special speed of 186,000 miles a second. In what frame of reference is it true that light goes at speed c? Why is that an equivalent question? Because one of the predictions of Maxwell's equations is that there should be electromagnetic waves, which we now know include light, and that those waves should travel at speed c. In a frame of reference in which Maxwell's equations are valid, that prediction will be borne out. It will be true. In a frame of reference in which Maxwell's equations are not valid, light will have a different speed. So the question, in what frame of reference does light go at speed c? is equivalent, logically equivalent, to the question, in what frame of reference are Maxwell's equations valid? We're going to ask that second question. In what frame of reference does light go at speed c?

Now, there's one easy possibility. You might say, well it's pretty obvious; it goes at speed c relative to its source. So if I have a light bulb and it's emitting light, then that light is traveling at speed c relative to the source of the light. And if I run with the source toward you, the light will be moving toward you faster than c, but it will be moving at speed c relative to the source.

Maybe Maxwell's equations apply to the source of light. That's a nice idea. It's not true. There's a lot of experimental confirmation that it's not true. But, perhaps, the easiest comes from observational evidence of double star systems. Astronomical observations of double star systems. About half the stars in our galaxy, it turns out, are double stars. There are two stars in close orbit or some orbit around each other like this. We don't happen to live in such a system or we'd have two suns.

So, here's a double star system. And it's going around like this. Two stars are going around each other. Now notice what's going on. Here's a time when this yellow star is coming toward you. Here's a time when this yellow star is going away from you. Now this star system may be a long ways away from us, thousands of light years or something, that means light takes thousands of years to get from it to us. If the speed of light depended on the speed of the source, the light leaving the yellow star when it's coming toward us is going to get to us a lot sooner than the light leaving the yellow star when it's going away from us. We may even see several images of the star as light from different positions arrives at the same time, because it took different amounts of time to travel and the star was in different places at different times. We certainly won't see this nice, simple motion, which, after all, is what Newton's law of gravity predicts. If the speed of light depended on the speed of the source, we would see very strange things with double star systems. We would be able to interpret what that double star system was actually doing—a nice, simple orbit—but we wouldn't see that.

Well, the fact is, when we look at double star systems, it's very, very clear that the speed of the light coming from the stars has no relation to the speed of the source. The light comes by us at speed c, regardless of what the source is moving. So, unfortunately, the simple answer, that light goes at speed c relative to the source is plain and simple wrong.

Now, in the nineteenth century, physicists gave a different answer to that question. They looked at other kinds of waves they knew about. Sound waves. They traveled about 700 miles an hour. With respect to what? With respect to the air. Because sound waves are disturbances of air. Earthquake waves travel through the solid Earth at speeds of several miles a second or more. Relative to what? Relative to the solid Earth, which is the thing which is being disturbed by the passage of those waves. Ocean waves travel along the ocean surface at characteristic speeds that depend, in the case of ocean waves, on how big the waves are. That makes them more complicated than electromagnetic waves by the way. But what determines that, what is that speed measured relative to? It's measured relative to the water. If the waves were in a flowing river, relative to the bank they would have the river speed added to the wave speed. It's very clear what the speed of a wave means when it's a wave that's a disturbance of the air, like a sound wave, or a disturbance of the water surface like an ocean wave, or a disturbance of the solid Earth like an earthquake wave.

Well, nineteenth century physicists were first of all very thrilled with the success of Newtonian physics in explaining everything in mechanical terms.

And they were still trying to think about electromagnetism as if it were mechanical. They were trying to think of light waves as being very much like sound waves—mechanical disturbances of some medium. So, they came up with the idea of a hypothetical medium. Medium being like the air or the water, the thing that the wave is traveling through that it's a disturbance of. And they called it the *ether*.

And the ether had to have unusual properties. It permeated all of space, because light waves, after all, travel in the so-called vacuum of empty space—for example, between the Earth and the moon, or the Earth and the sun. Sound waves don't travel there. Water waves don't travel there. There's no air. There's no water there. Earthquake waves—there's no Earth. But light waves do travel there. Electromagnetic waves. Radio waves. They travel through empty space, or seemingly empty space.

So the ether had to pervade all space. It got into the tiniest interstices of empty space, because we know that light travels in little vacuums we make in vacuum tubes and things like that. Light goes everywhere—at least if the material isn't opaque. So the ether had to be very able to get everywhere. It also had to be very tenuous, because, after all, the Earth goes around its orbit and has been doing so for billions of years, and it isn't feeling any drag, like a satellite does when we launch it in too low an orbit, and it feels drag from the Earth's atmosphere, and eventually spirals into the atmosphere, and crashes to Earth or burns up. Earth hasn't done that with respect to the sun. And it would if the ether presented much resistance. So, it had to be very tenuous.

Furthermore, it has to be very stiff, for a reason that's a little bit more subtle. It turns out, if you send a wave down something like a stretched string, a stretch of string, and wiggle it and send a wave down it, you'll find the wave goes faster if the string is stretched tighter. So the ether, to propagate waves, move waves at a very, very rapid speed of 186,000 miles a second, had to be very stiff. A property that's not really very compatible with being very tenuous. So, it was a strange substance that physicists of the nineteenth century proposed. But it seemed essential to those physicists, because they were steeped in this mechanical view.

Now, if you're saying to yourself, "Wait, this is a preposterous substance. Why on Earth did they make this? Why didn't they just say light waves go through empty space?" Well, if you say that, then you're back to the question, with respect to what does light go at speed c? Because empty space isn't a something, like a substance is. If I say, light goes at speed c relative to the ether, it's clear what I mean. But if I say, it goes at speed c relative to empty space; it's not clear what that means. And you're left with a deep philosophical question about your fundamental ideas of physics.

So, we're going to pursue the nineteenth century answer for a little while. The answer is light goes at speed c relative to this hypothetical substance— the ether. Now, there's a philosophical problem here if you accept this ether idea, and it rises in the form of a dichotomy in the two branches of physics I've just been describing. I've talked about the study of motion and how in the study of motion a relativity principle holds. The laws of motion are the same for anybody who cares to study them, as long as that individual is in uniform motion. There is no one special frame of reference where the laws of mechanics, the laws of motion, are valid. A relativity principle holds.

But now we've asked the question for electromagnetism, we said, in what frame of reference does light go at speed c? I argued that that was equivalent to saying, "In what frame of reference are the laws of Maxwell, the laws of electromagnetism, valid?" And the answer is: They're valid in this one special frame of reference, a frame of reference at rest with respect to the ether. So we have a dichotomy. One branch of physics obeys a relativity principle and the other doesn't. And that is, to say the least, philosophically unsatisfying. But I urge you to think about the alternative, and you'll quickly get mired in greater philosophical difficulties.

So that was where things stood in the nineteenth century. We have this dichotomy. Physicists believed that the laws of electromagnetism were valid in one special frame of reference, that the statement "I am moving" is meaning*ful* in electromagnetism. I am moving means I am moving with respect to the ether. Whereas, "I am moving" is a meaningless statement in mechanics. That means we ought to be able to do physical experiments involving electromagnetism that will tell us whether we're moving or not.

Now, remember back to the first lecture when I had you playing tennis. You played tennis on a cruise ship, on Venus, on this distant planet moving away from Earth at 80 percent of the speed of light. And you had no problem with that. Maybe you had a little more problem with the second thing I asked you to do, or it was a little more obscure. And that was heat a cup of tea in a microwave oven. That involved electromagnetism. That's why I had you do that. And if you were a believer in the status of these two branches of physics, as they were at the end of the nineteenth century, you would have told me, yes, I can play tennis on this distant planet moving at 80 percent of the speed of light, and it worked just the same as it does on Earth, because the relativity principle holds for the laws of motion. But you would say, no, the microwave oven isn't going to work at all. The microwaves really are going to get crashed to one side of the oven or left behind or something, because that oven is moving at 80 percent of the speed of light relative to Earth. And the laws of electromagnetism seem to work pretty well on Earth.

So, we have this dichotomy in physics. However, we're going to plow ahead with the philosophically unsatisfying dichotomy. We're going to ask another question. We're going to ask: Okay, there is this ether-like substance, how is Earth moving relative to the ether? Because if the ether exists, Earth must either be moving through it or not moving through it. And I want to take you through some important logic. And this logic is crucial, because if you follow this logic, you will be forced to believe in relativity when we get there. We're not there yet, but we will be soon, and you'll be forced to believe in it. Logically, you'll understand why it must be.

So, let's look a little bit at the question, Is Earth moving relative to ether? And there are really two possibilities. One possibility is that it isn't moving relative to the ether. If it isn't moving relative to the ether, then one possibility is Earth alone, among all the cosmos, is at rest relative to the ether. Now, that may be an absurd possibility, but maybe it's true. Or possibly the Earth is at rest relative to the ether because the ether that's locally in Earth's vicinity gets dragged along with it. Well, let's look at these two possibilities.

I think you can see that the first one probably isn't going to be very philosophically satisfying, regardless of its physical content—it isn't satisfying physically either because it violates the Copernican principle that the Earth isn't special.

Let's look at the second one in some degree of detail. Possibly the Earth drags the ether in its vicinity with it, and if the Earth does what Mars does too, and what Venus does, and that distant planet, and everybody sees themselves at rest with respect to the ether, even though the different ether is moving with respect to them, how could we find out if that's true or not? Of course, the other possibility is the Earth is moving relative to the ether. So we want to rule out the first possibility, if we can, or rule out the second possibility.

So, let's look at the two sub-possibilities—the possibility that the Earth drags the ether with it in its vicinity. Fortunately, there's an astronomical observation we can make that will answer this question of whether the Earth drags the ether in its vicinity with it, and presumably then whether the other planets and stars and astronomical objects do that as well. The phenomenon is called the *aberration of starlight* and it involves the question of whether we have to look in different directions to see the same star if we look at it at different times of year. Why would that happen? Why would we have to look in different directions? Well, it has to do with the motion of the Earth in relation to the motion of the light coming from the star. And the best way to explain that to you is to give you an analogy involving an umbrella.

So here's the analogy. Suppose it's raining, and raining vertically downward, and I want to keep dry. How do I do it? I put an umbrella directly over my head. I'm standing still in vertically falling rain and I want to keep driest; I put an umbrella directly over my head. There I go. I stay nice and dry. That's great.

But now what happens if I want to move somewhere—I want to run through the rain or walk rapidly through the rain. What should I do with the umbrella? Should I continue to hold it right over my head so it will keep my head dry, or should I do something different? Well, I'm going to convince you that you should do something different. Here's what happens if I run through the rain. I would argue that I have to tilt the umbrella at an angle. That angle is going to depend on how fast I'm moving relative to how fast the rain is falling. I'm going to have to tilt at an angle toward the direction I'm moving in order to stay driest.

Why does that work? Well, it works; you can understand that by looking at what the situation looks like in my frame of reference. Right now we're looking at a situation, first where I stood still, then if I run, but I'm still drawing that picture from the Earth's frame of reference, and I'm moving past the Earth and the rain is falling vertically. But in my frame of reference, it looks different. As I move toward the falling rain, the rain looks to me like it's coming at an angle. And that's the angle I have to tilt my umbrella at. And, again, that angle depends on basically how fast I'm moving relative to how fast the rain is falling. So, I've got to tilt my umbrella to stay dry. So here it goes. It's raining. I want to get from here to there. To do so, I tilt the umbrella. I tilt it at an angle that depends on how fast I'm going. If I'm walking slowly, maybe it's like that. If I'm running a bit faster, maybe it's more like that. And you might say, "How can that be? His head is exposed." Well, the rain that was going to hit my head is long passed by the time it gets there. The other rain has been intercepted by the umbrella. Rain that would have hit my head when I had reached this point has already hit the umbrella and has avoided me. So, I actually do stay driest if I tilt the umbrella in the direction I'm going, at a tilt angle that depends on how fast I'm moving relative to the falling rain.

Well, that's all well and good. But what happens if instead I drag some big volume of air with me? That would be analogous to the Earth dragging the ether with it—Mars, Venus, all the other astronomical objects dragging the ether with them. What happens if they do drag the ether with them? What happens to the angle I have to tilt a telescope to look at a star? What happens to the angle I have to tilt my umbrella if a large volume of air comes along with me? Actually, some volume of air actually does come along with me as I walk through the air. There's a small volume right next to my skin that isn't moving. It's been caught by my skin, and it moves along with me. But it's only a tiny, tiny fraction of a millimeter. So that's not important. But what if instead I drag a big volume of air, several times my size, big enough so that if the rain falls vertically in the air above that isn't moving with me, it gets into the air that is moving with me, it has time to get sped up by the motion of that air until it's moving along with me. What happens then? Well, the rain is going to fall. It's going to suddenly hit this volume of air that's moving with me. Temporarily, it's not going to be moving at that rate. But, it's going to feel a wind like force on it. It's going to speed up. Eventually, it's going to be moving with me. And then relative to me, it's going to be falling vertically. Relative to somebody standing still on Earth, it's going to now be moving at an angle within that huge blob of air.

Let's take a look at what that would look like. So, here I am. I'm running along. I've got this big volume of air I'm dragging with me. The rain is falling in my frame of reference at what looks like an angle. But once it gets into this blob of air, it's moving along with me, horizontally. And so it falls vertically again, and I do not need to change the angle at which I tilt my umbrella. So, if in fact I drag a big volume of air with me, then I walk along, don't tilt my umbrella. I walk in a different direction, I don't tilt my umbrella. Everything is fine.

How can I tell if I drag a volume of air with me? Simple. I do this experiment. I ask myself, "How do I stay driest?" I ask myself, "At what angle do I need to point my umbrella to stay driest?" If the answer is: I need to point it at some angle, and that angle changes if I change direction, because I always want to point it in the direction I'm going, then I've got good evidence that I don't drag a big volume of air with me.

On the other hand, as I move around, no matter which direction I'm going, I find I just hold the umbrella vertically and that will keep me driest. Then I conclude that I move a big volume of air with me, I drag a big volume of air with me.

Well, that's an analogy for this phenomenon of aberration of starlight that I mentioned. In the analogy, the umbrella becomes a telescope. We're going to look at a distant star with a telescope. So let me get a telescope. We're going to ask the question, do I need to point the telescope in different directions? Just like, did I need to tilt the umbrella? So what corresponds to the rain? Well, it's the thing that was falling in that case. Here, it's the thing that's coming to us from the star, namely, the starlight. So, in order to see the star, do I have to point the telescope in different directions depending on how I'm moving?

Now, I'm not going to run around the Earth with a big telescope. I couldn't run fast enough anyway. But what I am going to do is take account of the fact that the Earth, in its orbital motion, moves about 20 miles a second. So, at one time of year it's moving 20 miles a second this way. But we're in a circular orbit around the sun. So, six months later, I'm moving 20 miles a second in a different direction. And it's very much like the umbrella. Again, when I moved in this direction, if I wasn't dragging the air with me, I tilted the umbrella like this. When I get to the other end of the room here, and I want to go back, I don't simply hold the umbrella like that. That'd get me very wet. Instead, I tilt the umbrella in the other direction. So I'm still pointing it in the direction of my motion. And then I stay driest. The important point here is not so much the angle I had to tilt it at. The actual angle would be hard to measure accurately with starlight. What's important is the fact that if I move in a different direction, I've got to shift the actual direction the umbrella is pointing so that it's always going in the direction I'm moving. That's the crucial point.

So, if we go to the analogy again, that means if I look at the star at one time of year, and I point it in this direction, if I find out that six months later I have to switch the direction of the telescope, that is, it's still pointing at an angle, but before it was pointing this way, now it's pointing this way, then I conclude that aberration of starlight is occurring. The apparent position of the star is changing. And I conclude I don't drag a big volume of, not air but in this case in the analogy, ether, with me. So that is the phenomenon of aberration of starlight. And we can actually do that experiment.

Now remember what determined that angle. The angle in the umbrella case depended on how fast I was running relative to how fast the rain was falling. What's the analogy in the astronomical case? What's the analogy that gives me the angle at which I tilt the telescope? It's a question of how fast the Earth is moving, in its orbital motion relative to the speed of the stuff that's coming down from the star, namely, the speed of light. Well, the speed of light is a big speed. It's 186,000 miles a second. We don't know the absolute speed of the Earth relative to the ether, but we do know that it changes by about 40 miles a second, because one time we're going 40 miles a second this way, and six months later we're going 40 miles a second the other way. So, the change in angle is going to be a very small change, something like the ratio of 40 miles a second to 186,000 miles a second. It's going to be a tiny angle.

But we can measure it with astronomical telescopes, because what we do is take a picture of a field of stars. And we look at the positions of the stars on a photographic plate—or more currently on an electronic detector—and we see if those positions change with time of year. And we can measure that angle. And what happens when we do that experiment? What happens is we find that the positions do, indeed, shift or, equivalently, to get the star exactly the same place on our photographic plate, we have to tilt the telescope at a different angle; aberration of starlight occurs.

Back to our full picture here of the aberration of starlight problem. The analogy with the umbrella, we see that that rules out the possibility that we drag a big volume of the ether with us. So the Earth does not drag the ether with us. And that aberration of starlight is a convincing demonstration of that. So where does this leave us?

Let's go back to the logic we were talking about before. We had two possibilities. We asked the question: Is the Earth moving relative to the ether? One possibility, it isn't. The other possibility, it is. We want to rule out one of those two possibilities. Well, I think philosophical grounds alone are sufficient to rule out the idea that Earth, of all the things in the universe— Mars, Jupiter, Venus, the sun, the center of our galaxy, that planet in that distant galaxy moving away from us at 80 percent of the speed of light— that all those things are in motion relative to ether but Earth is not. That as Earth goes around the sun in its orbit, the entire ether throughout the whole universe moves to partake of that motion and keep Earth at rest relative to the ether. That's nonsense. That violates everything Copernicus was trying to tell us—that the Earth is not special, the Earth is not the center of things. Furthermore, if you don't like that philosophical argument for ruling out that possibility, there's plenty of observational evidence, from astronomy particularly, that rules out the possibility that Earth alone, amongst everything, is at rest relative to the ether. Just isn't true.

So, then there was the other possibility that would allow us to conclude that the Earth isn't moving relative to the ether. And that would be if the

Earth drags the ether in its vicinity with it, and the aberration of starlight observations allow us to conclude that that isn't the case. And, therefore, we have ruled out entirely the first possibility—that the Earth is not moving relative to the ether. It must be moving, relative to the ether. The second possibility is the case. And if it is moving relative to the ether, then there must be a wind of ether blowing past the Earth, because the Earth is moving relative to the ether. The ether, alternatively, is blowing past the Earth, just like if you stick your arm out the window of a moving car. Even on a calm day, you feel a wind. It's the wind, the apparent wind, caused by your motion through the air. Similarly, the Earth's motion through the ether should give rise to a wind of ether blowing past the Earth.

We should be able to determine the speed and direction of that wind. Why? Because we have a branch of physics, which is valid only in a frame of reference at rest with respect to the ether and will give different answers in a different frame of reference. And, in particular, will give different answers depending on how we're moving relative to the ether. So, we should be able to design an experiment that tries to answer the—we know logically, we've ruled out the possibility that we're not moving relative to the ether. We are moving relative to the ether. And we should, therefore, be able to devise some kind of experiment, if it's sensitive enough, that detects Earth's motion relative to ether and answers two questions: How fast are we moving relative to the ether? And in what direction?

It's going to have to be a sensitive experiment, because it's going to involve questions of the Earth's speed in relation to the speed of light, because we're going to ask electromagnetic questions. And we can't answer this question with mechanics, because in mechanics there's a relativity principle, and all frames of reference that are uniformly moving are equivalent. But we can answer it with electromagnetism, because we had this dichotomy. And at the end of the nineteenth century, physicists believed that the principles of electromagnetism were valid only in one frame of reference, the ether frame of reference. We've ruled out Earth being at rest in the ether frame of reference. Earth is, therefore, moving relative to that frame. And we're now going to set out and try to detect that motion. And one of the most famous and important experiments in all of physics did precisely that. And it's the experiment I'll be describing in the next lecture.

Earth and the Ether—A Crisis in Physics
Lecture 6

In the previous lecture we asked the question: In what frame of reference are the laws of motion valid? And we gave the same answer we gave back in Lecture 3. And the same answer you realized was true in Lecture 1, namely, they're valid for anybody in uniform motion.

Forced to the conclusion that Earth must be moving through the ether, physicists s+et out to detect and measure that motion. From our perspective on Earth, our motion through the ether should manifest itself as an "ether wind." As a result of that wind, the speed of light should be different in different directions.

In the 1880s, the American physicists Albert Michelson and Edward Morley conceived an experiment to detect the ether wind and answer the question "How is Earth moving relative to the ether?" Their experiment compared the travel times for light following two mutually perpendicular paths. It used interference of light to make a very precise comparison of the two travel times and was sensitive enough to detect motion much slower than that of Earth in its orbit around the Sun. Thus, the Michelson-Morley experiment could definitely determine Earth's motion through the ether.

The Michelson-Morley experiment used an ingenious system of mirrors to compare the speeds of light in two perpendicular directions. It takes longer to row a boat a given distance up a river and back than it does to row the same distance across the

The Michelson-Morley experiment used an ingenious system of mirrors to compare the speeds of light in two perpendicular directions.

river and back. Similarly, it should take light different times to make a round trip of the same length parallel and perpendicular to the ether wind. The Michelson-Morley experiment sought to detect this difference. The extreme sensitivity of the experiment resulted from its use of *interference* between the two light beams. The experiment did not actually measure speeds for the light beams, but rather was sensitive to *differences* in speed.

Michelson and Morley performed their experiment with the apparatus in many different orientations. They also performed it at different times of year, corresponding to different directions of Earth's orbital motion (relative to the hypothetical "ether wind"). But there was never any change in the interference pattern—meaning that Michelson and Morley could not detect any motion of Earth through the ether.

This "null result" gave rise to a serious contradiction. The Copernican paradigm and astronomical observations (especially aberration of starlight) rule out Earth's being at rest with respect to the ether. The Michelson-Morley experiment appears to rule out Earth's moving with respect to the ether.

In attempts to salvage the ether concept, the Irish physicist George Fitzgerald and the Dutch physicist Hendrik Lorentz independently proposed that objects shrink in the direction of their motion through the ether. This shrinkage was such that the Michelson-Morley light path parallel to the ether wind would be shorter by just the right amount to keep the travel time for the two perpendicular light beams the same—ensuring that the experiment could never detect motion through the ether. The Lorentz-Fitzgerald proposition had no physical or conceptual justification whatsoever; it was just an ad hoc assumption designed to explain away the Michelson-Morley result. ∎

Essential Reading

Hey and Walters, *Einstein's Mirror*, Chapter 2, pp. 38–45.

Hoffmann, *Relativity and Its Roots*, Chapter 4, pp. 75–80.

Suggested Reading

Einstein and Infeld, *The Evolution of Physics*, Chapter 3, pp. 160–186.

1. Did the Michelson-Morley experiment actually determine values for the speed of light in two different directions? If not, what did it measure?

2. Why was it necessary to repeat the Michelson-Morley experiment at different times of the year and with the apparatus in different orientations?

3. It is not necessary for the two arms of the Michelson-Morley apparatus to be the same length. Why not?

Wave Speed Depends on the Wind Speed and Direction

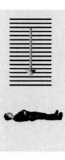

The air (represented by the gray square) is at rest relative to the viewer. The sound waves move toward the viewer at a speed of 700 mph relative to the air *and* the viewer.

The air moves toward the viewer at a speed of 50 mph. The sound, traveling through the medium of the air, is thus moving toward the viewer with a speed of 750 mph.

The air moves away from the viewer at a speed of 50 mph. The sound, traveling through the medium of the air, is thus moving toward the viewer at the reduced speed of 650 mph.

The Michelson-Morley Experiment

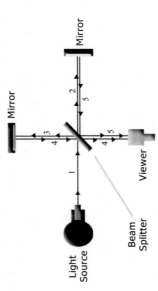

The Michelson-Morley experiment was an ingenious device for detecting the movement of an hypothesized "ether" medium with respect to the earth. With this elaborate device, a single beam of light (1), emanating from the light source (far left), is split by a beam splitter (center) into two beams (2 and 3) and through the use of mirrors (top and far right) is recombined into a single beam (4 and 5) aimed at a viewer (bottom).

The experiment relies not on precision timing of the separate light beams, but rather on the interference pattern created by the recombined light.

Wave Interference

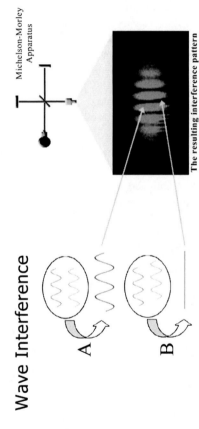

Michelson-Morley Apparatus

The resulting interference pattern

Wave interference, caused by two or more different waves combining, can have several possible outcomes, two of which are pictured here. In A, two light waves combine with wave troughs and wave crests coinciding, resulting in a wave with a higher amplitude. This is called **Constructive Interference**. In B, two light waves combine with troughs in the upper wave coinciding with wave crests in the bottom wave, resulting in a cancellation of the wave. This is called **Destructive Interference**.

Earth and the Ether—A Crisis in Physics
Lecture 6—Transcript

In the previous lecture we asked the question: In what frame of reference are the laws of motion valid? And we gave the same answer we gave back in Lecture Three. And the same answer you realized was true in Lecture One, namely, they're valid for anybody in uniform motion.

Then we went on and asked a similar question for the laws of electromagnetism that we had just learned about, including their prediction that there were electromagnetic waves, and the electromagnetic waves moved at a certain speed, the speed c, the speed of light. So we asked the question: In what frame of reference are those laws valid? We found that was equivalent to the question: With respect to what does light go at speed c? And then, I introduced the nineteenth century answer to that question—with respect to the ether, a hypothetical medium that permeated all of the universe and was the medium in which light waves are a disturbance. Similarly to the way that the air is the medium in which sound waves here on Earth are a disturbance, or the ocean is the medium in which water waves are a disturbance.

And then we asked the logical question: How is Earth moving relative to the ether? And we've been through a series of very rigorous, logical steps in an attempt to answer that question. And we're partway there now, and we're going to continue that logic today. And I want to do this with considerable rigor. Because at the end of today's lecture, we're going to be in real crisis in physics, and I want you to understand how we got to that crisis and how difficult it's going to be to get out of it.

So, let me just review briefly the logic that got us to where we are now. We asked the question: Is Earth moving relative to the ether? There are two possibilities. Either it isn't, or it is. We looked last time in considerable detail at the possibility that it isn't moving relative to the ether. There were two possibilities then. Either Earth alone among everything in the cosmos was at rest relative to the ether. That's an absurd conclusion given the Copernican principle that the Earth is not a special place in the universe. It's particularly absurd in light of what we know from modern cosmology, namely, that there are places in the universe, distant galaxies in particular, that are moving away from us at speeds that are very close to the speed of light—80, 90 percent of the speed of light. It's absurd to imagine that everything in the universe is pinned to the Earth, when there's such a wide range of speeds relative to Earth throughout the universe. So, we've ruled out the first possibility, that Earth alone among everything in the universe is at rest with respect to the ether. We ruled that out, basically, philosophically, on Copernican grounds. I also guarantee you, it's also ruled out on grounds from observational astronomy and other measurements we can make. But it suffices to rule it out on this philosophical ground.

The second possibility was that the Earth drags the ether in its immediate vicinity with it. And, presumably, so do all the other astronomical bodies. We looked at the possibility that that was true. I gave you an example involving an umbrella and trying to keep dry in the rain as I moved through the rain. Analogous to the Earth moving through the ether and having to point a telescope in different directions. And astronomical observations convince us, on that basis, that the Earth does not drag the ether in its immediate vicinity with us. So, we've ruled out that possibility as well. Together then, we've ruled out the possibility that the Earth is not moving relative to the ether.

So we're forced to the logical conclusion that the Earth must be moving relative to the ether. If it is moving relative to the ether, then just as a rider in a convertible car is feeling a wind going past them, even though the air may be still because of the motion of the car, similarly, if Earth is moving through the ether, there must be an ether wind blowing past the Earth. And we ought to be able to detect that wind somehow. We can't detect it by a mechanical means, by the laws of motion, because, you know, they're the same in all frames of reference. They don't care about the motion of ether. But we should be able to detect it with some kind of electromagnetic experiment. Some kind of experiment involving electricity and magnetism. Because we believe in the ether paradigm that the laws of electricity and magnetism are valid exactly only in the ether's frame of reference, and if we're moving relative to ether, we should be able to detect some subtle differences.

So, the other possibility is that the Earth is moving relative to the ether. That's the only possibility we have left. And if that's true, there must be this ether wind blowing past Earth. We should be able to detect that ether wind, and, in particular, to measure both its speed and direction. And that's what we want to try to do.

Let me carry this a little bit further and suggest how we might do that. If you have a wave of any kind, and let me talk first about a sound wave here in air. If you have a wave that travels at a certain speed with respect to that, to its medium, in the case of a sound wave, that medium is air. Sound waves go about 700 miles an hour with respect to air. That's, incidentally, why jet planes travel no more than about 600 or so miles an hour, because they don't want to go supersonic. It requires a whole different engineering to do that. So that kind of sets an upper limit to the speed of everyday airplanes.

The speed of sound is about, anyway, 700 miles an hour, about a thousand feet per second. That's why when you hear, when you see lightening and count the seconds until thunder, every five seconds represents about a mile— because the speed of light is almost instantaneous, and the speed of sound is about a thousand feet a second. Anyway, the speed of sound is about 700 miles an hour. If there's a wind blowing at, say, 50 miles an hour in the direction the sound waves are going, the sound waves will be going relative to the Earth and relative to you, if you're standing still on Earth, at 750 miles an hour.

So what's that look like? The wave speed—that's the important point—the speed of these sound waves, depends on the wind speed and on the direction of the wind. I'm showing a picture here, which shows you standing at rest on Earth. A blob of, at this point, air surrounds you. And the air is at rest. And there is a series of sound waves coming at you at 700 miles an hour relative to the air. And they're also moving at 700 miles an hour relative to you, because the air is at rest relative to you. The speed of sound is 700 miles an hour relative to the air. The air is not moving with respect to you. Therefore, this sound is going 700 miles an hour with respect to you.

If, on the other hand, the wind is blowing toward you, and I've put a little arrow here on the blob of air to represent the fact that the whole air is moving, the air is carrying the sound waves. The sound waves are moving at 700 miles an hour relative to the air. Let's say the wind is blowing at 50 miles an hour. Then the sound waves are coming at 750 miles an hour relative to you. So that the wave speed that you measure, the speed of sound that you measure, is different than it would be if there weren't a wind blowing. In this case, with the wind blowing in the direction the sound waves are moving toward you, you will get a speed that is greater—750 miles an hour.

If, on the other hand, the wind is blowing away from you but the sound is still coming toward you, then the speed of sound will be reduced for you. It's still 700 miles an hour relative to the air, but because the air is blowing away from you at, say, 50 miles an hour, the sound waves are going past you at only 650 miles an hour. So, the wave speed depends on the wind speed and the direction.

Incidentally, you might speculate about what would happen if the wind speed were exactly the speed of sound. Those sound waves would never get to you.

Well, let's re-do this example. But now, let's talk about light and the ether. If we talk about light and the ether, my blob of air, the medium in which sound waves propagate, now represents the ether. The sound waves now become light waves. Light waves travel at a known speed—the speed c, 186,000 miles a second, 300,000 kilometers a second, one foot per nanosecond—relative to the ether. That's the nineteenth century hypothesis. So, if we're at rest with respect to the ether, we will measure c for the speed of sound. And we will measure c for the speed of sound no matter which—or speed of light rather—we will measure c for the speed of light. And we'll measure c for the speed of light, no matter which way the light is moving, because the ether is still with respect to us.

But we've ruled out the possibility that the ether is at rest with respect to Earth, or Earth is at rest with respect to the ether. We know the ether must be moving with respect to Earth or Earth is moving through the ether. There must, therefore, be an ether wind blowing past the Earth. And therefore, in some directions, depending on whether light is moving with the ether wind or against the ether wind, or maybe sideways to the ether wind, the speed of light should have different values. And if we could build a sensitive enough experiment and send light in different directions, we should be able to detect that difference. And we should be able to use that difference to calculate the speed of the Earth and the direction of the Earth relative to the ether. That's the experiment physicists wanted to do in the nineteenth century.

There were a number of attempts to do that experiment. Early attempts failed, for a number of reasons—some of them fairly theoretical and some of them because they simply weren't sensitive enough. The problem is, we already know that even though Earth must be moving relative to the ether—that's what our logic told us—that motion can't be very, very fast. If that motion were, say, 90 percent of the speed of light, it would be very obvious to us that Maxwell's laws of electromagnetism worked differently in some directions than others. It would be very obvious that the speed of light was different in different directions.

If we are moving relative to the ether, which logic tells us we are, that motion is not real fast, compared to the speed of light. That already should raise up your hackles a little bit. Because if there are people out there on planets or in distant galaxies moving away from us at 80 percent of the speed of light, they must be moving pretty fast relative to the ether. Because, remember, they don't drag the ether with them. But we are moving relatively slowly compared to the ether, and we want to find out just how fast that is.

We can't move very fast at all on Earth relative to the speed of light. But we do have a faster speed available to us, and that's the speed of the Earth in its orbit around the sun. As I mentioned before, that speed is about 20 miles a second.

Now, we don't know how fast the whole sun and the solar system are moving relative to the ether. But we do know this: We do know that at one point in the year, if we're moving at 20 miles a second this way, six months later we're moving at 20 miles a second the other way. And that's a difference in speed of 40 miles a second—from going 20 miles a second this way, to 20 miles a second that way. And we have that difference in speeds to play with. And 20 miles a second is not big compared to the speed of light. But at least it's not absolutely negligible.

By the way, for you mathematically inclined types, these experiments depend not on the ratio of the speed of the object through the ether, to the speed of light, but on the square of that ratio, which is an even smaller number. Take 20 miles a second, or 40 miles a second, over 186,000 miles a second. That's a tiny number. Square it, it's even smaller. So these experiments have to be very sensitive.

I'm not going to describe the earlier attempts. But I'm going to describe in some detail the attempt that was carried out in the 1880s by the American physicists Michelson and Morley at Case Institute in Cleveland, Ohio. And this was an experiment they started in the early 1880s. By 1887, the experiment was sensitive enough to detect motion 10 times slower than the known motion of the Earth in its orbit. So it was clearly sensitive enough to detect the motion of the Earth through the ether. No question about that at all. I want to spend a good bit of time describing this experiment, because it really locks us into this crisis we're going to get to by the end of this lecture.

So let's take a look at the Michelson-Morley apparatus. The Michelson-Morley experiment consists, schematically, of several parts. It has a light source. It has two mirrors. It has a viewer. And it has a device called a beam splitter. And the beam splitter is nothing but a lousy mirror. It's a piece of glass that has had silver coated on it or aluminum coated on it, but not thick enough to make it a really good mirror. It reflects about half the light that hits it. But about half the light gets through. Today, we use slightly more sophisticated devices. But that was what was used in Michelson-Morley's day. And it works quite satisfactorily. So the beam splitter, think of it as a mirror—half the light that hits that mirror goes through and half the light is reflected.

So let's see what happens when we turn on the light source in the Michelson-Morley apparatus and send a light beam out. Out comes the light beam and it hits the half-silvered mirror. Some of it goes straight through and hits a distant mirror. Some of it reflects off the half silvered mirror, bends at a right angle, because the mirror is placed diagonally, and hits another mirror.

So, what we've established here are basically two arms. There's a path from the half-silvered mirror to the upper mirror, and there's a path for light to travel horizontally from the half-silvered mirror to the right hand mirror. The light has been split, and it's going to travel on two different paths. It then bounces off these flat mirrors and comes back. So, back comes one light beam. Back comes the other light beam. They join at the half-silvered mirror. Some of the light goes straight through, some of the light is reflected in each case, and the important point is, there is light that's been recombined. The original light left the source, traveled to the half-silvered mirror, split into two beams that went out in two mutually perpendicular directions, came back and joined, and goes into a viewer. This is the experiment with which we're going to try to detect Earth's motion through the ether.

How are we going to do it? Well, we have in this Michelson-Morley apparatus these two paths for the light to travel. These two arms. The one that goes vertically, the one, in this case, that goes horizontally, although the whole apparatus was actually laid out horizontally on the ground.

Let me just give you a little historical description of this apparatus before I go on. We can, today, build something that looks very much like this on an optical table, which is just a big, heavy hunk of metal that doesn't vibrate much. About this big. And we can build this apparatus and pretty much carry out an experiment like this using lasers and modern optical components. Michelson and Morley couldn't do that. They actually floated their apparatus on a 13-foot diameter slab of stone floating in liquid mercury. Sensitive optical experiments have to be isolated from vibration. And it took a lot of effort to do that. And, furthermore, they had lots of mirrors, and the light bounced back and forth many times to make the light paths longer. But in principle, this is what the apparatus looked like.

What we're going to try to do is detect differences in the travel time for the light on these two paths. For now, think of those two paths as being identical in length—but I'm going to assure you that isn't at all crucial, and I'll show you why in just a minute.

Now, it's not good enough simply to time the travel time of the light and the two paths with a stopwatch. We just don't have stopwatches accurate enough. We couldn't even do that—well, in today's technology, we might just barely be able to actually time the travel time. But in the 1800s that was impossible. Instead, Michelson and Morley used a much cleverer and more sensitive method to try to detect a difference in travel time for the two light paths. They weren't concerned about the actual value of the speed of light in two different directions. That didn't matter. All that mattered was whether the speed of light was different in two different directions. That's enough, knowing how different it is and knowing the different directions, to tell you something about the speed of the Earth relative to the ether.

Why do you expect a difference in travel time? Suppose there's an ether wind blowing in this apparatus. Suppose the wind is blowing horizontally, from right to left, for example. I assert that it will take light longer to travel up this path and back than it will to travel along this path and back at right angles to the wind. You might say, well, why is that. Why should it take any longer to travel upstream and back?

Think about a boat rowing in a river. If I were rowing upstream and downstream, would it take me just as long if there were no current or if there were a current? You might say, well, if you row upstream you're slowed down, but if you rowed downstream you sped up and it all comes out in the wash. But that's not true. And the reason it's not true is very simple. If I row the boat upstream, because the river is trying to push me backwards, I'm spending a lot more time on the slow leg of the trip. I'm spending a lot more time when I'm fighting against the river than when I come back and the river is speeding me up. And the result of spending more time going slowly, means the overall time takes longer. It actually also takes longer to go across the river and back, because I've got to angle myself a little ways upstream to land on the opposite shore. But that effect is much less because I'm not working directly against the current. You can work out the mathematics of this if you want using the Pythagorean theorem. But you don't need to. Take my word for it, it takes longer to go upstream and back than it does to go across the wind and back again.

So, if there is an ether wind blowing past the Earth, and we orient this apparatus so one leg is in the direction of the wind and the other leg is perpendicular to it, then we should be able to see a difference in travel times for the light beams.

How are we going to detect that difference? Well, that was the genius of the Michelson-Morley experiment. And, incidentally, this design is used to this very day as a sensitive measurement of distances or optical properties, because the clever idea that Michelson-Morley had works so well and is so very, very sensitive.

So let's take a look at how it works. To do that, we need to understand the phenomenon of *wave interference*. So the picture I'm showing here has a little picture of the Michelson-Morley apparatus in the upper right corner to remind you what we're talking about. Wave interference is the following thing: When two waves combine, as they do in the Michelson-Morley experiment after they've gone their separate paths and come back together again, they may be in step with each other, as I show for two waves here. These two waves are exactly in step. This is what would happen if they had traveled identical paths in identical times. They would be in step, and when they combined they would make a wave that was bigger. That's one possibility. That's called *constructive interference*. The waves combine constructively to make a bigger wave. We're going to see more about interference later, when we get to quantum physics. So, that's one possibility for interference.

On the other hand, what if the path lengths are different or the times that are taken are different because of the ether wind, and the two waves arrive slightly out of step? In fact, what if they're out of step by just the right amount that the crests, the peak of one wave, line up with the troughs of the other? Then when those waves combine, we get nothing. Darkness. No light. That's wave interference. What's happening in the Michelson-Morley experiment is we're combining two waves that have traveled on two different paths. If the path lengths are identical and the speeds of light are identical, they're going to come back in step. We're going to get constructive interference. If the path lengths are identical but the speeds are different, we're going to get different interference. It may be destructive. It may be partially destructive.

In fact, the situation is a little bit more complicated than that. The light beam is not perfectly straight, but it spreads out a little bit and parts of the light travel at slight angles. And they travel slightly different paths. And the result of all this is, what an observer looking in the viewer sees is something like this. This is an actual photograph taken by putting a camera at the viewer of a Michelson-Morley experiment. And what you see are a band of what are called *interference fringes*. You see light and dark and light and dark and light and dark. This particular one was done with a red laser light. The same kind of red lasers you see in supermarket checkouts. That's why we have that color.

So, we see alternating bands of light and dark. The light bands, the bright bands, are due to waves that traveled the two paths and came back interfering constructively. The dark regions are due to waves that came back and interfered destructively. And again, because light is leaving the source at slightly different angles and traveling just slightly different distances, we get this whole range of bands. Here is why it doesn't matter that we build the apparatus with exactly the same length arms; I don't care if the arm lengths are different. That will just make the interference pattern shift slightly right to left. But once I build the apparatus, that interference pattern will sit there and it will look like that unless something changes.

And here's the key to how the experiment works. Michelson-Morley said, let's do this experiment now, when the Earth is going this way—relative in its orbit around the sun. We'll get a certain interference pattern. Let's do the experiment again six months later when the Earth is moving in a different direction. And if the interference pattern shifts—and that's all that matters—if the interference pattern shifts, then we know that the change in the orientation of the apparatus relative to the ether wind caused a change in the relative times it took the light to travel the two different paths. We don't care absolutely how long it took to travel those paths, we don't care that the paths were the same length. All we want to know is, is there a shift in the interference fringes.

Not only do we do the experiment at different times of year, but we also do the experiment at different orientations. And that's why Michelson and Morley floated their apparatus in a great big bath of liquid mercury. OSHA would have a big problem with all that liquid mercury and mercury fumes today. But they floated the whole thing on a big stone slab—which floats in mercury because stone is less dense than liquid mercury—and they were able to rotate the entire apparatus.

Why did they need to do that? Well, several reasons. Suppose it happened that they started out in a situation where the apparatus happened to be aligned with one of its paths along the ether wind. So here's the apparatus, once again, I'm showing now a big picture of the whole apparatus with the light beams. And here comes the ether wind. So the ether wind, in this picture, is blowing from right to left. The whole apparatus is immersed in this ether wind that's blowing past the Earth. And in this configuration, we expect the travel time for light, on what in the picture is the horizontal path—remember, the whole apparatus, though, is lying horizontally on the Earth—we expect a longer travel time on this path than we do on this path.

Now, we're not going to know that just by looking at one interference pattern. We're just going to see one pattern of interference looking in the viewer—a sequence of light and dark bands. We don't know which path is longer at this point and which path is shorter, because we don't know if the two paths have exactly the same length, and so on. That's not important. What's important is what happens if we say, okay now let's rotate the apparatus to different angles. I mean, it might have been, we don't know which way the ether wind is blowing. We might not have had this configuration. Let's rotate the apparatus to a different orientation. There it is in a different orientation. If this picture is correct, and the ether wind is blowing at us from the right, then in this case the two paths are going to take the same amount of time if they have the same length. Whether they have the same length or not, that's not crucial.

What's crucial is the interference pattern we get, in this case, is going to be shifted somewhat from what it was before. Because the light path, the light traveling the two different paths is going to take different amounts of time than it did before. And consequently, we're going to see a shift in the interference pattern as we rotate the apparatus. Let's rotate it another 45 degrees until it's now 90 degrees from where it was before.

Now, this beam, which originally was going against and with the ether wind, is now going across it. So now the travel time on this path is shorter. And the travel time on this path, which was originally the path that was perpendicular to the wind, this path is now parallel to the wind and its travel time is longer. And that's going to result, again, in a shift of those interference fringes. And

I want to emphasize this about this experiment; this is not an experiment to measure the speed of light in two mutually perpendicular directions. It's an experiment to determine whether the speed of light differs in two mutually different, mutually perpendicular, directions. And the way we detect that difference is by looking at the interference pattern. In one case, we learn nothing from that single case. We rotate the apparatus and we see if the interference pattern shifts. And the amount of shift tells us something about how different the times were along those two paths.

So you say, well, maybe Michelson and Morley just could have done this experiment in one day just sitting there rotating the apparatus and seeing how much the interference pattern shifts. But they had a worry about that. They were worried, what if, by some unlucky chance, on the day they chose to do the experiment, the Earth in its orbital motion happened, at that moment, to coincide with the ether. So at that moment the Earth happened to be at rest with respect to the ether.

Now, that doesn't bother us so much. That doesn't violate the Copernican paradigm so much, because it's okay instantaneously, by chance, for the Earth to happen to be at rest relative to ether. Pretty soon it's not going to be. And, in fact, six months later, it's going to be moving in the opposite direction in its 20-mile-a-second orbit around the sun. That'll give it a 40-mile-a-second speed difference from when it was at rest with the ether. And, so, if we do the experiment at different times of year, we should be able definitely to detect this change in the interference pattern and, therefore, the motion of the Earth through the ether.

Well, what happened when the experiment was done in 1887? What happened was, there was never, never, in any orientation, at any time of year, there was never any shift in the interference pattern. None. No shift. No fringe shift. Nothing. What's the implication? Here's an experiment that was done to measure the speed of the Earth and direction of the Earth's motion through the ether. It was an experiment that was 10 times more sensitive than it needed to be. It could have detected speeds as low as two miles a second, instead of the known 20 miles a second that the Earth has in its orbital motion around the sun. It didn't detect it.

What's the conclusion from the Michelson-Morley experiment? Well, the conservative conclusion is, there's no fringe shift. But what's the implication of that conclusion? The implication of that conclusion is the Earth is not moving relative to the ether. If the Earth were moving relative to the ether, the path lengths, the times to travel the two paths in the Michelson-Morley apparatus would have differed, either at different times of year or in different orientations. We would have detected that as a shift in the interference pattern. A shift that was easily measurable for speeds much less than the known speed of the Earth in its orbit. And we would have seen that shift. No shift. So-called null result. One of the most famous experiments in physics did not give the result it was intended to give.

Michelson and Morley confidently expected their experiment would measure the direction and speed of the Earth's motion through the ether. Result, zero. No interference pattern shift. No fringe shift. No motion through the ether.

Now, think back to the logic that got us to this point. We argued that the Earth is either at rest with respect to the ether or it's not at rest with respect to the ether. We ruled out it being at rest with respect to the ether, because we ruled out Earth alone among all the things in the universe being at rest—that totally violated the Copernican paradigm. We ruled out the possibility that there was a blob of ether surrounding the Earth that was being dragged along with Earth. So, locally, the Earth was taking the ether in its vicinity with us, and, locally, the Earth was at rest with respect to the ether. We ruled that out from the observations of aberration of starlight. That ruled out the possibility that Earth is not moving relative to the ether. The only possibility left is the possibility that Earth is moving with respect to ether. That has got to be true. We're forced to that conclusion by the beginning of this lecture.

Now, we've gone through the Michelson-Morley experiment. Null result. No fringe shift. No detection of any change in the speed of light in different directions. Conclusion, the speed of light is the same in all directions. How could that be? That could be only if the Earth is at rest with respect to the ether, because as I described earlier with that sound example, if the Earth is moving through the ether, then the speed light will differ in different directions. And that's exactly what the Michelson-Morley experiment was designed to detect.

So we have a real serious contradiction. We've ruled out the possibility that the Earth is at rest with respect to the ether. We confidently do an experiment that's designed to show us exactly how fast and in what direction Earth is moving relative to the ether. The experiment is sensitive enough to measure that. And it gets the answer, zero. It says the Earth is not moving with respect to the ether. And that's exactly the possibility we ruled out.

That's a deep and serious contradiction that hits physics in the 1800s, particularly in 1887. How can that be? Well, it took a lot of physicists thinking about that. What could be going wrong? Is there something wrong with the basic hypothesis of the Michelson-Morley experiment? Is the experiment not sensitive enough? No. The experiment is sensitive enough. Does wave interference not really occur? No. Wave interference had been discovered, for light, in the early 1800s and was a well-known phenomenon. Light was known to be a wave. It was known to interfere. There's nothing new in all that. What's new is that Michelson and Morley are masters of experimental technique and were able to put together an experiment that was clearly and demonstrably able to measure the effect they were looking for.

So, what could be wrong? Well, there were some hypotheses developed. The Irish physicist Fitzgerald and a Dutch physicist Lorentz, both pretty much independently came up with one hypothesis. And these two physicists deserve some of the credit for the development of relativity, because their hypothesis was, in one sense, correct and in another sense not correct. They assumed that somehow the motion of the experiment through the ether caused the whole apparatus to compress in the direction it was moving. As though the push against the ether—which is strange, because the ether doesn't exert any resistance to objects like planets moving through it—somehow the push against the ether or something, caused the apparatus to shrink. If the apparatus shrank, let's look at that apparatus again, in the direction of its motion, then this leg, the leg running parallel to the ether wind, the leg on which light was going to take a longer time to travel because it had to go upstream and then come back, if that length were shortened by just the right amount, that would compensate for the longer travel time, the slower speed of light going against the ether wind. And Fitzgerald and Lorentz proposed that that's what happened. They called this, it's now called the Lorentz-Fitzgerald contraction hypothesis, sometimes just called the Lorentz contraction, they calculated how much that contraction would have to be in terms of the speed of light and the speed of the apparatus relative to the ether, and they found there was an answer they could get that would allow that effect to go away; it would allow the experiment to give no fringe shift, the experiment to be unable to measure the speed of the Earth relative to ether because the whole apparatus had shrunk. It's a strange hypothesis. It's an ad hoc hypothesis. We'll see later that it has seeds of truth in it, but it has absolutely no theoretical basis. And it still assumes that there is an absolute frame of reference in which, if we put this apparatus, it has the right length, and if you move it relative to the ether it somehow gets squished. A very ad hoc assumption no physics behind it, no theoretical basis. Very unsatisfying.

This is now the late 1800s. The stage is set for 1905 when Albert Einstein is going to come on the scene, and Einstein and his resolution of this contradiction will be the topic of our next lecture.

Einstein to the Rescue
Lecture 7

It's now the year 1905, and physicists have been puzzling for several decades over the null result of the Michelson-Morley experiment. Why was there no fringe shift? Why didn't we detect the Earth's motion relative to the ether? The best we have going is the Lorenz-Fitzgerald contraction hypothesis suggesting the apparatus got compressed.

In 1905, after 10 years of pondering the nature of light and of time, Einstein resolved the contradiction with his *special theory of relativity.* As early as age 16, Einstein had puzzled about what a light beam would look like if one ran alongside it at speed *c*—a puzzle that he eventually solved with relativity. Einstein was 26 and a young father at the time of his 1905 paper on special relativity. He was working in the Swiss patent office because he had not been able to secure an academic position. In that same year, he published three other scientific papers, two of which were also seminal works. One provided the final convincing evidence for the existence of atoms, and the other helped lay the groundwork for quantum physics.

As early as age 16, Einstein had puzzled about what a light beam would look like if one ran alongside it at speed *c*—a puzzle that he eventually solved with relativity.

Even though I have stressed the logical dilemma posed by the Michelson-Morley result, historians of science debate whether Einstein even considered this result in developing his theory of relativity. Einstein's reasoning was based at least as much on how he felt nature should be as it was on experimental results.

Einstein declared the ether to be a fiction in his paper "On the Electrodynamics of Moving Bodies." Instead, he asserted the principle of relativity for all of physics, electromagnetism as well as mechanics. Thus, the essence of Einstein's theory is summed up in a single statement, the *principle of special relativity*: *The laws of physics are the same for all observers in uniform motion.* What's *special* about special relativity is the restriction to the case of uniform motion (later, we will take up the general theory, which removes this restriction).

(As a historical note, Einstein actually proposed two postulates as the basis of his theory: the principle of relativity and the constancy of the speed of light. A more modern approach is to consider the latter a consequence of the former.)

Einstein's relativity is both radical and conservative—conservative because it asserts for electromagnetism what had long been true in mechanics, namely that motion doesn't matter (recall Galilean relativity from Lecture 3). Relativity is radical because it radically alters our notions of time and space. From a modern perspective, Einstein's relativity should come as no surprise. You already agreed to the principle of relativity in Lecture 1, when you recognized that both tennis (mechanics) and microwave ovens (electromagnetism) should work the same in any uniformly moving reference frame. Special relativity is really more than a theory; it has been verified and is unlikely to be refuted.

Why did it take a genius like Einstein to recognize the truth of relativity? Because relativity requires us to relinquish deeply ingrained ideas about space and time. The principle of relativity means that the laws of physics are exactly the same for all observers in uniform motion. In particular, the predictions of Maxwell's electromagnetic equations will be the same for all observers.

One such prediction is the existence of electromagnetic waves—including light—that travel with speed c. Therefore, all observers will measure the same value c for the speed of light—*even though different observers are moving with respect to (or relative to) each other*!

To see what this means, imagine I'm standing by the roadside, equipped with a device for measuring the speed of light. The device uses a very fast clock to time the passage of light over a known distance of exactly 1 meter. You drive by in a car at 70 mph that is equipped with an identical apparatus. Down the road, a traffic signal flashes, and we both measure the speed of the light as it passes us. Despite the fact that we're in relative motion, we both get exactly the same speed for the light! We get the same results if we repeat the experiment with you going past in a jet plane at 600 mph or even in a spacecraft at half the speed of light! How can this be? It's possible only if our measures of time and space are different. Time and space are not absolute, but are relative to a particular observer. ■

Essential Reading

Hoffmann, *Relativity and Its Roots*, Chapter 5, pp. 81–95.

Mook and Vargish, *Inside Relativity*, Chapter 3, Sections 1–6.

Suggested Reading

Einstein and Infeld, *The Evolution of Physics*, Chapter 3, pp. 187–209.

Questions to Consider

1. The speed of light is the same for anyone who cares to measure it. Explain how this fact follows from the principle of relativity.

2. The observer standing by the roadside in our example measures c for the speed of the light from the traffic signal. If you didn't know about relativity, what speeds would you infer for the light as measured by observers in the car, the airplane, and the spaceship?

3. What's wrong with the following explanation of the fact that all observers in our example get the same value for the speed of light? "Strange things happen to time and space when you move, so the moving observers' clocks and meter sticks are distorted in just such a way that they all get the same value for the speed of light as does the stationary observer."

Measuring the Speed of Light

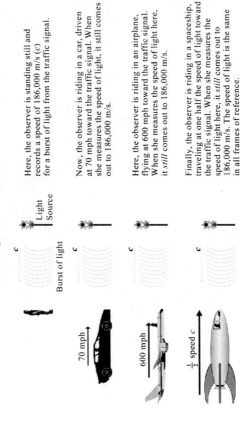

Here, the observer is standing still and records a speed of 186,000 m/s (c) for a burst of light from the traffic signal.

Now, the observer is riding in a car, driven at 70 mph toward the traffic signal. When she measures the speed of light, it still comes out to 186,000 m/s.

Here, the observer is riding in an airplane, flying at 600 mph toward the traffic signal. When she measures the speed of light here, it *still* comes out to 186,000 m/s.

Finally, the observer is riding in a spaceship, traveling at one half the speed of light toward the traffic signal. When she measures the speed of light here, it *still* comes out to 186,000 m/s. The speed of light is the same in all frames of reference.

Einstein to the Rescue
Lecture 7—Transcript

It's now the year 1905, and physicists have been puzzling for several decades over the null result of the Michelson-Morley experiment. Why was there no fringe shift? Why didn't we detect the Earth's motion relative to the ether? The best we have going is the Lorenz-Fitzgerald contraction hypothesis suggesting the apparatus got compressed. The idea being that all objects perhaps get compressed in the direction of their motion relative to the ether. Physicists are reluctant to give up this notion of the ether for reasons that will be very obvious in just a moment. You may think we can give it up but hold on.

Let me begin with a few historical touches here about Einstein and the year 1905 and what Einstein was like. One hears a lot about Einstein. One hears, for example, that he was actually rather dull as a child. That's not true. But what he was in school was very rebellious. And some of the trouble he got into in several schools, especially a very rigorous German school he attended, was that he simply couldn't put up with rigorous discipline. He wanted to study and work on his own, and this led him to problems throughout his academic career. But he was not dull. He was noted for his intelligence, although he didn't necessarily do well in all his courses. In particular, he didn't do particularly well in some mathematics courses. And one of his mathematics professors, named Herman Minkowski, was very distressed about this. And Minkowski, nevertheless, later went on to become a major interpreter of relativity and mathematical terms. So Einstein was a rebel, but he wasn't dull.

There's also a rumor that he talked rather late. And this is probably true. And Einstein, himself, probably embellished this a little bit. He claimed at some point that he didn't talk until he was three years old. And, perhaps, his talking late had something to do with a brain in which he could visualize abstract ideas better than most of us can. I don't know about that.

Einstein was an interesting character. Although he published his paper on the special theory of relativity in 1905 and wrote it in a matter of only about six weeks after he got the key insight that led to the idea, he had been thinking about this problem for many, many years—when he was 16, for about 10 years, in fact. When he was 16, he asked the question, "What would a light wave look like if I ran along side it at the speed of light?" That's an interesting question. It may have occurred to some of you also. Why is it an interesting question? Well, it's a particularly interesting question because if you believe Maxwell's theory of electromagnetism, you know that that theory predicts the existence of light waves, and light waves should move at speed c. If you run along a light wave at speed c, you see a structure that looks like a light wave—electric and magnetic fields, in the particular configuration of a light wave—but it's not moving relative to you. That's not a solution of Maxwell's equations.

If you believe in the ether paradigm, you simply say, "Well, I'm not in the ether frame of reference so I don't expect Maxwell's equations to be correct. I don't expect the laws of electromagnetism to work in this frame of reference." But Einstein puzzled about this. He really worried, what would a light beam look like if I ran along side it?

Another thing that bothered Einstein came out of shaving. He's sitting there looking in the mirror, and he wondered, "What if I started to run with this mirror in front of me? Walk along. Still continue to shave. No problem. What if I run faster? What if I run at the speed of light? The light leaving my face that's going to bounce off the mirror is never going to get to the mirror, because the mirror's moving away at the speed of light. So as I reach the speed of light, will the mirror suddenly go dark?" By the way, *Einstein's Mirror* is the title of one of the more recent books I've put in my bibliography for this course. You might be interested in reading it. It draws its name from this particular pondering of Einstein's.

"What would happen if I ran at the speed of light? The light wouldn't be able to catch the mirror. The mirror would go dark." Well, those are things Einstein pondered about, before 1905. When he reached 1905, he was still a young man. He was only 26 years old. Most of us think of Einstein as in the picture I'm showing here, as a very old, rather interesting looking character with this grizzled face and this wild, white hair. We forget that Einstein did some of his most creative work, in fact, basically all his most creative work, as a very young man. And in 1905, the year he formulated the special theory of relativity, Einstein was a young father. Twenty-six years old. That's four years beyond college graduation age. He had a young child and a young wife who was a fellow physics student of his. And he was working in a position as a patent clerk in the Swiss patent office. He actually enjoyed that position because he was working with patents and technological devices. So, that was interesting to him. He also had plenty of spare time to work on his own ideas in physics, including the theory of relativity.

Why was he a patent clerk rather than a physics professor? Because of all the classmates of his from his graduate school class, he was the only one who was not recommended for a professorship upon graduation. He had trouble finding an academic position. Again, not because he wasn't smart, but because he walked to his own drummer. He was a rebel. He was difficult to work with in an institutional setting. He simply was not recommended for a professorship. And it was a while before he achieved a professorship. Once his relativity fame became established, it was no problem for him.

But in 1905, he is working in the Swiss patent office. He's watching a lot of trains and clock towers in Switzerland. Thinking a lot about time and space. And he's writing papers. And he actually published a number of papers in 1905—four significant papers and some smaller ones. And of the four substantial papers, three of them were really seminal in the history of physics. One of them, the one I'm going to talk about today, is the paper that introduced to the world special relativity.

The second paper was a paper that was on the subject of Brownian motion, which is the behavior of very small particles—little dust motes, or little bacteria, or things like that—that you might see through a microscope as they're jostled around by the invisible particles that make up, say, a solution like water. And Einstein published a paper on Brownian motion. And that was the paper that convinced the few remaining skeptics—in 1905 there were still some skeptics who didn't believe there were atoms, they didn't believe in the reality of atoms—and Einstein's paper on Brownian motion convinced the remaining skeptics of the reality of atoms. So, that was a seminal paper in physics.

The third important paper is one I'll have a lot more to say about later in the course when we deal with quantum physics, because Einstein interpreted the results of the so-called *photoelectric effect* experiment—which, again, we'll describe in detail—to give an interpretation in terms of the new quantum ideas. He was the second person to introduce the idea of the quantum. He was the first person to apply it to the nature of light. And I'll have a lot more to say about that. That was a really crucial and important paper.

And the fourth paper of the batch of four significant papers was not particularly interesting. It was a run of the mill paper. Later in that year, he also published a paper in which he introduced the seeds of the idea behind the equation $E=mc^2$. And I'll have a lot more to say about that. But at this point, suffice it to say that $E=mc^2$, which you may think of as the essence of relativity, was not even in the original paper on relativity that was published in 1905.

So, what did Einstein have to say? He published a paper, which was not called the special theory of relativity, it was called, "On the Electrodynamics of Moving Bodies." And I want to emphasize that title, because I argued, when I introduced you to electromagnetism, that electromagnetism was crucial in understanding the evolution of relativity theory. Relativity is, really, an outgrowth of electromagnetism. And you now know why. It's because electromagnetism raised the questions, with respect to what does light go at speed c, in what frame of reference are the laws of electromagnetism valid? A question that had already been answered for mechanics, for the study of motion, way back in Galileo's time, and that now begged for an answer in terms of electromagnetism. And we've been looking for that answer. And we got some logic that told us that the answer should be: There is some motion relative to the ether. And then we did the Michelson-Morley experiment and found there was no answer to that question. We are still left with that question—in what frame of reference are the laws of electromagnetism valid? And Einstein's paper, a direct outgrowth of electromagnetism, was called, "On the Electrodynamics of Moving Bodies."

By the way, I have to give you one other historical aside here, which I find fascinating and a real reflection of Einstein's genius. There is, among historians of science today, still some debate about how much the Michelson-

Morley result influenced Einstein's thinking. There have been some people who've argued in the past that he may not even have known about the Michelson-Morley result. That's probably not the case. He makes reference in his paper to "unsuccessful attempts to find the light medium." He's surely referring there to some of the early experiments at trying to detect Earth's motion through the ether. And he may well be referring to the Michelson-Morley experiment. But how much the Michelson-Morley experiment influenced his thought, we don't know. And it may not have been very much. For us, learning about relativity, it's crucial to understand the Michelson-Morley experiment because we need that absolute hard and fast logic that tells us inexorably we have to get to relativity.

Einstein, who thought about how the world ought to be, and said, "I think the world ought to be simple, I think it ought to behave this way," he may not even have thought much about the Michelson-Morley experiment. He may have thought, instead, that this is just the way the world has to be. I need a simpler world than the one I see. And here's the theory that describes that.

So what did Einstein do? Well, basically, he declared the ether to be a fiction. Something you may have had at the back of your mind in the last few lectures. Why are we going to all this trouble about the ether if we're going to have trouble finding the Earth's motion through it? The problem is, once you give up the idea of ether, you're left with that question, with respect to what does light go at speed c? In what frame of reference are the laws of electromagnetism valid?

So what did Einstein do? He declared the ether as a fiction. No more ether. And he asserted, instead, the principle of relativity; the same principle of relativity that Galileo had talked about. The principle of relativity that in Galileo's time said the laws of physics are the same for all observers in uniform motion. And it said that because in Galileo's time the laws of motion were basically all there was to physics pretty much. We've modified that since to say, Galilean relativity states the laws of motion, the laws of mechanics, are valid in all uniformly moving frames of reference. And then we introduce this new branch of physics—electromagnetism. And we ask the same question for electromagnetism, in what frame of reference are the laws of electromagnetism valid, with respect to what frame of reference does light go at speed c?

Well, Einstein declared the ether was a fiction. And he asserted the principle of relativity again, for all of physics. In that sense, relativity is a profoundly conservative step. What had been known for hundreds of years—that the laws of physics or the laws of mechanics, at least, didn't depend on your state of motion as long as that motion was uniform—that had been known for hundreds of years. That's the Galilean principle of relativity. Einstein simply reasserted that relativity was true for all of physics. It's that simple.

And I introduced you to this idea in the very first lecture. And I introduced you to it through the microwave oven and the tennis match. The tennis match represents motion—mechanics. The microwave oven represents electromagnetism. And I think you drew the conclusion in your mind that it made eminent sense, that both those things, the game of tennis, a physics experiment in mechanics, and the microwave oven, a physics experiment in electromagnetism, would work the same on the cruise ship, on Venus, and even on that distant planet in a distant galaxy moving away from Earth at 80 percent of the speed of light.

So to you, this probably makes very good sense. In fact, it probably makes more sense to us at the turn of the 21st century than it did to people at the turn of the twentieth century who were steeped in the idea that there had to be an ether. To us we know about the existence of distant galaxies moving away at 80 percent of the speed of light. And it makes eminent sense that microwave ovens should work the same there as they do here. So, I think you can accept what Einstein had to say, at least at present, in a sort of gut way that maybe wouldn't have been possible around 100 years ago. You can accept the principle that laws of physics are simply the same, all laws of physics, for anybody who is in uniform motion, in any reference frame, uniform motion. And that is all Einstein said. That is it.

Let's take a look at that statement. It's an important statement. It's special relativity. The statement of special relativity is, simply, the statement that the laws of physics are the same for all observers in uniform motion. And that's it.

I want to make a couple of comments about this statement. I mean, you can assert that to your friends and say, I know everything there is to know about Einstein's special theory of relativity. Because that is everything there is to say about it. The laws of physics, all the laws of physics, are the same for everybody in uniform motion. Any frame of reference in uniform motion is an equally good place for doing physics. Period. And that's it.

A couple of comments though. First of all, if you read Einstein's original paper, and you read older textbooks that talk about this, you'll find that Einstein didn't quite put it this simply. He had what he called two postulates. One postulate was this one—that the laws of physics are the same for all observers. And the other postulate, which he said seemed to be in contradiction with this but wasn't, is that the speed of light is independent of the speed of its source. I gave you some observational evidence earlier why that was the case. Why we know that's the case that the speed of light doesn't depend on the speed of the source, that evidence involving double star systems. Einstein asserted that as a second postulate. I think a more modern way of looking at relativity is to simply state this principle: That the laws of physics are the same for all observers in uniform motion. All frames of reference in uniform motion are equally good places for doing physics—that includes experiments in electromagnetism and measurements of the speed of light. And we'll get to that in just a minute.

So this suffices. One statement. The laws of physics are the same for everybody as long as they're in uniform motion. There is no preferred frame of reference. There is no preferred state of motion. The Earth is not special. That distant planet in that galaxy moving with respect to Earth at 80 percent of the speed of light, that is not special. No place, no state of motion is special, at least as long as it's uniform motion.

That gets to the second point I want to make about the statement of special relativity. You hear the word "special" you might think this is kind of neat. And you know, really super. Special doesn't mean that in this context. Special means this is a specialized theory. It applies only to a limited, special case. It applies to the case of uniform motion only. In the context of special relativity, nonuniform motion doesn't count. The laws of physics are not the same, at least in the paradigm of special relativity, in frames of reference that are undergoing non-uniform motion, that are accelerating, that are changing speed, that are rotating, that are airplanes going through turbulence, that are cruise ships in rough seas, that are planets in circular motion around their stars, maybe, that one's a subtle one, and we'll get to that later. In fact, as I argued a little earlier, it's sometimes difficult to decide whether we have a frame of reference in uniform motion. And all of special relativity has a slightly slippery foundation for that reason.

There is a general theory of relativity that generalizes this statement to all motion. That takes away the in uniform motion clause. But it's a much more difficult theory. We'll spend less time on it. But we will get to it later on in the course.

So, for now, we are talking about special relativity. And special relativity means specialized to the case of uniform motion. It doesn't say everybody is equal as far as the laws of physics are concerned. It does say everybody who is in uniform motion is equal as far as the laws of physics are concerned.

If you can find one frame of reference that's in uniform motion and the Earth, although it's not quite, is moving uniformly enough that special relativity applies to the Earth to a very good approximation. Not perfectly. But to a very good approximation. If you can find one frame of reference that is essentially in uniform motion, then any other frame of reference that is moving at a constant speed in a fixed direction relative to that one must also be in uniform motion. So, if you can find one, you can find others. And we at least can find approximate uniformly moving frames of reference, such as the Earth and the other planets, and spaceships moving at constant speed in space and relative to, say, the Earth, and airplanes moving through calm air, and so on. So we have frames of reference in which this is true. But again, we're going to get further a little bit later.

It's going to be important though to remember that special relativity works only in uniformly moving frames of reference. Because we are going to come across a situation in the next few lectures where we do have some nonuniform motion. And you have to remember that, because that nonuniform motion, the laws of physics do not work when the motion is nonuniform.

Now, I asserted that Einstein's relativity is a very conservative statement. It's a conservative statement because it reasserted, for all of physics, particularly for electromagnetism, what had already been true for the physics of motion—for Newton's laws, for the classical physics that came before the study of electromagnetism. Einstein reasserted relativity. He's very conservative. He says what Galileo already knew is now true for all of physics. But it's not only conservative, it's also radical. And it's radical because, as you will see in the rest of this lecture, it does very, very strange things to our notions of space and time. And that's what most people think of when they think of relativity. They think of $E=mc^2$. But then they immediately think of things like the twin example that I started this course with—an example that we will revisit in a couple lectures in some detail and understand exactly why it has to be.

Relativity does strange things to space and time. I didn't want to emphasize that up until now, because I want you to understand that relativity is much simpler than that. Relativity is the statement that the laws of physics are the same for all observers in uniform motion. And I want you to have two feelings about that statement. One is it makes good gut sense. Why is Earth special? Why should Earth or any other frame of reference be special? In an era when we know about distant galaxies moving away from Earth at 80 percent of the speed of light, it makes no sense to think that the laws of physics work here on Earth or in some frame of reference that Earth's not moving very fast relative to, but then they certainly don't work in that distant galaxy. What's so special about us? Nothing. So, it makes good gut sense.

I also want you to understand the inexorable logic that led us to it. The question: In what frame of reference are the laws of electromagnetism valid? The attempts to answer that question in nineteenth century physic's terms with the ether, the contradiction that said the Earth is not at rest with respect to the ether. Then the Michelson-Morley experiment coming along and said but the Earth is at rest with respect to the ether. That contradiction. We were driven inexorably to that.

The reason I want you to think both about your gut feel and that logic is because in the next few lectures we are going to be doing those very strange things to space and time. And you will find those things unacceptable. You will want to fight against them intellectually. But they must be the case. And they must be the case if you believe this simple statement that the laws of physics, all laws of physics, including electromagnetism, are the same for all observers in uniform motion. If you believe that, and you have the logic leading you to it, and you have the gut feel for a universe which is so vast and full of objects moving at such high speeds, that it ought to make sense. So this is the simple statement of relativity. This is special relativity. The laws of physics are the same for all observers in uniform motion.

By the way, although this is often called the theory of relativity, I really like to avoid the word theory, because it conjures up in most people's minds the idea this is just a theory. You know, evolution is just a theory. We shouldn't be teaching it in schools, because it's just a theory. We don't know if it's true. Well, that's hogwash. It's hogwash about evolution, and it's hogwash about relativity. Relativity, special relativity particularly, is one of the most solidly verified "theories" in all of physics. It's much more correctly verified than Newton's laws, which turn out only to be an approximation. Special relativity is true to many, many decimal places of accurate measurement as far as we can tell. It is a very accurate description of physical reality, at least when we are dealing with frames of reference in uniform motion. And we're avoiding questions of gravity. We'll get to those and general relativity.

So, I don't like to think of special relativity as a theory. Sure, it's a theory. Could it be proved wrong? Very, very, very unlikely. Could it be modified slightly? Probably not. It may be expanded upon, with the general theory of relativity, and maybe something beyond that, but it's not wrong. It's not *just* a theory. It is really one of the foundations of our understanding of physical reality. And along with it go these very strange things that happen to space and time.

Now, let me get to those things now. Why, after all, did it take Einstein's genius to come up with special relativity? Here it is, a simple statement. You can go out and make this statement to your friends. I know relativity. The laws of physics are the same for all observers in uniform motion. And that is really it. There is nothing else to it than that statement. There is no other foundation. From that, all else follows. And if you ever get stuck in relativity, just go back to that statement. It's all there. Why was that so hard? Well,

Einstein puzzled about it for 10 years. And six weeks, roughly, before the paper on special relativity was finalized, Einstein said, "I suddenly got the answer." The key was time. Einstein was able to give up his commonsense notion of time and accept something very strange and different about the way time behaves. And then he was easily able to come up with the special theory of relativity. And everything went ahead from there.

To see why it took an Einstein to do that, though, let's take a little bit more logic from this statement. Okay, here's the special theory of relativity or the special relativity—the laws of physics are the same for all observers in uniform motion. Einstein's statement in 1905. This includes electromagnetism. For the first time, Einstein brought electromagnetism under the same relativity principle that had already governed mechanics. And part of electromagnetism—and this is why I spent a whole lecture on electromagnetism and showed you Maxwell's equations and all the mathematics of them, because I wanted you to see that they led inexorably to the prediction of electromagnetic waves and that those electromagnetic waves move at a known, calculable speed c, 186,000 miles a second, etc.—includes that prediction.

In what frame of reference are the laws of electromagnetism valid? That's the question we've been asking for two lectures. Now we have Einstein's answer. And it's a very simple answer. For any observer in uniform motion, in any frame of reference in uniform motion, the laws of electromagnetism are valid. Any frame of reference in uniform motion. What that means is all predictions of electromagnetic theory are valid in any frame of reference in uniform motion, including the prediction that electromagnetic waves travel at speed c. And what that means is the answer to our second question: With respect to what does light go at speed c? Which we saw was equivalent to the question: In what frame of reference are the laws of electromagnetism valid? The answer to that question must also be the speed of light is c with respect to any uniformly moving frame of reference. Any uniformly moving observer will measure c for the speed of light. So, therefore, all observers in uniform motion will measure the same value for the speed of light.

So far so good. That is a direct consequence of our belief in the fact that the laws of physics are the same in all frames of reference in uniform motion. All laws of physics, including, as Einstein added, the laws of electromagnetism, with their prediction that there are light waves and those light waves move at the speed we call c. That prediction must be true in all uniformly moving frames of reference. Therefore, anybody who's in uniform motion measures the speed of some light, and they get the answer c for that speed. And here's where it gets difficult for you—even if they are moving relative to each other. And I believe that was difficult for Einstein. I believe it's difficult for any of us, in our gut, to accept this. But it has to be true if you accept the principle of relativity. If you accept the principle of relativity, then it must be true that all observers, as long as they're moving uniformly, will measure the same value for the speed of light. Even if they're moving relative to each other.

I want to give you an example that illustrates that. Let me begin with a picture of the example. We're going to try to measure the speed of light. So here you are or here I am standing by the roadside. And I'm looking at a traffic light. And that traffic light is going to flash and emit some light waves. And those light waves are going to go by me at speed c. And I'm going to measure that speed by taking some simple measuring apparatus. I'm going to take a meter stick, a stick that is exactly one meter long. And I bought it from a reputable company so I know it's exactly one meter long. And I'm going to take a beautiful, perfect clock, which can measure very, very precise time intervals. And what I'm going to do is I'm going to notice with my clock when the light reaches the front end of my meter stick and when it passes the back end of my meter stick. I'm going to measure on my clock the time that takes. I'm going to take speed equals distance divided by time. And from that I'm going to calculate a speed for light.

Now, that's not actually how we calculate the speed of light, of course. This clock would never do. This distance, well, it's a little short, but actually we could measure the speed of light over this distance. My undergraduates at Middlebury College, by the way, in their sophomore year regularly measure the speed of light over a distance of about 30 feet. They send a laser beam down to a mirror about 30 feet away. It bounces back. And they use a high-speed clock—it's a device called an oscilloscope—that notices when the light beam leaves the laser and notices when the light beam returns and makes a little blip on a screen. And the distance between those blips is a measure of how long the distance is, the time is, between the departure and the return of the light beam. And they can measure the speed of light to an accuracy of about a tenth of a percent. In a simple undergraduate laboratory. Thanks to modern technology, particularly a laser that we can switch on and off in about a nanosecond, in about a billionth of a second that, by the way, was built by another undergraduate of mine.

So this is not a difficult thing to do. And with a little more fancy technology, but certainly within the realm of what we have, one could easily measure the speed of light over distances as short as a meter. One simply couldn't do it with this clock. But I'm going to pretend we can, or I'm going to pretend that this clock has the capabilities to measure the speed of light, to measure times as short as the time it takes light to go a meter very, very accurately. By the way, since the speed of light is 300,000 kilometers a second, that's 300 million meters a second, that time is one 300-millionth of a second. That's roughly the time a fast computer does a calculation. So it's not an unheard of kind of time. It's about three or four nanoseconds, because light travels one foot per nanosecond, and that's about three feet. A meter is about three feet. So it's not an immeasurable time. We regularly measure times a million times shorter than that in physics laboratories. So that's a reasonable measurement to try to make.

Now, here's what I'm going to do. So, I'm equipped with this meter stick and this clock. And I'm going to measure the speed of that light as it goes by me. Now, you're going to go past in a car at 70 miles an hour. Before you got in the car, I equipped you with an identical meter stick and an identical clock. And again, we bought them from the same reputable company. They came of the same manufacturing run. They are identical. Absolutely identical. You're going by in that car. You take your meter stick. You hold it up. You notice when the light passes the front of your meter stick. You notice when the light passes the back of your meter stick. You do the calculation of velocity equals distance divided by time. And you come up with an answer for the speed of light. You are moving relative to me at 70 miles an hour. If that picture I showed you of waves, speed, and the ether worked, you would measure a speed for c of 186,000 miles a second plus 70 miles an hour. Whatever that comes out to be. You don't. According to the principle of relativity, you measure exactly the same speed I do. Exactly. Well, I'll give you the number. It's 299,798,458 meters per second. That's the exact value for the speed of light. It's an exactly defined value. It's so well known that it's now used to set the standard measure of length. The meter is now defined in terms of the speed of light rather than the other way around.

And we get exactly the same value. Exactly. Even though you are moving relative to me at 70 miles an hour. How can that be? Worse, I'll put you instead in a jet plane. Again equipping you with the same clock and the same meter stick. Now you're going 600 miles an hour relative to me. You're in uniform motion. The laws of physics are equally valid for both of us. And you measure the same speed for the light. Exactly 299,792,458 meters per second. Same answer. Same light, also.

Let's make it even worse. I'm going to put you in a high-speed spaceship. And it's going to be going past me at half the speed of light. Are you going to get one-and-a-half c for the speed of light you measure? If you think you do, then you are thinking nonrelativistic. They say, "Oh, it's only on Earth that the speed is really c." There's something special about Earth. It's the special frame of reference. Nonsense. There is no special frame of reference. All frames of reference in uniform motion are equally valid. And both those observers will measure c for the speed of light.

And to make that clearer, let me take away the ground, which doesn't matter at all, and take away the traffic light—just some light going by and we're both going to measure it. And I'm—this is now from my point of view—I say, "Oh, I'm at rest, light's going by me at c, you're going by me at one-half c, and yet you measure the same speed for light that I do." You are equally equipped to say, or equally justified in saying, "Oh, I'm at rest in my spaceship and that other observer is going past me in the other direction at half c, because no frame of reference is preferred. All frames of reference in uniform motion are equally valid."

How can this possibly be? It must be that something funny has happened to clocks and meter sticks. But you're not justified in saying, "Oh, something funny happened to the spaceship measuring instruments, because the spaceship is moving." No, there is nothing special about your frame or the spaceship's frame. Something funny must have happened, not so much to the clocks and meter sticks, but to space and time themselves. And that is what we will be exploring in the next few lectures.

Uncommon Sense—Stretching Time
Lecture 8

> How can it be that if you move past me at half the speed of light, and we measure the speed for the same light beam, that we get the same value when you're moving relative to me at such a high speed? It violates everything we know from our common sense. How can it possibly be?

The simple statement of relativity—that the laws of physics are the same for all observers in uniform motion—leads directly to absurd-seeming situations that violate our commonsense notions of space and time. But common sense is built of our limited experience, in which we never move at speeds anywhere near c relative to things in our immediate environment.

Our common sense isn't wrong; it's just an approximation that works when relative speeds are small compared with c. We simply don't move at speeds anywhere near c relative to the objects with which we interact. (We do move at speeds near c relative to distant galaxies and cosmic rays, but most of us aren't directly aware of that.) To describe the universe accurately and to understand what happens at high relative speeds, we need to abandon common sense and instead embrace the principle of relativity. When we do, we find that measures of time and space differ in different frames of reference.

It is all too easy to speak in a way that violates the principle of relativity; even many books on the subject do. We need to make an effort to be *relativistically correct* (RC) in describing what happens to space and time. In the traffic light example, for instance, I can't dismiss your strange clock readings because "you're moving." That's because the principle of relativity denies the concept of absolute motion. Both of us are equally positioned for doing physics, and your measurements in the spaceship zooming past Earth at $0.5c$ are every bit as good as mine on Earth. It could be that the different clocks keep different time.

Let's consider the concept of "time dilation" or the stretching of time. A "light clock" consists of a box containing a light source at one end and a mirror at the other. A flash of light leaves the source, bounces off the mirror, and returns to the source. This process repeats, so the round-trip travel time for the light constitutes the basic "ticking" of this clock. We want to examine the length of this "ticking"—that is, the time between the event of light leaving the source and returning to the source—in two different frames of reference.

First let's define "event": An event is something that happens at a time and a place. To an observer at rest with respect to the light clock, the light makes a round-trip journey twice the long dimension of the box. To an observer relative to whom the box is moving, the light takes a longer path, following two diagonals that are each longer than the long dimension of the box. Here's where relativity comes in: The speed of the light is the *same* for both observers. Because the path lengths for the light are different, however, so must be the times between the emission of the light and its return to the source!

The phenomenon of different times for different observers of the light clock is called *time dilation*. There's nothing special about the light clock; it's just a convenient device for visualizing why time dilation occurs. Times on ordinary clocks would show the same discrepancy. Time dilation is not about light or light clocks; it's about *time itself*. Measures of time are simply different for different observers in motion relative to each other—a consequence of the principle of relativity.

Time dilation is often described by saying that "Moving clocks run slow," but this is very poor wording and not RC. Why not? Pause a minute and think about that question before proceeding. In particular, an observer moving with the light clock would feel *nothing unusual whatsoever*! Things wouldn't seem to be happening in slow motion nor would anything else seem strange. The frame of reference of the light clock is just as good as any other uniformly moving frame, so all physical events happen perfectly normally.

What time dilation really says is this: Suppose there are two events that occur at different places in some frame of reference; in that reference frame, the time between the events is measured by a pair of clocks at the two different places. If another clock moves so that it is present at both events, then the time between the events as measured on that single clock will be less than that measured by the pair of clocks.

Let's try a quantitative example. Let t' be the time measured by the observer at rest with respect to the light clock, and t, the time measured by the observer for whom the box is moving. Then the two times are related by $t' = t\sqrt{1-v^2}$ where v is the speed of the light clock relative to the observer for whom it's moving, with v given as a fraction of the speed of light (i.e., $v = 0.5$ is half the speed of light). Anyone who remembers the Pythagorean theorem and who is fluent in high school algebra can derive this result, but we'll leave that as an exercise for those who want to try it. ∎

Essential Reading

Hey and Walters, *Einstein's Mirror*, Chapter 3, pp. 46–55 (but watch out for relativistically incorrect wording in one heading!).

Hoffmann, *Relativity and Its Roots*, Chapter 5, pp. 96–106.

Mook and Vargish, *Inside Relativity*, Chapter 3, Sections 7–10.

Suggested Reading

Thorne, *Black Holes and Time Warps: Einstein's Outrageous Legacy*, Chapter 1.

Questions to Consider

1. What's wrong with the statement "Moving clocks run slow"? Can you find this or a similar "relativistically incorrect" statement in a book on relativity?

2. (For the mathematically courageous; you need not be able to do this to understand relativity!) Using high-school algebra and the Pythagorean theorem, derive the time-dilation equation $t' = t\sqrt{1-v^2}$.

Time Dilation : A "Light Clock"

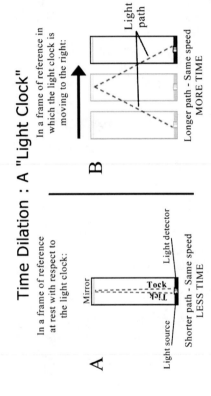

A

In a frame of reference at rest with respect to the light clock:

Mirror

Tick
Tock

Light source — Light detector

Shorter path - Same speed
LESS TIME

B

In a frame of reference in which the light clock is moving to the right:

Light path

Longer path - Same speed
MORE TIME

The light clock is a device that clearly shows how time dilation occurs. In this device, light is produced from a light source at the bottom of the clock, travels up the length of the light clock, bounces off a mirror, and returns to hit a light detector on the bottom of the clock. In A, a frame of reference in which the light clock is at rest with respect to the viewer, the path of the light is exactly equal to twice the height of the clock. In B, a frame of reference in which the light clock is moving to the right with respect to the viewer, the light path is longer. *But the speed of light has the same value for both observers* and thus the time as measured in frame B is longer than the time recorded by clock A.

TIME DILATION

~~"Moving clocks run slow"~~ — ***Bad Wording : Not RC !***

The time between two events is shorter when measured by a single clock present at both events than when measured by two different clocks located at the two events.

In this diagram, time dilation is explained in terms of "real" clocks rather than light clocks. Box One sets the stage for the experiment. Two clocks (B_1 and B_2) are located at two different places. Clock A is a clock whose state of motion is such that it will be present at B_1 and then later will be present at B_2. The two **events** that we are interested in are the concurrence of A with B_1 and the concurrence of A with B_2. Time dilation states that the time elapsed as recorded by clock A which is present at both events will be less than the elapsed time as recorded by the two other clocks. Box Two shows the results: t' is the elapsed time recorded by A. t is the elapsed time as recorded by the two B clocks. The equation in the center of Box Two describes the mathematical relationship between t' and t.

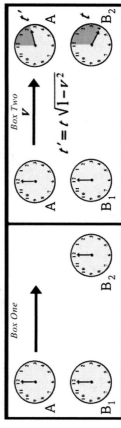

$$t' = t\sqrt{1 - v^2}$$

Uncommon Sense—Stretching Time
Lecture 8—Transcript

In the preceding lecture I equipped you with this very high precision clock and this precision meter stick. And I equipped myself with a pair of them also—from the same manufacturer; they're identical. I sat on the roadside. And I put you, first, in a car moving at 70 miles an hour, and then I put you in an airplane moving at 600 miles an hour, and then I put you in a spacecraft moving past me at half the speed of light. And we applied the principle of relativity to this situation. And we asked, what would happen if each of us measured the speed of the same pulse of light that happened to come from a traffic signal, but could have come from anywhere else, and we arrived at the conclusion that seems absurd to us—that we, in all those situations, measured identically the same speed for light. Even though you and I were moving relative to each other, at one point at half the speed of light. Nevertheless, with our identical clocks and meter sticks, we reached the same conclusion for the speed of light—299,792,458 meters per second. Period. No discrepancies.

The fact of your motion relative to me made absolutely no difference. Why? Because we agree that the laws of physics are the same in all uniformly moving reference frames. That's all special relativity says. And if that's true, one consequence of the laws of physics, the laws of electromagnetism in particular, is that there should be electromagnetic waves and they should move at speed c. That's what light waves are, and, therefore, we all have to get the same value for the speed of light.

But, how can it possibly be? How can it be that if you move past me at half the speed of light, and we measure the speed for the same light beam, that we get the same value when you're moving relative to me at such a high speed? It violates everything we know from our common sense. How can it possibly be?

Well, it seems absurd, actually, for one particular reason—because it violates our common sense. And now I want to insult all of my audience. The Teaching Company audiences tend to be a fairly sophisticated lot. But you're actually very provincial. You may think yourself a world traveler and have been all over the place, but you've spent all your time basically crawling around at relatively slow speeds. The fastest you've ever gone is 600 miles an hour in a commercial jet. Maybe a few of you have gone supersonically in the Concorde or a military plane or some thing. The fastest you've gone, maybe, at most, is a couple thousand miles an hour, relative to this hunk of rock that we call planet Earth. That is our home.

You have very little or no experience of moving at high speeds relative to anything that seems to be significant to your life. So you have a very limited provincial experience of the range of possible activities that are in this universe, including moving relative to things that are important to you at high speeds. Now, I have to put in a little parenthetical remark here. You actually do move at speeds close to the speed of light relative to some things. They are just things that aren't particularly important in your life. For example, we've been talking throughout the course about that distant planet in that distant galaxy that happens to be moving away from us at 80 percent of the speed of light. Relative to that, you are moving at 80 percent of the speed of light. And if you had business to communicate with the beings on that planet, you would certainly have to take relativity into account. You move at the speed of light relative to cosmic rays, which are coming toward Earth at 0.99999 the speed of light. They don't play a major role in your life. You hardly know about them. But, if they did, you would have to take in to account relativity. And it would be much more obvious to you.

When you're sitting watching this course on video, for example, the electrons that paint the picture on your TV screen are coming at you from the back of the TV tube at about 30 percent of the speed of light. Relative to them, you're moving at 30 percent of the speed of light. That's not a huge speed, compared to light. But it's significant. It doesn't affect you because you don't think about those electrons when you watch TV, but the engineer who designed your TV picture tube sure had to take relativity into account or otherwise the TV picture would be out of focus.

So there are people in the world for whom motion at speeds near that of light is important. An astrophysicist studying that distant galaxy does have to take relativity into account. An engineer designing that TV tube does have to take relativity into account. A cosmic-ray physicist does have to take relativity into account.

Now, where did our common sense come from? Well, think about how we developed our notions of time and space. When you were a baby and were crawling around, you began to get a sense of what space meant. What it meant to get from one place to another. When you were a little kid and you were sitting in the back of the car on a long trip and saying, "Daddy, are we there yet?", you were beginning to develop a sense of time. And that sense of time and space that you developed were coming out of that very limited, provincial experience of, as I say, crawling. And I use that term to mean moving slowly compared to the speed of light. Even 2,000 miles an hour in a supersonic airplane is far, far, far slower than the speed of light. You crawl around on this hunk of rock, and that is your limited experience.

I would assert that if you had grown up able to crawl at 90 percent of the speed of light, relativity would have been obvious to you. You wouldn't need this course. Nobody would need to go to school to study relativity, it would be intuitively obvious to us. If we were beings that were somehow based in interstellar gas clouds or something, as a science fiction novel by Fred Hoyle once described, and we tended to move around the universe at speeds near the speed of light all the time, then relativity would be intuitively obvious to us. It would be our common sense. The reason it's not common sense is simply because it's not in our everyday experience.

In order to appreciate relativity and to appreciate the richness of a universe in which the relativity principle holds, I'm going to have to ask you to suspend common sense. It's a very difficult thing to do. I'm not going to ask you to give up common sense entirely, although I'm going to ask you to expand on it. To enlarge the possibilities. That's what relativity asks us to do. And I think it's remarkable that this very simple principle—that the laws of physics are the same in all uniformly moving frames of reference, it sounds very benign—is so immediately at such odds with our common sense. But it is. And it is for the reason, again, that our common sense is built out of very limited experience. If you can abandon or at least agree to stretch your notion of common sense, then you're going to encounter the richer universe that relativity offers us.

Now let me make very clear, common sense isn't wrong. If it were wrong, common sense wouldn't tell us how to throw a tennis ball, or drive a car, or do many of the other "physics experiments" that we do here on planet Earth all the time and do very successfully. It's not that our common sense is wrong. It's that our common sense is limited. Newtonian physics, if you will, is the physics of common sense. It was built on our common sense notions of space and time. Those notions proved to be limited. Relativity tells us that, and the example from the last lecture of you whizzing by me in a spaceship at half the speed of light and are still both measuring the same speed for that light, that tells us that our common sense notions of space and time simply have to be wrong.

But when the relative speeds involved in our lives are small compared with c, small compared with the speed of light, then our common sense notions are good enough. They're not perfect. Even me pacing very slowly around this room, there are slight differences in space and time for me than for you because of my motion relative to you. These effects are so small that they're essentially negligible in our everyday lives. Even when you're traveling in a jet airplane at 600 miles an hour, or in a supersonic plane at 2,000 miles an hour, or even in an orbiting satellite going at 25,000 miles an hour, 17,000 miles an hour, 25,000 miles an hour on the way to the moon, these effects are very, very, very small. They're measurable. And there are a few instances—one of which is the global positioning system of satellites, which I'll talk about a little bit later in the course—where relativistic effects, even at the kind of speeds we use in everyday space travel or even airplane travel, have measurable, relativistic effects. But for most of us, those are completely negligible.

So there's nothing wrong with our common sense. It's just very limited. In fact, mathematically, if you take the equations of relativity and you reduce all velocities to something much, much smaller than c, then those equations reduce to the ordinary equations of physics that we knew about before relativity. So common sense isn't wrong. It's just limited. And what I'm doing here is asking you to expand your common sense.

I'm going to also ask you to watch your language. And this is a plea, not only to you in my audience, but also to some of my fellow teachers of relativity, because it's very easy to make statements that violate the spirit of the relativity principle. You've heard about political correctness on our campuses. You have to make statements that are in tune with the prevailing political norms. Well, I don't want to get into that. But I do want to ask you, and ask myself, and ask my fellow teachers of relativity to be relativistically correct in our language. And what that means is when you're describing what happens to space and time, be very careful to avoid any statements that might suggest there is a preferred frame of reference. A preferred state of motion. Why? Because there isn't a preferred frame of reference, and there isn't a preferred state of motion.

The Copernican principle has been expanded, through special relativity, to say that all uniformly moving frames of reference are equally valid places, situations to do physics. So, avoid any statements that would suggest anything to the contrary.

For example, in the last lecture, when we put you in first the car, then the airplane, and then the spaceship, and we said, well, we both measure the same speed for light, and we said how could that possibly be? And I simply hinted at the end of the lecture that it must mean that something is different about our measures of space and time.

You might jump to the conclusion that, well, maybe the meter stick in the moving vehicle shrinks. Maybe something funny happens to the clock in the moving vehicle because the vehicle is moving. That's sort of the philosophy that Lorentz and Fitzgerald had when they suggested, maybe the apparatus, the Michelson-Morley apparatus, contracts in the direction of motion. It's really this long, but because it's moving, it got contracted.

What's wrong with a statement like that? What's wrong with a statement like that is the sense that somebody is moving and somebody isn't. Somebody, probably you think the person on Earth, has the right to assert that they are at rest, and funny things are happening to clocks and meter sticks and space and time for the people who are moving. Nonsense. You, in the moving spacecraft moving relative to me; you, in the airplane moving past me at 600 miles an hour; you, in the car moving past me at 70 miles an hour—because those motions are uniform, you have every bit as much right to claim that you're a place for doing physics. You have no more right to claim you're the right place than I do. We both have equal claim on doing physics and doing physics correctly. That's what the principle of relativity says.

So don't think funny things happen to clocks and meter sticks or space and time because things move. The phrase, because things move, is a relativistically incorrect phrase. You're allowed to say, I am moving relative to you. Or, that clock is moving relative to this clock that I'm looking at. But you're not allowed to say that clock is moving and this one isn't and, therefore, funny things happen to that clock. That violates, directly, the spirit of the principle of relativity, which says all uniformly moving frames of reference are equally good places to do physics.

And that means any uniformly moving clock keeps time just as well as any other uniformly moving clock. It's just as accurate. Clocks don't become inaccurate "because they move." Because no uniformly moving clock has the right to say, "I'm at rest and all the other ones are moving." What may be the case, and, in fact, is the case, is that different clocks moving relative to each other keep different time. But, there's not one of them that can say, I keep the right time and all the rest of you are wrong, because you're "moving." Because nobody has the right to say, I am moving.

I will even slip occasionally and use phrases that are not correct. If I ever say the moving clock, or this is moving, what I really mean is, this is moving relative to that. That's why it's a relativity theory. It says only relative motion matters. Lecture Three we went over this with the Galilean relativity principle. Only relative motion matters. Statements about absolute motion are meaningless. They were in mechanics all along. They became so in all of physics with Einstein's special relativity. So everybody who's in uniform motion is in equally good motion to do physics.

So let's now move on and see what actually happens to space and time in special relativity. We'll start with time. We're going to be talking a lot in the next few lectures about events. I just want to tell you what an event is so we make it very clear. An event is something specified by giving both its place and its time. For example, your birth is an event. It occurred at a specific place, and it occurred at a specific time. Just to give the place, to say, New

York City, for example, is not sufficient to specify your birth. Just to give a year, 1960, is not sufficient to specify your birth. New York, 1960, that does it. An event is a place and a time. And events are the key things in relativity. So we're going to talk a lot about events.

What I'm going to do now is give you a somewhat artificial seeming example that I hope will convince you of what exactly happens to time because of the principle of relativity being true. I'm going to imagine a device, which I'll call a light clock. And here's a picture of a light clock. And you audio people have a picture of the light clock in your booklet. The light clock consists of a box. Just a simple box. At one end of the box is a light source. And at the other end of the box is a mirror. And we're going to imagine a pulse of light leaves the light source. Travels to the mirror. Reflects off the mirror. And comes back. And we're going to be interested in the time interval between two events. And those two events are the departure of the light pulse from the light source and the return of the light pulse back to the light source after it's bounced off the mirror.

If you will, this device is a clock, and that's the fundamental ticking. Tick as the light goes up, tock as it comes back. There's one period of the clock. And it keeps doing that. But we're just going to focus on one tick-tock of the clock. One outgoing light pulse hitting the mirror, coming back to the light source again. And the two events are: light pulse leaves the source, light pulse returns to the source. We want to know what the time interval is between those two events.

So, here's our light clock. Now, whenever you see pictures in relativity books or in my course or in anything dealing with relativity, it's crucial to ask yourself, and it's crucial to understand the answer to the question: In what frame of reference is this picture being drawn? Many relativity books fall down, I think, because they fail to tell you that. Or sometimes worse, they'll draw pictures that are ambiguous. They're partly from one frame of reference and partly from another. And in relativity, things tend to look different in different frames of reference. And it's important to catch that difference. And I'll try to do the best I can. Sometimes I'll slip, but I want to tell you, whenever possible, what frame of reference I'm dealing with.

So the picture I'm showing you right now on the left side of the screen is a picture drawn in a frame of reference at rest with respect to the light clock. What that means is we're in a frame of reference, we put this light box or light clock here, and it's at rest with respect to us. That's what I mean by saying this is a picture drawn in a frame of reference in which the light clock is at rest or, equivalently, a frame of reference at rest with respect to the light clock. And I'm assuming that I've got a uniformly moving frame of reference.

You might say, well, maybe it's at rest. Well, maybe it is at rest, but being at rest is meaningless. Maybe it's at rest on Earth. That's a particular state of uniform motion. Maybe it's at rest on Mars. Maybe it's whizzing through space at 99 percent of the speed of light relative to Earth. But as long as it's moving uniformly, that's an okay frame of reference. So, I don't know exactly what the light clock is doing relative to Earth. Doesn't matter. The point is I'm drawing a picture here from a frame of reference in which the light clock is at rest.

And in that frame of reference, here's what's going to happen. The light is going to leave the source. It's going to go up and hit the mirror. It's going to return to the source. It's all that happens. And the light is going to travel a total distance, which is twice the length of the box. And that's all we need to know at this point. What's the length of the light path for that light? It leaves the source. It travels up to the mirror. It reflects off the mirror; it comes back and hits the source. And we know how far it traveled. It traveled twice the length of the box. No problem.

Now, there are certainly plenty of other frames of reference that aren't at rest with respect to the light clock or light box equivalently through which the light clock is moving. If I am at rest with respect to the light clock and you go by me at 70 miles an hour in the car, then relative to you, the light clock is, in fact, moving. In fact, suppose you go by me in that direction. Then, judged from your frame of reference, the light clock is moving in the other direction relative to you. And you might say, oh, but it's really at rest. Uh-uh. That's not relativistically correct. Both of those frames of reference are equally valid because they're both uniformly moving frames of reference. And no one has the right to claim they are at rest or not.

So, is the light clock at rest or not? It is in one frame of reference. It isn't in another. And that's just the way it is. And that's O.K.

So let's look at that second frame of reference and see what happens. We'll look at the same situation. The light clock, the emission of the light—that's the first event—bounces off the mirror, comes back to the source—that's the second event. We're going to look at that sequence of events, at those two events in a frame of reference now in which the light clock is moving to the right. That means, relative to the frame in which the light clock is at rest. That frame is moving to the left. And, therefore, the light clock looks like it's moving to the right. It is moving to the right in that frame.

By the way, a word to avoid. You might want to say, "But isn't it really at rest?" Avoid the word "really" in relativity. It's a very, very dangerous word. Because really implies there's some special case, some special frame of reference in which things are really the case, which they're really right. The time is really this time. The length is really this length. This event really happened before this event. Uh-uh. To use the word "really" is to sort of pin a specialness on one frame of reference. So let's avoid that.

So here we are, simply in a different frame of reference, a different place. A different state of motion that is equally valid for doing physics. It's a frame of reference, which happens to be moving to the left, relative to the first frame of reference I discussed—the one in which the light clock was at rest. And so, in this new frame of reference, the light clock itself is moving to the right.

So it starts out. There it is. And I'll make a little arrow to show that it's moving to the right at some speed. And to make this make any sense, that speed is probably pretty significant compared to the speed of light. We already expect that because we know that these relativistic effects aren't important at very slow speeds. You won't see these effects if you do this experiment and move the light clock at 70 miles an hour or 600 miles an hour. But if you move it at 80 percent of the speed of light, this will be a big effect.

So, what happens? The light flash leaves the source and it travels straight to the mirror. But in this frame of reference straight to the mirror means it travels like that. At an angle. At the time the light pulse hits the mirror, the mirror is in a different location. The light pulse still hits the middle of the mirror. The same physical things have to happen. In one frame of reference, the light pulse left the source, it hit the middle of the mirror, and it came back. That's got to happen in the other frame of reference. There's no question about what happens. It's a question of interpretation, about how long the intervals are between what happens, and so on. So, here we are half way through the process. The light pulse has hit the upper mirror. It's in the process of reflecting off it. It comes back. And by the time it returns to the source, the light box is in a different place. And there is nothing funny and relativistic yet. I could have described this all in old-fashioned classical terms. The light clock could be in a car, which happens to be moving to the right. And the first frame of reference is the frame of reference of the car. And the second frame of reference is the frame of reference of an observer on the road who sees the car go by. And there's nothing unusual about this. These are two different descriptions of the same sequence of events from two different points of view. Two different frames of reference.

Here's the crucial point. In the second frame of reference, the one in which the light box is moving, notice the light path. The light path is longer. Why is it longer? It's longer because the light travels diagonally to get to the mirror, reflects, and comes back diagonally. It's a longer path. It still has the up and downness of the length of the box. But, in addition, it has a component of its motion in the direction the light box is going.

In classical physics, in old fashioned pre-relativistic physics, we say, oh well, that's okay because the light is actually moving a bit faster in the right-hand frame of reference, because it has the motion of the light box imposed on it. And we could describe that in terms of the ether and so on if we wanted.

But, now we believe in the principle of relativity. The principle of relativity says the laws of physics are the same for all observers in uniform motion, all laws of physics, including the laws of electromagnetism, including their predictions there are light waves, including the predictions that those light waves go at speed c. So, what has to be the case? The speed of the light has to be the same in both frames of reference. And, yet, in one frame of reference, the light is moving further than in the other frame of reference. This is where relativity comes in. Relativity comes in, in our assertion that the light moves at the same speed in both these frames of reference. That's what's going to give us all the trouble with time. That's what gave us the trouble in the experiment of measuring the speed of the light flash from the traffic light too, the assertion that all these observers had to measure the same speed. Observers in both these frames of reference measure the same speed for light. But in the left hand frame, in the frame of reference at rest with respect to the light clock, the path length is shorter. And we can conclude, in the left frame we have a shorter path; we've got the same speed; and, therefore, the time between the events is shorter.

In the frame of reference in which the light clock is moving to the right, we have a longer path. But by the principle of relativity we have exactly the same speed c for the light. And, therefore, it takes more time. And that's simply the way it has to be.

If you accept the principle of relativity—the very simple, benign-sounding principle that says the laws of physics, including electromagnetism and predictions about light, are the same in all frames of reference in uniform motion—then this conclusion has to follow: The time intervals between those two events are different in the two different frames of reference.

Now, you might say, oh, well, okay, this is just a peculiarity of this funny light clock we're using here. This is something that if I used a real clock, real clock, like this clock, that result wouldn't happen. But that's nonsense. We've made a clock. What's a clock? A clock is a device that accurately, repeatedly, undergoes some kind of happening that is periodic, that occurs over and over and over again. In an old pendulum clock, that could be the swing of a pendulum, once a second. In this light clock, it's the bouncing of the light pulse from the source back to the source. In this modern quartz clock, it's the fundamental vibration of a quartz crystal that vibrates 32,768 times per second. That's a power of two from one second. And then when you divide that down with electronic circuits you get a ticking every one second. Exactly 1.0000 seconds. A clock is simply a device that undergoes a repetitive, periodic motion or change of some kind. And that light clock is a perfectly good clock.

And what we are doing here is discovering, not something about the light clock, or not something even about light, but something about time. The phenomenon we've described here is called *time dilation*. Time is different in different frames of reference. Clocks do not keep the same time when they are moving relative to each other. They're still perfectly good clocks. There is nothing wrong with the light box when it's moving. It's the same one. It's just looked at from a different frame of reference. It isn't keeping bad time "because it's moving." It is keeping perfectly good time in either frame of reference. It's just keeping different time.

Time dilation isn't about light. It isn't about light clocks. It's about time itself. Measures of time are simply different for different observers in relative motion. The time interval between two events is not an absolute quantity. It's something that depends on your point of view. We're going to have to give up the absoluteness of time, and soon, the absoluteness of space if we accept the simple principle that the laws of physics are the same for all observers in uniform motion.

When you read books about relativity, you often hear time dilation described by saying, "moving clocks run slow." There are two things wrong with that. It's probably grammatically incorrect. It should say slowly. But much more profoundly, it's simply not relativistically correct. And I'd like you to pause right now; turn off your VCR; turn off your cassette player for just a minute and think about what is wrong with the statement: Moving clocks run slow. Why is that statement not relativistically correct? Think about that for a few minutes. When you're done thinking about it, start up again and we'll continue.

So, what's wrong with the statement: Moving clocks run slow? It implies there's something special about motion, special about movement, that there's somebody who's not moving, for whom the clock is right and doesn't run slow, and somebody else for whom, because they're moving, the clock runs slow. Nonsense. Nobody can say, I'm moving and you're not, or, you're moving and I'm not. That's the whole point of relativity. Moving clocks run slow is an absurd statement. It's nevertheless used very commonly, including in physics books or books for non-science people about relativity, to describe this phenomenon of time dilation.

One problem with the idea of moving clocks run slow is you might imagine yourself in a spacecraft whizzing past Earth with a clock. That's a moving clock. Right? And the moving clock runs slow. So you might imagine, in your frame of reference everything seems to have slowed down. You watch the hands of your clock creeping slowly. And your heart beats slowly. Nonsense. That frame of reference is just as good a frame of reference for doing physics in as any other frame of reference. Things have to seem perfectly normal in that frame of reference. If they didn't seem perfectly normal in that frame of reference, physics would be somehow different there. And the whole point of relativity is the laws that govern physical reality don't depend on your state of motion. They are the same for all observers in uniform motion. Things can't seem strange even if you're in a spacecraft whizzing along at 0.99999 the speed of light. Everything in the context of that spacecraft, in your frame of reference—your heartbeat, your aging—all that seems perfectly normal. Time, for you, may be running differently than it does on Earth. We'll get to that in the next lecture. But everything seems perfectly normal.

So, what does time dilation really say? Let me describe what it says. Here's what it really says. It doesn't say moving clocks run slow. Instead it says something a little more complicated. It says, if I have two events, and I measure the time between them on a clock that is present at both events—remember the light clock when it was in the frame of reference where it was at rest, that light clock, that same box, was present at the emission and the return of the light flash. We are going to now replace that light box with a couple of real clocks. And so what does it say? It says something like this: It says, first of all we are not going to say moving clocks run slow anymore. That's bad. Relativistically incorrect. We're going to say, a little more sophisticatedly, the time between two events is shorter when you measure it with one single clock that's present at both events, rather than measured by two different clocks.

So here's the picture. This is the picture that should come to your mind when you think about time dilation. I've got two clocks—clock B-1 and clock B-2.

And they're located at two different places. And I've got another clock—clock A—whose state of motion relative to the other two clocks is such that A will be present at clock B-1, and then later it will be present at clock B-2. And the two events we're interested in are the first event, which is shown here, when clock A passes clock B-1, and the second event, which will be when clock A passes clock B-2.

Now you may be very tempted to say, oh, A is the clock that's moving. And I've got an arrow, showing that, indeed, A is the clock that is moving. And later, clock A will be above clock B-2. And in that sense, it is moving—in the frame of reference of the clocks B-1 and B-2. And that's the frame of reference from which this picture is drawn. Remember, I asked you to always think about what frame of reference is the picture shown in. That frame of reference, clock A is indeed moving. But that doesn't say clock A is absolutely moving. It simply says clock A is moving relative to the frame of reference in which the two clocks B are at rest. They are at rest relative to some frame of reference. The one in which they're at rest. They are at rest relative to each other. I put them that way. I've glued them to the wall in that frame of reference. Clock A is moving relative to that frame of reference, but clock A isn't moving absolutely.

Clock A has every right to say, "Oh, the clocks B are moving relative to me, I'm not moving." So, that's why I don't want to call clock A the moving clock. Clock A is simply a clock that happens to be present at the two events. It's a single clock. Clock A is present at the event of clock A and clock B-1 coinciding. And clock A is present at the event of clock A and clock B-2 coinciding. Neither clock B1 nor clock B-2 is present at both those events.

And what does time dilation say? Time dilation says that the time elapsed on clock A between those two events will be less than the time as measured by the two other clocks. And that is crucial. It takes two clocks in one of the frames of reference, and only one clock in the other frame of reference to do this measurement—to determine the time interval between the two events.

Now, this is not a mathematically oriented course, but I do want to tell you the actual result of doing this measurement and calculating what the time differences should be. There is a little formula. And we will use this occasionally. And any of you who have had high school algebra and can use the Pythagorean theorem could easily go back to the light clock picture and look at those diagonals. Hint: Diagonals have to do with the Pythagorean theorem. And you could calculate the time in the one frame versus the time in the other frame.

And you would find that the time, which I will call t prime (t'), that's the little thing after the t for time on the left side of that equation. The thing I call t prime, that's the time measured on clock A, the one clock that's present at both events. That time is the time measured on the other two clocks multiplied by a factor 1 minus v squared, square rooted, where v is the speed measured in terms in the speed of light. So if you're going half the speed of light, v is a half. And that number is always smaller than one, and, therefore, t prime is always less than t. And we're left with the conclusion that time is simply not an absolute quantity. The time interval between two events is relative. It depends on your frame of reference.

Muons and Time-Traveling Twins
Lecture 9

Measures of time between two events simply are not absolute. Now, I grant that you probably have great difficulty accepting this intuitively, because all your experience suggests that to say this class is one hour long, or this lecture is a half hour long, or this airplane flight is two hours long, that that has an absolute meaning. It doesn't.

Experiments with subatomic particles called muons dramatically confirm that time dilation really occurs. These particles are produced in the upper atmosphere and travel downward at nearly the speed of light. But they're radioactive, and they decay in such a short time that they shouldn't last long enough to make it to the ground. Yet they do—showing that time runs slower in the muons' frame of reference by just the amount expected from time dilation.

Muons are subatomic particles produced by cosmic rays high in Earth's atmosphere. They rain down on Earth at a steady rate and can, therefore, be considered "clocks." The muons are radioactive and decay in a time so short that very few should be expected to reach sea level.

In the 1950s, physicists measured the number of muons arriving each hour atop Mount Washington in New Hampshire; it was about 600 per hour. These muons were moving at $0.994c$ and, even at that high speed, their radioactive decay meant that only about 25 should survive to sea level. But a measurement at sea level revealed about 400 muons each hour. This is consistent with time in the muons' frame passing at about 1/9 the rate it does on Earth. Work out the quantity $\sqrt{1-0.994^2}$ and you will see that it is just about 1/9. So time dilation really happens!

Here is another example: Suppose we had a spaceship capable of $0.8c$ relative to Earth. It sets out on a trip to a star 10 light-years distant (one light-year is the distance light travels in a year, so the speed of light is simply 1 light-year/year). From the Earth's point of view, the ship is going 10 light-years at 0.8 light-years/year. Because distance

= speed × time, the time this takes is $t = 10 \text{ ly}/0.8 \text{ ly per year} = 12.5$ years. But according to time dilation, the time on the ship is: $t' = (12.5 \text{ years}) \times \sqrt{1 - 0.8^2} = (12.5 \text{ years}) \times (0.6) = 7.5 \text{ years}$

Similar to this example, the "Twins Paradox" is a famous seeming paradox of relativity. Imagine the same star trip, but now the ship turns around once it reaches the star. The return trip is just like the outbound trip, so it takes 12.5 years according to observers on Earth and 7.5 years in the ship. When the traveling twin returns, she is 15 years older, but her brother is 25 years older!

Why can't the traveling twin consider that she's at rest and that Earth goes away on a 10-light-year journey at $0.8c$, in which case she should conclude that her brother will be younger? The answer is that the situation is not symmetric.

The Earthbound twin stays all the time in a single, uniformly moving reference frame (here we neglect the nonuniform motion associated with Earth's rotation and orbital motion; because these are slow compared with c, they don't have a significant effect on the results).

The traveling twin occupies two different uniformly moving reference frames, going in opposite directions. They're separated by the ship's turnaround. The traveling twin feels that turnaround; Earth doesn't. The situation really is different for the two twins, so there's no paradox.

The twins' experiment actually has been done, but with an atomic clock flown around the Earth.

Think about what's *special* about special relativity: It is restricted to frames of reference in *uniform* motion. The statement "I am moving" is meaningless according to the principle of relativity, but the statement "My motion changed" is meaningful. Here, the traveling twin can rightly say, "My motion changed," but the Earthbound twin cannot.

By going faster, the traveling twin could make her trip time arbitrarily small; as v approaches the speed of light c (or v approaches 1 in our formula), the time-dilation factor $\sqrt{1-v^2}$ approaches 0. If she goes at very nearly c, the trip will take just barely over 25 years' Earth time, but a negligibly small amount of ship time—so she will return 25 years younger than her twin.

By going farther, the traveling twin can go further into the future. Suppose she goes to the Andromeda galaxy, 2 million light-years distant, at nearly c. The shortest Earth time a round trip can take is just over 4 million years, but the ship time can be arbitrarily small. So she can return to Earth 4 million years in the future. But this is a one-way trip! If the traveling twin does not like what she finds, there is no going back! Time travel to the past is not permitted.

Can we hope to duplicate the twin's trip? Not with today's spacecraft. Scientists have sent an atomic clock on a round-the-world airplane trip. On return, it read less time—by some 300 nanoseconds (300 billionths of a second)—than its stay-behind twin. The effect is clearly measurable but hardly dramatic, because the airplane's speed is so much less than that of light. This experiment actually involves effects of both special and general relativity, as we'll see in subsequent lectures. ■

Essential Reading

Hey and Walters, *Einstein's Mirror*, Chapter 3, pp. 64–66.

Hoffmann, *Relativity and Its Roots*, Chapter 5, pp. 109–110.

Mook and Vargish, *Inside Relativity*, Chapter 4, pp. 110–111.

Suggested Reading

Davies, *About Time: Einstein's Unfinished Revolution*, Chapter 2.

Fritzsch, *An Equation that Changed the World*, Chapter 11.

Hafele and Keating, "Around-the-World Atomic Clocks: Predicted Relativistic Time Gains" and "Around-the-World Atomic Clocks: Observed Relativistic Time Gains," *Science*, vol. 177, pp. 166–170 (July 1972).

Moore, *A Traveler's Guide to Spacetime*, Sections 4.7.3 and 5.8.

Taylor and Wheeler, *Spacetime Physics*, 2nd edition, Chapter 4.

Questions to Consider

1. Suppose two triplets leave Earth at the same time and undertake round trip space journeys of identical length and at the same speed but in opposite directions. When they return, will they be the same age or will one be older? How will their ages compare with their third sibling, who stayed at home on Earth?

2. In 1999, scientists discovered a planetary system orbiting a star 44 light-years from Earth. How far into the future could you travel by taking a high-speed trip to this star and returning immediately back to Earth? Under what conditions would you achieve this maximum future travel? How long would you judge the trip to take?

3. Suppose the twin in the spaceship traveled at $0.6c$ instead of $0.8c$. By how much would the twins' ages differ when the traveling twin returns to Earth?

Time Dilation : An Experiment with Muons

In this experiment subatomic particles called muons, which have a fixed rate of decay, stream down from high in the atmosphere toward earth at a speed of 0.994c. The experiment has two stations, one at the top and one at the bottom of the mountain, that measure the number of muons passing by.

In the first box, we see the results that would be expected if time dilation did not occur. As the muons stream down toward earth, many are detected at the top of the mountain in the first station. But since muons decay at such a rapid rate, the expected count of muons passing the second station at the bottom of the mountain is dramatically lower.

In the second box we see the results as they actually do occur, with time dilation affecting the results. Since the muons are traveling at a significant fraction of the speed of light (0.994c), their measure of elapsed time is significantly slower than the time recorded by the two stations. Thus, with the muons' time running more slowly, most of them do indeed reach the second station.

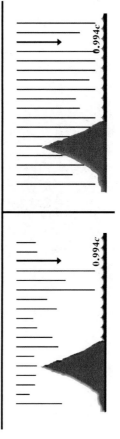

157

A Little Math...

Distance = speed × time, so in Earth-star frame of reference, trip takes:

$$t = \frac{d}{v} = \frac{10 \text{ light-years}}{0.8 \text{ light-years / year}} = 12.5 \text{ years}$$

Time dilation: $t' = t\sqrt{1-v^2}$, so in ship frame, trip takes:

$$t' = (12.5 \text{ years})\sqrt{1-0.8^2}$$

$$= (12.5 \text{ years})(0.6) = 7.5 \text{ years}$$

Star Trip!

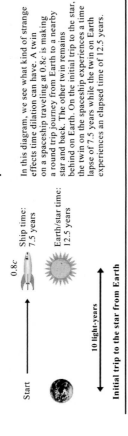

0.8*c*

Ship time:
7.5 years

Earth/star time:
12.5 years

Start

10 light-years

Initial trip to the star from Earth

In this diagram, we see what kind of strange effects time dilation can have. A twin on a spaceship traveling at 0.8*c* is making a round trip journey from Earth to a nearby star and back. The other twin remains behind on Earth. On the initial trip to the star, the twin on the spaceship experiences a time lapse of 7.5 years while the twin on Earth experiences an elapsed time of 12.5 years.

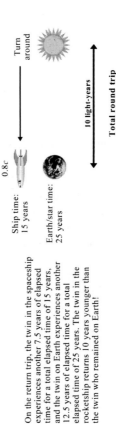

0.8*c*

Turn around

Ship time:
15 years

Earth/star time:
25 years

10 light-years

Total round trip

On the return trip, the twin in the spaceship experiences another 7.5 years of elapsed time for a total elapsed time of 15 years, and the twin on Earth experiences another 12.5 years of elapsed time for a total elapsed time of 25 years. The twin in the rocketship returns 10 years younger than the twin who remained on Earth!

Muons and Time-Traveling Twins
Lecture 9—Transcript

Welcome to Lecture Nine: "Muons and Time-Traveling Twins." In the previous lecture, I introduced you to this notion of time dilation. The idea that the time interval between two events depends on your frame of reference. Time is no longer an absolute. And we did the time dilation experiment by thinking about a special device, a light clock. But we concluded from that, that this was really a property of time itself. Time is simply not an absolute quantity. Different observers measure different time intervals between the same pairs of events. And we described in some detail—with a pair of clocks in one frame of reference and another clock that moved between one clock and the other—exactly what it meant to say that time was different in one frame of reference than another frame of reference.

Measures of time between two events simply are not absolute. Now, I grant that you probably have great difficulty accepting this intuitively, because all your experience suggests that to say this class is one hour long, or this lecture is a half hour long, or this airplane flight is two hours long, that that has an absolute meaning. It doesn't. To say that the time between your birth and your sitting here is 40 years or whatever, that is not an absolute statement either. In different frames of reference, different observers will judge different time intervals for the time difference between those two events. Time is simply not an absolute.

If you have trouble accepting that, on the other hand I hope you don't have trouble accepting Einstein's principle that the laws of physics are the same for all observers in uniform motion. That all frames of reference in uniform motion are equally good places for doing physics. If you accept that—the logic I applied last time showing that this light path from the light being emitted by a source in that light clock to the return to that source, that the path was different in two different frames of reference—but if you believe relativity, you believe that light goes at speed c in all reference frames and that led us immediately to the conclusion that the time intervals between those events were different in different frames of reference.

And I can't emphasize enough that this is not a property that is peculiar to that device I used in the previous lecture. The light clock. It's not even peculiar to clocks at all. All measures of time, all processes, physiological processes, biological processes, chemical reactions, physical processes, all processes that measure time in any way are affected in exactly the same way, because this is not something that happens and drags back the hands of clocks and slows them down. This is time itself that's different in different frames of reference. Don't look for some mechanical thing that slows down the clocks because they're "moving." That would be a relativistically incorrect statement to make. It's simply that measures of time are different in different frames of reference. Time is not an absolute.

Some things are absolutes, by the way. Relativity is not the theory that says everything is relative, nothing has any absoluteness. Nonsense. There are things that are absolute. The laws of physics are the same for all observers in all frames of reference. The speed of light, as a consequence, is the same for all observers in all frames of reference.

But some things that we used to think of as absolutes and applying to everybody, particularly measures of time and, as we'll see soon, of space, are not absolutes.

Well, does this really happen? Does time dilation really occur? We sure don't notice it when we drive around. If I jump in my car in New York and drive to San Francisco and I compare my clock when I get to San Francisco with the clock that's there, that was synchronized with the clock in New York when I left, I'll hardly notice the difference. In fact, I won't notice the difference unless it's a super accurate atomic clock. And then I might notice a few billionths of a second difference. These effects are negligible in our everyday lives.

But there are situations in which they become completely obvious. And there are situations, eventually, maybe involving interstellar space travel, where we could put them to advantage. Let me give you an example of an experiment. This experiment was done in the 1950s with the particular goal of, not only testing, but more, demonstrating the principle of time dilation. Demonstrating that special relativity is correct. By the time this experiment was done, there was no question about special relativity being correct. I mentioned before it's one of the most tested theories in all of physics. It's at the foundation of our understanding of physical reality. So, the experiment I'm going to describe is not the clincher that says, "Oh, Einstein was right." But it's an experiment that demonstrates, quite straightforwardly, that it had to be correct.

And the experiment involves some clocks. Now, what's a clock again? A clock is anything that regularly measures time intervals. And the clocks in this case are tiny subatomic particles called *muons*. And these muons are being created all the time in the upper atmosphere as high energy cosmic rays—which are themselves subatomic particles that originate sometimes in the sun and sometimes in the vastness of the cosmos in our galaxy, or maybe even outside it—and they come barreling through the universe at very, very high speeds. And when they slam into Earth's atmosphere, they cause all kinds of nuclear reactions to occur in the nuclei of the atoms they hit, and they produce different particles. And showers of particles are coming down through the upper atmosphere of the Earth all the time as these cosmic rays hit the upper atmosphere.

And among these particles are these subatomic particles called muons. And the muons are radioactive, as are many elementary particles. Not really elementary, because they break up into other particles. We'll say more about that much later in the course. But these particles are radioactive. And they decay. And they decay at a certain rate. And the rate of their decay is what we're going to use as a clock in this experiment. And we're going to demonstrate that these clocks, these muons, when they're moving relative to us at high speed, seem to decay more slowly than they would if they were at rest relative with us.

Now, "seem" is not quite a good word, because it implies there's something funny about their frame of reference. So let me change that to "do decay at a rate which is different than if they were at rest relative to us." So let me describe for you this experiment. The experiment was done on Mount Washington in New Hampshire. Mount Washington sticks up about 6,000 feet in to the atmosphere from sea level. And there is a rain of muons coming down due to these cosmic ray interactions high in the atmosphere. And the particular muons that were singled out for this experiment were muons that were moving at about 0.994 of the speed of light. That's 99.4 percent of the speed of light. That's pretty fast. Relativistic effects will be quite dramatic at speeds like that.

So here come these muons. And the scientists measure how many muons per hour were coming into a detector. A muon detector. A device that gave off a little flash of light whenever a muon came into that detector and disintegrated. They could tell how many muons were coming in. And the number of muons coming in to the detector was something like 600 every hour. At the top of the mountain there were a lot of muons coming into the detector.

Now, scientists can measure the lifetime of these muons. And when they stop one of the muons, or stop a bunch of the muons in their detector and they watch them decay, they can determine the rate at which the muon's internal clock is running. And they can conclude that even though these muons are traveling very, very rapidly—99.4 percent of the speed of light—they decay so fast that there should be very few of them left when you get to sea level. The prediction was that there might be about 25 or so when you get to sea level.

So here's the experiment. Muons come down on top of Mount Washington. You measure how many are coming down every hour into your muon detector. They're coming about 600 every hour. Pretty big number. You study these muons and see how fast they decay. Or maybe you just look up that information in a book, because the radioactive decay rates of these elementary particles are pretty well known. So you look up that quantity. Or you measure it by watching the muons decay and detecting them with Geiger counters and other radiation detectors. And you ask yourself, "Okay, if they're decaying at that rate, and they're moving at 0.994c, how many of them will decay between the top of the mountain and sea level?" And the answer comes out: Most of them will decay. And you'll expect, on the average, only about 25 to survive every hour.

So that's the essence of the experiment. It's kind of a statistical experiment. You count the muons present at the top of Mount Washington. You're then going to go down to sea level. There's a wonderful movie, by the way, made of this experiment. An old 1950s movie. And you see them moving all their apparatus down Mount Washington in leased trucks and setting it up again in Cambridge, Massachusetts, at MIT, essentially at sea level, to do the experiment all over again down there. And they carefully compensate for the thickness of the atmosphere, which affects the muons and things like that. And they get all that stuff right. And they set up the counters. And they expect to find 25 muons every hour if they don't believe in relativity. And, instead, what do they find? Well, they expect very few muons at sea level; what actually happens, actually most of the muons reach sea level. It's over 400 an hour.

Most of the muons have survived. Why? Their internal clocks that caused them to decay radioactively are keeping time at a certain rate. The muons decay at a certain rate. You expect it in the relatively short time—it's a few microseconds, a few millionths of a second—that it would take them to get from the top of Mount Washington to sea level, most of them would have decayed. How do you explain the fact that most of them didn't decay? Well, their internal clocks must have been keeping time at a different rate than the clocks in the laboratory—the clocks at rest with respect to Mount Washington, the clocks at rest with respect to the laboratory at sea level. That's the explanation for what's going on.

You'll remember I introduced the time dilation factor near the end of the last lecture. And this is about as mathematical as we're getting, and you don't need to worry about this too much. But, in fact, the time measured on the "moving clock of the muon," if I may use that expression—again, it's a relativistically incorrect expression—the time measured on the muon clock, that was my t prime. That's given by the time measured in the laboratory, by the clocks at rest with respect to Mount Washington, multiplied by the square root of 1 minus the speed squared, where the speed is given as a fraction of the speed of light. Well, 0.994 squared is about 0.99. And 1 minus that is about 0.01. So that's a very, very tiny number. It comes out be about a ninth. And what that says is moving at $0.994c$, the muons should be keeping time at one-ninth the rate of the clocks on Earth. And that is, in fact, completely consistent with the survival of about 400 muons to reach sea level instead of the 25 you would expect if the clocks, on the muons—their internal, radioactive, time-keeping mechanism—if that were running at the same rate of speed as the clocks in the laboratory.

So this is a fairly dramatic confirmation of time dilation. It really does occur. It's not the kind of satisfying demonstration you might like where I pick up a clock and run past you and show you time dilation that way. We can't do that yet. We can't move macroscopic objects at $0.994c$. But we have plenty of subatomic particles moving around at those speeds, and they very dramatically show time dilation. And this example illustrates that beautifully.

So there's an experiment that verifies that time dilation occurs. It really does happen. And it really would happen if we could pick up that clock and move it at speeds approaching that of light. In fact, it really does happen even if we move that clock in a jet airplane, as I'll describe later, it's just a rather insignificant effect.

But let me give you now an example of what would happen if we could move macroscopic, that is, big-sized objects like people in spaceships at speeds substantially like that of light. So, we're going to do an experiment, or "thought experiment" like Galileo and Einstein have done before us, we're going to do a thought experiment about a star trip. A trip to a distant star. And we're going to imagine we have a spaceship capable of traveling at say, 80 percent of the speed of light.

Now, it may be technologically not possible to do this, at least at this point in humanity's history. But there's nothing in the principles of physics that rules out what I'm about to describe. That's why it's a valid thought experiment. It's something we could imagine doing, and we believe we know what the outcome of the experiment would be because we understand the laws of physics.

So, we're going to do a star trip. And we're going to go from Earth to a star, which I'm going to say is 10 light years away from Earth. Now let me take a minute and explain what a light year is. I'm going to be measuring distances in the next few lectures in light years. Sometimes in light minutes. Sometimes in light seconds. This is a measure of distance, not of time, even though it's got the word year in it. It's a measure of distance.

What's a light year? A light year is the distance that light travels in one year. How big is that? It's trillions of miles. But I don't really care. It's 186,000 miles a second, times the number of seconds in a year, which is about 3.15 times—31 million, roughly. So multiply those out you'll figure out what it is. But I don't care what it is in real distance—it's a quarter of the way to the nearest star, approximately, if you want to get a feel for it. It's much bigger than the solar system, for example. That's not the point. The point is, it's a convenient measure of distance when you're talking about relativity and speeds near the speed of light, because what's the speed of light? Well, we know it is 186,000 miles a second. We know it is one foot per a billionth of a second. One foot per nanosecond. We know it is 300 million meters a second. But, what is it in light years per year. Think about that a second. A light year, by definition, is the distance light travels in one year. So what's the speed of light? Real simple. It's one light year per year. That makes all the mathematics, all the arithmetic, very easy.

So I'm going to deal with distances—not in miles or meters or centimeters or nanometers—I'm going to deal with distances in light years. Sometimes I'll talk about light minutes. For example, the sun is about eight light minutes from Earth. What does that mean? It means light from the sun times the time it leaves the sun until the time it gets to Earth takes about eight minutes. Jupiter, at least at some points in its orbit, is about half a light hour from Earth. What does that mean? It means it takes light 30 minutes—the duration of one of these lectures—to get from Earth to Jupiter; or a radio signal, which is also an electromagnetic wave like light, from the Galileo space probe out at Jupiter now going in orbit around that planet, takes about 30 minutes for the radio signals from that spacecraft to come back. The moon. The moon is about one and a quarter light seconds from Earth. Round trip transit time to talk to an astronaut on the moon is about two-and-a-half seconds.

You do a telephone conversation by a satellite and the satellites that carry those conversations are 22,000 miles up. There's a reason for that. They orbit the Earth in exactly 24 hours. The same as the Earth goes around. And they sit at one place above the Earth for that reason. That round trip is about 40,000 miles or a fraction of a light second. A foot is one light nanosecond. You get the idea.

We're going to measure distances in light time units. Light times time. And the light year is the appropriate unit for measuring interstellar distances. A light year is simply the distance light travels in one year.

So here we have this star trip. And going at the speed of light it would take, therefore, 10 years to get from the Earth to this star. We're going to go in a spacecraft, which I've hypothesized is technologically capable of achieving speeds of 80 percent of the speed of light—$0.8c$. So here we go. We're going to start out on this star trip at $0.8c$, and we're going to do a little math to figure out how we get there. Now, hold on. This is just arithmetic. It's the fifth grade formula, distance equals speed times time. And I want to apply it to that situation. No relativity here at all. Distance equals speed times time.

We know the distance. It's 10 light years. We know the speed. It's $0.8c$ or 0.8 light years per year, because the speed of light is one light year per year. So, distance—10 light years—equals speed—0.8 light years per year—times the time. We can figure out how long it's going to take.

By the way, I don't believe in doing math problems unless I sort of know the answer already. So I'm just going to think a little bit about the answer. I'm going at 0.8 the speed of light, 80 percent of the speed of light. If I were going at the speed of light, since that star is 10 light years away, it would take me 10 years. I'm going a little slower, so I expect an answer a little more than 10 years.

So let's do the math. Time is going to be the distance divided by the speed. That's 10 light years divided by 0.8 light years per year. And it all comes out nicely at 12-and-a-half years. So that's how long I'm going to say, from Earth, that that trip is going to take. It's going to take 12.5 years as I measure time on Earth.

But what about in the spaceship? Well, the situation is very similar to our time dilation experiments, especially the last visual I showed you in the last lecture where we had two clocks at rest with respect to each other, and another clock that moved from one to the other. The spaceship is carrying a clock—that's like the clock that's present at both events. The spaceship's clock is there when the spaceship passes Earth; it's there when the spaceship passes the star. We can imagine there are clocks on the Earth and star that are themselves synchronized together. And we've got a perfect time dilation situation. Exactly like we set up in the last lecture. And so, we can argue that the time on the spaceship is given by the time on Earth multiplied by that relativistic factor, the square root of one minus the speed squared, where the speed is measured in a fraction of the speed of light.

Again, if you don't like the arithmetic, don't worry about it. But it easily computes to give us the answer. Time dilation says t prime—the time in the ship's frame of reference—is the time in the Earth's frame of reference. And again, there's nothing special here about Earth. It's the fact that we have the two clocks in the Earth-star frame of reference. We're assuming the Earth and the star are pretty much at rest with respect to each other. Those two clocks versus the one clock carried by the ship.

So, we're going to just apply that formula, t, the Earth time, is 12.5 years. We're going to multiply it by the square root of one minus 0.8 squared. I chose those numbers for a particular reason—0.8 squared is 0.64. One minus 0.64—if you had change from it from a dollar for 64 cents, you'd get back 36 cents—0.36. The square root of 0.36 is 0.6. So that factor is 0.6, and it all comes out to 7.5 years.

So, if we look at the star trip, we start out at $0.8c$. We travel to the star. Earth time that has elapsed is 12.5 years. Ship time that has elapsed is 7.5 years.

That's just the way it is. It's just like the muons. Here we're keeping time in the spaceship at a smaller rate—about 0.6 the rate we were on Earth. In the muon case, it was more dramatic, because they were going not at $0.8c$, but $0.994c$. So that's what would happen if we could do this experiment. Just exactly like the muon situation. Just exactly like the situations I set up in the last lecture.

You might ask me, does 7.5 years really elapse in the spaceship. Isn't it really 12 years? Suppose you're the pilot of the spaceship, what will you have experienced when you get to the star? Won't 12 years—really 12.5 years—won't you really be 12.5 years older? No, 7.5 years will have elapsed for you. Remember, we're not talking about clocks here. We're not talking about something dragging on the hands and slowing the clocks back because we're in a bad frame of reference. All these frames of reference are good.

What we're talking about is the measures of time are different in different reference frames—7.5 years really have elapsed for the time-traveling twin, or the traveling twin there in the spaceship.

So let's go back now and revisit that twin idea that I described in the very first lecture. It doesn't take much to do the twin experiment from where we are already. Already we've put one twin in a spaceship. She's traveled the 10 light year distance to the star. The Earth time that's elapsed is 12.5 years. In her ship, according to her ship clock, according to her physiological aging processes, according to everything else that involves time, 7.5 years have elapsed.

However, she's now 10 light years away, so it's a little bit hard to sort of look at her and compare. I mean, if she sends back a radio signal or a TV picture of herself, well we got to worry about the travel time for that light. And things get a little bit complicated. Let's turn her around, and let's bring her back right here on Earth where you, her twin, are sitting. And let's see what things look like then.

Then there's no ambiguity. You and the twin are right side by side. No question. Don't have to worry about time for light to get between you or radio signals or whatever. So let's turn this ship around. And let's head it back at $0.8c$. Now, this is exactly the situation we just had, except instead of leaving Earth, we're leaving the star.

So we're going to head back at $0.8c$, and how much time is going to elapse? Well, it's exactly the same situation we just had. The spaceship is carrying one clock moving between a clock on the star and a clock on the Earth. And, of course, it doesn't really matter whether there are clocks there or not. All manifestations of time exhibit this effect. So the ship is going to go back to Earth. There it is. And how much time has elapsed? Well, Earth-star time it was 12.5 years one way, so it's 25 years round trip. What about in the ship? Well, it was 7.5 years one way, so it's 15 years coming back. The twin hops out of the spaceship and he's 10 years younger than you are.

Can it be? Isn't this just something that happened to the clocks? But she's really experienced the same physiological wear and, tear on her body. No. She has experienced 15 years of time. And you have experienced 25 years of time. And you're standing right next to each other. And you're different. And you were twins. You still are twins. And yet you're 10 years apart.

Could it be? Is it possible? Yes, it is. And it could be. The reason this bothers a lot of people has to do with, first of all, it bothers you because you don't want time to do strange things like that. But even if you accept that time does strange things, then you may ask a more subtle question. You may say, wait a minute. You've been teaching me about everything being relative. Well, why can't the twin who jumps in the spaceship say, oh, here's what I see. I'm sitting in my spaceship. I see the Earth's going away at 80 percent of the speed of light, and that's a perfectly legitimate thing for her to say. And then I see the Earth turn around, and come back at 80 percent of the speed of light. And so why isn't my brother 10 years younger than I am? Why don't 25 years elapse for me and 15 years for my brother? We can't have it both ways, because we're standing there side by side. And there is some objective truth to either, we're the same age, or one's older, or the other's older. So, who's right? Why can't I simply reciprocate? Why can't I do the same picture, except draw it from my point of view—the spaceship at rest, Earth zooms out 10 light years away, comes back, and the traveler on Earth should be younger. Why not?

This is what bothers people. Once they get beyond the notion that time gets messed up a little bit, you begin to accept that strange things are going to happen to time. Why isn't this reciprocal? Why doesn't it happen like that? Well, there's a subtlety here. And the subtlety is best understood if you remember what's special about the special theory of relativity. The special theory of relativity, special relativity, is special because it applies only to uniform motion. It says the statements about "I am moving" are meaningless. It's meaningless for me on Earth to say, oh, I'm at rest, but the twin in the spaceship is traveling. It's similarly meaningless, for the spaceship twin to say, I'm at rest and the Earth is traveling. Those are both equally meaningful and equally meaningless statements. Nobody has any authority to say, I'm at rest and you're moving.

However, in special relativity at least, although motion doesn't matter, uniform motion doesn't matter, changes in motion do matter, because special relativity applies only to uniformly moving reference frames. Remember what uniform motion is, way back from Galileo and Newton, it's motion in a straight line at constant speed. If you do any thing else—slow down and stop, speed up, change direction, turn around, go in a circle, turn around, that's what happened in this experiment—then that motion is no longer uniform. There really is an asymmetry in this situation. The two twins are not in the same situation, as far as special relativity is concerned. One twin, the one on Earth, is in uniform motion the entire time. The other twin, the one in the spaceship, is not in uniform motion the entire time. There's a small interval when she's turning around. During that interval, she feels great big forces. Just like if you turn a car very suddenly on a high-speed turn around a curve, you feel yourself thrown to the outside of the car. Or you sit in an airplane while it's accelerating down the runway and you feel yourself pushed back in the seat. The traveling twin feels forces like that when her spaceship turns around. Back on Earth, they don't feel those forces.

As far as special relativity is concerned, her motion has changed. And that's a real thing. That's an absolute. She can't say, I was moving and now I'm not moving. But she can say, my motion changed. And Earth's motion did not change. She's in two different frames of reference in this trip, she's in one frame of reference going out, and she's in a different frame of reference, a different state of motion, going back. There is no symmetry between the two. And so there is no problem with the fact that one twin comes back younger.

Now, you might think about this a little bit more, and I'm just going to give you hints of two answers to questions you might want to ask that I don't have time to go into in detail. This whole process can be treated from the point of view of general relativity, which treats completely arbitrary kinds of motion. And it gives the same result. You might ask, what happens if the twins try to communicate with each other as they're going? Well, basically, what happens is you will find that they can keep in touch with each other, and they can count up the number of signals each one sends back. Suppose one sends back a pulse or a message every year. You can count up those messages. And it all works out. And they come out to having spent exactly the amounts of time we've calculated—15 years in one case, and 25 years in the other case. This process really does occur, and it's happening to all manifestations of time. The twin's body really ages 15 years in the process of her trip. Your body really ages 25 years. It's not about clocks. It's about time itself and all processes, including aging, that involve time.

Well, I talked in the beginning about time travel. My first lecture was called "Time Travel, Tunneling, Tennis, and Tea." Time travel. In a way, your twin sister, in this experiment, has traveled into the future. She's traveled 10 years into your future. She's now only 15 years older, and you're 25 years older. Could she do better than that? The answer is yes. And to see that, I want to look very quickly, once again, at the time dilation formula that we developed. Again, this is something that, if you've had high school algebra, you could derive yourself. So it's not very mysterious. And it simply says the time as measured in the spaceship's clock is the time as measured in the Earth clock times this factor: $1-v^2$ square rooted. And again, nothing special here about Earth and spaceship. This doesn't happen because you're in space or because you're in a spaceship. It happens because you have one clock moving to be present at two events versus two clocks that are at rest with respect to each other at those separate events.

Look what happens if v gets close to 1. In our units, $v = 1$ means the speed of light. That square root gets close to zero. And the square root of zero is zero. So that number, t prime, can be made as small as you want. In the case of the Earth's star trip, if the twin goes faster and faster, she can make the time elapse in her clock, not 15 years, but one year, or one hour, or one second if she goes fast enough. If she goes very fast, she'll be going at almost the speed of light, and the trip will take just a hair over 10 years out, as far as Earth is concerned, and just a hair over 10 years back, for just a hair over 20 years. So, for that Earth-star trip, the minimum time that could take in the Earth-star frame of reference is just over 20 years for speeds that are very close to the speed of light. But she can make that time as short as she wants, as close to zero as she wants by going as close as possible to the speed of light.

Remarkable. In that case, for a trip to a star 10 light years distance, she could come back almost, but not quite, 20 years younger. With the 0.8 speed of light, it was 15 years younger. It could be 20 years younger. She can do better than that though.

Suppose she wanted to go to the Andromeda galaxy. The Andromeda galaxy is the only other galaxy than our own that is just barely visible to the naked eye. It's sort of to the right of the constellation Cassiopeia. On a clear night sky you can see it as a fuzzy little blob. It's 2 million light years away.

So, suppose she goes to the Andromeda, 2 million light years distant. And you might say, she's never going to make it. How can you go 2 million light years when the human life span is only 80 years or something? But she can because she can go as fast as she wants. And she can make t prime as short as she wants. So she can get there in a year, say, or a day, or whatever she wants, going as fast as she wants. If she's going almost the speed of light—the galaxy is 2 million light years away—at the speed of light, that would take 2 million years, judged by somebody in the Earth and galaxy frame of reference, they're essentially at rest with respect to each other.

So, the outgoing trip takes a minimum of 2 million years. Two million years coming back. As far as somebody in Earth is concerned, that trip takes a minimum of 4 million years. But she can make it take as close to zero time as she wants. And so, she'll come back to Earth. And she'll be 4 million years in the future. Her twin brother is long gone. Civilization, as we know it, is probably long gone. She's traveled 4 million years into the future. And how much time has elapsed for her, and her body, and her clocks and her spaceship, and all those things? An hour? A day? Twenty-five years? Whatever it was, depending on how close that v is to the speed of light. How close the v is to 1 in our equation.

Remarkable. She can travel into the future. There's a problem. If she gets there and doesn't like what she finds, maybe Earth is gone, maybe civilization has made a runaway greenhouse effect, like on Venus, and the planet is no longer habitable—all kinds of things could happen. If she doesn't like it, there's no going back. She can't get back to her time. She can leap frog further into the future with another trip. In fact, she could get even further into the future, she could get 2 billion years into the future by going to a galaxy a billion light years away at such a high speed that her time is negligible. Remarkable. Time travel, into the future, is possible. But you can't go back.

Well, this sounds absurd. Does it really happen? It has happened, and it's been demonstrated in an experiment, which I list, incidentally, in the bibliography for this course. Some articles in *Science* magazine from a couple of decades ago took very sensitive atomic clocks, put them in airplanes, and flew them around the Earth. One clock was left behind. Other clocks went around in the aircraft. They went in a circular path—that's nonuniform motion. They came back. The two clocks were compared and they differed, in that case, by about 300 nanoseconds—300 billionths of a second. Not a big amount of time but easily measurable by today's standards.

In that experiment we had the effects of special relativity. We also had some effects from general relativity, because of changes in gravity with the airplane's altitude. And we'll get to those later. But the fact is, those experiments confirm, remarkably, although with fairly trivial change in time, the time dilation formula. The twin event really does occur. The traveling clock really did come back younger—in that case by only 300 billionths of a second. But it did come back younger. Time travel to the future really can occur, and it really is possible.

Escaping Contradiction—Simultaneity Is Relative
Lecture 10

Most of the contradictions are resolved by a fact that's spelled out in the second half of my title, namely, that simultaneity is relative. What it means to say that two events are simultaneous—occur at the same time—actually depends on your frame of reference. And we're going to explore that in some detail in this lecture.

There seems to be a big problem with time dilation: "Moving clocks run slow," but who's to say which clock is moving? If clock B sees clock A move by and concludes that clock A is "running slow," why can't clock A claim to see clock B go by and conclude that B is "running slow"? It can! There's no contradiction, because of another remarkable implication of the principle of relativity: Two events that are simultaneous (i.e., that occur at the same time) in one frame of reference are not simultaneous in another frame that is moving relative to the first. It is this relativity of simultaneity that allows two observers in relative motion to see each other's clocks "run slow" without contradiction.

Recall that we arrived at time dilation by considering *one* clock that moves past *two* others that are at rest relative to each other and that are *synchronized*—meaning the events of their hands pointing to a given time are simultaneous. Those events are simultaneous in the clocks' own frame. Remarkably, as we'll see shortly, relativity implies that events simultaneous in one frame of reference are not simultaneous in another frame.

They are not simultaneous in the frame of the "moving" clock. In fact, the right-hand clock reads a later time, and that is why the elapsed time on the two clocks is longer than on the "moving" clock, even though the "moving" clock judges the other two to be "running slow." There is no contradiction; observers in each frame judge the other's clocks to be "running slow," and each is correct. In general, the relativity of simultaneity explains many of the apparent paradoxes that arise in relativity. They are only paradoxical if one insists that simultaneity is an absolute concept—and it isn't.

To understand why simultaneity is relative (or why events that are simultaneous in one frame of reference are not simultaneous in another frame of reference), we will revisiting the star trip example that we discussed earlier. This example reveals the phenomenon of *length contraction*. Earth–star distance was 10 light-years in the Earth–star frame. But the spaceship made the trip in 7.5 years, going at 0.8c. Therefore, in the ship frame, the distance must have been (0.8 ly/y) ✕ (7.5 years) = 6 light-years.

In general, the length of an object is longest in a frame in which the object is at rest. (Here, the "object" is the Earth–star system, 10 light-years long to an observer at rest in that frame.) An object is shorter when measured in a frame of reference in which it is moving, with the contraction given by the same factor $\sqrt{1 - v^2}$ that arises in time dilation. This is called "Lorentz contraction" (or sometimes "Fitzgerald contraction"). Recall from Lecture 6 that they came up with the right idea for the wrong reason.

> **Remarkably, as we'll see shortly, relativity implies that events simultaneous in one frame of reference are not simultaneous in another frame.**

An example in which length contraction is important is the Stanford Linear Accelerator, which is 2 miles long as measured on Earth, but only about 3 feet long to the electrons moving down the accelerator at 0.9999995c. Even a TV picture tube, in which electrons attain about 0.3c, would not focus correctly if engineers did not take length contraction into account.

If we are being relativistically correct, then isn't the distance between Earth and star really 10 light-years and the length of the Stanford Linear Accelerator really 2 miles? No! To claim so is to give special status to one frame of reference, and that is precisely what relativity precludes. Be careful of using the word "really" when talking about relativity.

To examine the relativity of simultaneity further, consider two very high-speed airplanes passing each other in opposite directions. There is some frame of reference in which both have the same speed and, therefore, the same length (although it is shorter than their length at rest). In this frame, the left ends of the two planes coincide at the same time that the right ends coincide; the events of the two ends coinciding are, therefore, simultaneous.

In a frame in which the upper airplane is at rest, the right ends of the two planes coincide before the left ends, so the events aren't simultaneous. In a frame in which the lower airplane is at rest, the time order of the two events is reversed. Continuing this line of reasoning, would it be possible for your birth and death to occur at the same time?

Is there a problem with causality? After all, if the time order of events can be reversed in different reference frames, there might be. But there is no problem; we'll see why in the next lecture. ∎

Essential Reading

Mook and Vargish, *Inside Relativity*, Chapter 4, pp. 111–113.

Suggested Reading

Moore, *A Traveler's Guide to Spacetime*, Chapter 3.

Questions to Consider

1. Devise an experiment that observers in the reference frame of clock A could do to verify that, as measured in the upper clock's frame, both the clocks in the lower frame are "running slow."

2. A famous "paradox" of relativity is the following: A high-speed runner carries a 10-foot-long pole toward a barn that is 10 feet long and has doors open at both ends. The runner is going so fast that, from the point of view of the farmer who owns the barn, the pole is only 5 feet long. Clearly, the farmer can close both barn doors and trap the runner in the barn. But to the runner, the pole is 10 feet long and the barn, rushing toward the runner, is only 5 feet long. So clearly the runner can't be in the barn with both doors closed. Can you resolve the paradox, using the fact that events simultaneous in one reference frame aren't simultaneous in another? (By the way, the speed required here is $0.866c$.)

Who's Moving? Who's "Running Slow"?

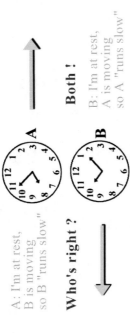

A: I'm at rest,
B is moving
so B "runs slow"

Both !

B: I'm at rest,
A is moving
so A "runs slow"

Who's right ?

This diagram illustrates what initially seems to be a contradiction created by the Principle of Relativity. Because the laws of physics are valid in all frames of reference in uniform motion, observers in the reference frame of each clock can draw the same conclusions about the other clock. Thus the following statements hold:

#1. From a frame of reference in which Clock A is at rest, Clock B is in motion and thus is "running slow" compared to Clock A.
#2. From a frame of reference in which clock B is at rest, Clock A is in motion and thus is "running slow" compared to Clock B.

The two statements seem to be completely contradictory, but relativity insists that both are correct. The explanation lies in the fact that simultaneity itself is actually relative, as can be seen in Dr. Wolfson's later arguments.

The Situation in Different Frames of Reference

This diagram illustrates the same experiment, viewed from two separate frames of reference. In the experiment, clock A moves from clock B_1 to clock B_2, similar to the experiment in Lecture 8.

Box One details the experiment as seen from the frame of reference of clocks B_1 and B_2. In this frame, the two clocks are synchronized and an observer in the B-clock frame can conclude that clock A "runs slow" as it moves between B_1 and B_2.

Box Two details the exact same experiment from the frame of reference of Clock A. To observers in this frame, the B clocks are running slow.

The problem of contradiction is resolved through the concept of relative simultaneity. While the B clocks are synchronized in *their* frame of reference, they are *not* synchronized in A's frame of reference. Although they're "running slow," clock B_2 is ahead and thus the elapsed time $(B_2 - B_1)$ is still longer than that measured on clock A.

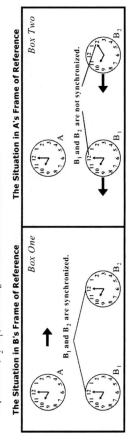

The Situation in B's Frame of Reference

Box One

B_1 and B_2 are synchronized.

The Situation in A's Frame of Reference

Box Two

B_1 and B_2 are not synchronized.

Star Trip, Revisited

In spaceship's frame of reference:
Earth-star trip takes 7.5 years at $0.8c$
Therefore distance is:
distance = speed × time
$\quad\quad\quad$ = (0.8 ly/y) × (7.5 years)
$\quad\quad\quad$ = 6 light-years

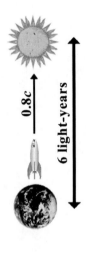

0.8*c*

6 light-years

Two Identical Airplanes Pass

In a frame where both are moving, their ends coincide at the same time:

In a frame where the upper plane is at rest, their right ends coincide first:

In a frame where the lower plane is at rest, their left ends coincide first:

Escaping Contradiction—Simultaneity Is Relative
Lecture 10—Transcript

Welcome to Lecture Ten. The first part of my title for Lecture Ten is "Escaping Contradiction." People who study relativity at first seem to be stuck with the idea that there are an awful lot of contradictions here. There seem to be a lot of things that could not really be true. How can we possibly have time dilation? How can we possibly have the twin being younger? And, yet, the other twin isn't younger. And these seem to be contradictions.

Most of the contradictions are resolved by a fact that's spelled out in the second half of my title, namely, that simultaneity is relative. What it means to say that two events are simultaneous—occur at the same time—actually depends on your frame of reference. And we're going to explore that in some detail in this lecture.

This is about the most technical look we're going to have at special relativity. Not mathematical. But we're getting pretty much into the thick of it and drawing conclusions thick and fast from, ultimately, the principle of relativity. If you're feeling a little bit lost at this point, you might want to hit rewind and go back to the last couple of lectures and convince yourself that you see how all this must follow from nothing more than the principle of relativity—that the laws of physics are the same for all observers in uniform motion. There is nothing else coming in here. There is nothing new. Nothing beyond that.

We're just exploring the implications of that one simple fact. And I'd like to look at a seeming contradiction that may, or may not, have occurred to you yet but is implicit in our discussion of time dilation, and that rather poor phrasing of it by the phrase "moving clocks run slow." What's wrong with that phrase, of course, and I asked you to think about this at one point, what's wrong with that phrase is neither clock has the right to say the other is moving and I'm at rest. Each frame of reference is an equally good frame of reference for doing physics. Each is entitled to say, "Oh, I'm here and you're moving relative to me." The other frame can say, "Yeah, well, but you're moving relative to me." And they're both right.

And I want to explore the implications of that briefly for this statement that "moving clocks run slow." And I think I'll show you in the process why that's such a bad phrase to use in describing time dilation.

Here is your clock. You're at rest, with respect to that clock, and I'm going to go by you with my clock. I'm moving, relative to you. And if you wanted to set up one of those time dilation experiments, you would say, "Oh, Wolfson's clock is the moving the clock, mine's the clock at rest. His clock runs slow."

Now the way you would set that experiment up, remember, is to have a second clock and compare the time when my clock passes your first clock and the second clock, and so on. But the essence of the statement you would arrive at is, "Oh, Wolfson's clock is running slow, because he is moving relative to me." Now, naively, you might think, well, if my clock is moving slow compared to yours, yours must be moving fast compared to mine. But that violates the spirit of relativity, because the spirit of relativity says no frame of reference is special. My frame of reference is just as good a frame of reference for doing physics as is your frame of reference. Provided, at least, I'm moving uniformly.

Remember, it was that non-uniformity of the twin's motion when she turned around at the distant star that broke that symmetry. But at this point I'm simply going to be moving past you in uniform motion, and my frame of reference is just as good a frame of reference as yours on Earth. Nothing special about Earth or Earth's state of motion. I'm in just as good a frame of reference as you are for doing physics. And therefore, I should draw exactly the same conclusions you do, in exactly the same situation.

So what conclusions do I draw? Well, I'm at rest in my frame of reference. And I see your clock going by me. Your clock's moving relative to me. And therefore, your clock must run slow. And if I wanted to set this up very carefully, I would take a second clock in my frame of reference, and I would watch the time when my first clock passed yours, and my second, and so on. And I'd do the time dilation experiment, exactly as we did it in the preceding two lectures. And I would come to exactly the same conclusion that your clock moves slow compared to mine. Runs slow compared to mine.

How can that possibly be? There's one of the seeming contradictions of relativity. Two clocks are in relative motion. Each one must conclude that the other runs slow. If that weren't the case, then we would have an asymmetric situation, and one of them would be able to say, oh, that one's moving fast and this one's moving slower. That one's at rest, and this one isn't at rest. And that's exactly what relativity precludes us from doing, because relativity says all the laws of physics, all the conclusions you draw from the laws of physics, are the same in all uniformly moving frames of reference.

How can that be though? Well, the resolution lies, as I said, in the second half of my title, "Simultaneity is Relative." And we'll come to that in a minute. But before we do, let's look in a little bit more detail about exactly how we do this time dilation measurement. Here we've got two clocks—clock A and clock B. And we're raising the question, who's moving? And who's running slow? Here's what clock B thinks. Clock B says, "I'm at rest, A is moving, so A runs slow." But A says, "No, I'm at rest, B is moving, so B runs slow." That's exactly the situation I just demonstrated for you. So which one of these clocks is right?

Well, the answer is both are right. How are we going to resolve this? How are we going to understand that this is not a contradiction? Well, let's look at the situation in the reference frame of clock B and how clock B would really set up an experiment to prove, or show, that clock A is "running slow."

Here's how clock B would do it. It would do exactly what we've done in the previous couple of lectures. We would have a couple of clocks in some frame of reference. That's clock B's frame of reference. Those clocks are at rest with respect to each other. And clock A is going to move relative to that frame of reference in such a way that it goes from clock B-1 to clock B-2. And we're going to compare the elapsed time on clock A with the time that elapses on clocks B-1 and B-2. And we already know the answer. The time elapsed in clock A is going to be shorter by that square rootish factor. That doesn't matter. It's going to be shorter. That's the important point. That's the situation in B's frame of reference.

Now we have to look at one other thing that I didn't mention before when we set this experiment up. B has two clocks. And in order for this experiment to have any meaning, in order for the time interval measured on those two clocks to make sense, B has got to make sure those clocks are synchronized. That is, they both point to 12 noon or 10 o'clock, as they're shown in the picture, at exactly the same time. Otherwise, they're not keeping time right. I mean, they may be running at the same rate, but unless they're also synchronized—they point to the same time at the same instant—we've got a problem. And we won't be able to do the measurement carefully or accurately, because one of the clocks won't have been set right. No problem. You know, you start some operation and say "synchronized watches, folks." Well, everybody's watch has got to read the same thing if we're going to be talking about the same time.

The same is true here. To do this experiment, clock B-1 and clock B-2 have to be synchronized. They have to be running at the same rate. Well, they are, because they're perfect clocks and they're in the same frame of reference. That's no problem. But they also have to be set to the same time at the same instant. And I didn't stress that before, because it was so obvious. But, in fact, it's not quite so obvious. It's a bit subtle. And now I want to stress it. And I want to raise the question, how, in fact, would one synchronize those clocks, because they are, after all, located at different places. One thing we know we cannot do is bring them together. Set them, you know, something like this. Bring them together, set them until they read exactly the same, and then run this one over here. Because if I do that, this clock has spent some time in a different frame of reference. And then it's keeping time at a different rate. And then they have failed to be synchronized. That will not work.

There are ways to synchronize the two clocks, though. For example, if I know how far apart they are. Let's say they're one light year apart. Pretty far, but it's a good example. Let's say they're one light year apart. Or make it one light hour apart, that'll be easier. Make them one light hour apart. Okay, what do I do? At noon, I set clock B-1 to noon, and at the same instant I send out a light flash, and I say, hey, B-2, when you get that light flash, set yourself to—what should it be? Not noon, but one o'clock, because we know it's taken one hour for the light flash to go one light hour. That will assure that those clocks are synchronized.

But notice it's a little bit subtle because the way I can communicate between the two clocks, the fastest way I know how is by sending a light signal, or a radio wave, or some other electromagnetic wave. And that takes time because that wave travels at speed c. I cannot simply bring the clocks together and then separate them because then one of them at least has to go in a different frame of reference.

There's a more symmetrical way of synchronizing. If I'd like, I could go into the position in my frame of reference that's midway between the two clocks. And I could send out a flash of light. And I could say, okay folks, when you get this flash of light, both of you set yourself to time zero, or some specified time, whatever it is. And whether or not we get the absolute time right, in that frame of reference, we at least will have the two clocks saying the same time at the same instant, if, in fact, we send out this light flash from a point midway in between them. So, it is possible to synchronize two clocks that are at rest with respect to each other in a frame of reference. But it is a little bit subtle. But it's entirely possible to do. But we do have to take in to account the finite speed of light.

So, we assume we've achieved that synchronization, and we've achieved it in this frame of reference. So here's the situation. Clocks B-1 and B-2 are synchronized. And again, that was so obvious I didn't even mention it before. But now it's crucial to an understanding of what's going on.

Back to the second half of my title. First half—"Escaping Contradiction." Here we have this seeming contradiction. Clock A says clock B is running slow. Clock B says clock A is running slow. How can it be true for both of them? I asserted it had to be true for both of them. It is true for both of them. How is it not a contradiction? "Escaping Contradiction, Simultaneity is Relative." Let me say that in great big letters. Events that are simultaneous in one frame of reference are not simultaneous in another frame of reference. Two events that occur at the same time in one frame of reference, do not, cannot occur at the same time in another frame of reference. I'm going to ask you to accept that statement for the next five or ten minutes, but then I'm going to prove it to you before the end of this lecture. I'm going to show you why that has to be. Events that are simultaneous in one frame of reference are not simultaneous in another frame of reference.

There is an exception to that I just ought to mention in case you get picky about this. The exception is two events that occur at exactly the same place and the same time—those two events can be simultaneous. But they're really, essentially, the same event then. If two events happen, if I snap this finger and snap this finger at the same time and the same place—literally the same place—those two events are simultaneous by anyone who cares to observe it, because they take place right at the same place. But the minute they're separated in space—I snap my fingers simultaneously—there are observers for whom those events are not simultaneous. There are observers for whom this snapping occurs before this one. And there are observers for whom this snapping occurs before this one. And you say to yourself, "Yeah, but you really snapped them at the same time." And if you say that, you're saying Wolfson's frame of reference is special. It's the only frame of reference where things are really true. I don't believe in relativity. I can throw this whole course out the window, because I don't accept the principle of relativity. If you say, they're really simultaneous in my frame, you're singling out a special frame of reference in which the laws of physics are right and they aren't right in all the other frames. And that simply isn't what relativity is about.

What relativity is about is the statement that the laws of physics are the same in all uniformly moving frames of reference. So if there is another uniformly moving frame of reference in which these two events are not simultaneous, then that is just as legitimate a picture of reality. And indeed, in that frame of reference the events are not simultaneous and there is no "really" about it. To me they are simultaneous; to an observer in another frame of reference they're not.

We'll see why that is by the end of the lecture. But let's see what that does to the problem, the contradiction, we seem to have with the time dilation situation in which clock B says, "Oh, clock A is moving relative to me. It's running slower." Clock A says, "Oh, clock B is moving relative to me. It's running slower." And I say they're both right. They are both right. Here's how.

We're going to look at the situation next in clock A's frame of reference. We've been looking all the while in clock B's frame. We've been drawing the picture of the two clocks in B's frame of reference and the one clock moving between them. Let's now look at the situation in clock A's frame of reference. And remember, events that are simultaneous in one frame of reference are not simultaneous in another. I'm going to prove that to you. Hold on to that for just a little bit.

What are the simultaneous events we're talking about in this case? Well, the simultaneous events were the setting of the two clocks. In B's frame of reference, clock B-1 and B-2 were set to the same time at the same instant. In my picture, they were both set to read 10 o'clock at the same instant. How? Well, there was this synchronizing process. B-1 sent out a light signal, and B-2 set itself to the time B-1 told it—the time B-1 was reading when it sent out the light signal plus the time it took the light signal to travel. And then they were synchronized. Or we put somebody midway in between them, and they sent out a light flash in both directions. And the two clocks set themselves to the same time when they received the light flash. And then they were synchronized. So, they've been synchronized in B's frame of reference, in the frame of reference in which those two clocks are at rest.

But clock A is moving past, from B-1 toward B-2, relative to that frame of reference. And I just told you—again not yet with proof, but I will prove it soon—I've just told you that events that are simultaneous in one frame of reference cannot be simultaneous in another frame of reference. May I take that back a little bit—in another frame of reference moving on a line between the two events. They can't be simultaneous. If it were moving sideways, might be. Events that are simultaneous in one frame of reference are not simultaneous in another frame of reference.

Well, here we have clock A moving on a line between clocks B-1 and B-2. The events of setting the two clocks in B's frame to, say, 10 o'clock— which is what they show in the picture—those events are simultaneous in B's frame of reference. Therefore, they cannot be simultaneous in A's frame of reference. Therefore, from A's point of view, the two clocks do not read 10 o'clock at the same time. That's what it means to say the events are not simultaneous. What are the events? The events are the events of the two clocks reading 10 o'clock. Those events are simultaneous in B's frame. We set the two clocks together to 10 o'clock. They, therefore, cannot be simultaneous in A's frame. Therefore, as far as A is concerned, the two clocks cannot point to 10 o'clock at the same instant.

And so, what does the situation look like in A's frame of reference? Here's the situation in A's frame of reference. It sees clock B-1 and clock B-2 going past it to the left. Why to the left? Because in B's frame of reference, we had clock A moving past to the right. So, clock A sees clock B-1 and B-2 moving to the left. And I haven't, again, proved rigorously why it happens this particular direction. But, in fact, it's the case that clock B-2, in the set up of this situation in which B-1 and B-2 are moving past A to the left, and B-2 is further to the right, B-2 will be reading a later time at the instant that B-1 reads 10 o'clock. So, that's the situation from A's point of view.

Now, remember what occurred in this whole time dilation thing. The elapsed time on A was shorter than on B-1 and B-2, as measured for the event of A passing B-1 and then A passing B-2. From A's point of view, it's B-1 passes A and then B-2 passes A. A measures an elapsed time, which is short. A says, "Wait a minute, clocks B-1 and B-2 should be running slow. So they should measure a shorter elapsed time." Well, they sort of do. But the problem is clock B-2 is ahead. And so even though clocks B-1 and B-2 are "running slower," as far as A is concerned—because it's completely reciprocal—because clock B-2 is ahead, when A compares the time on clock B-1 and the time on clock B-2, and when they pass, it will see that clock B-2 is further advanced. Why? Because B-2 is already ahead of the game. So what if it's running slower, it's already ahead in time, because the events of the two clocks reading 10 o'clock were not simultaneous.

So there's the key to that contradiction. It is really true that clock B, the B clocks, say clock A is running slow. It's also true that clock A says the B clocks are running slow. Everybody agrees that the difference in time between clock B-2 and clock B-1, for the events of the clock A passing B-1 and then B-2, everybody agrees that that interval is longer than the interval elapsed on clock A between those two events. Everyone agrees. But the interpretation is different. The interpretation in B's case is, "Oh, A is running slower. So, of course, less time elapses."

The interpretation A is a little more complicated. A says, "Well, the clocks in B are running slow. If they were synchronized, then less time would elapse in B. But they're not synchronized. Clock B-2, the right-most clock, is further ahead. So the elapsed time I get by subtracting the measurement in B-1 from the measurement in B-2 is still longer than the time on clock A. Even though those clocks are running slow." Now, A's interpretation is, those clocks were not synchronized. That's a bit to think about.

You might want to go back and review that a few times. But there is the way out of the contradiction. It really is true that if a clock goes by me and I say, quote, it's running slow, it can say my clocks are running slow, and there is no contradiction. And the lack of contradiction comes about because events that are simultaneous in one frame of reference are not simultaneous in another. Well, why is that?

Let's revisit that star trip we took in the last lecture. Here's the star trip. We had a trip of what seemed to be 10 light years distance. We were going at $0.8c$. I'm going to assert now, that that trip is not 10 light years distance as measured in the spaceship's frame of reference. And we can easily see why that is. The spaceship, after all, is moving at $0.8c$, and it goes some distance, which we thought was 10 light years. But I'm going to show you it isn't 10 light years now. It's going at $0.8c$, 0.8 light years per year. We know that it takes 7.5 years. We figured that out. So distance equals speed times time, which is 0.8 of a light year per year times 7.5 years, and that comes out six light years. It must be the case with the distance between the Earth and star, in the ship's frame of reference, is not 10 light years, but six light years. It would be a contradiction if that weren't true. Distance equals speed times time wouldn't be true. It is true. And the answer is that the distance is different. The spatial interval is different.

That shouldn't surprise you. We've already seen that temporal intervals— the interval between two events in time—is different in different frames of reference. Now, I'm telling you the interval between two events in space, the interval between the event of the ship passing Earth and the event of the ship passing the star is different in a different frame of reference. And, in this case, the difference is given by the original distance in the frame in which the two objects were at rest, multiplied by that factor, $1-v^2$.

So if I take this meter stick, this is true of any object, if I take this meter stick and I move it along in my frame of reference, it's a perfectly good meter stick. It's a meter long. But if you measure it, and I'm coming toward you, say, at 0.8c, you'll measure this meter stick to be 60 centimeters long. There's that square root of 1-0.8² factor coming in again. It's really a meter long. Well, you can try to say that. But if you say that, you're saying the frame of reference where it's at rest is the special frame. In a way, that's a special frame for the meter stick. But, if somebody else is in another frame of reference and they measure it and they get 60 centimeters, 0.6 meters, they're right, too.

Measures of space are not absolute any more than measures of time are. Not a surprise. Space and time are the things we had to give up the absoluteness of when we accepted the absoluteness of the laws of physics. The fact that the laws of physics are the same for all observers in uniform motion. We gave up space and time being absolute when we first did that first experiment where I handed you a clock and a meter stick, and I put you in a car or an airplane or a spaceship, and we determined that you measured the same speed for light that I did. How could that be? Well, I said, "Well, something strange must have happened to your meter stick or your clock or maybe both. So that despite the fact that you're moving relative to me, you get the same answer for the speed of light that I did."

Well, something, "strange" has happened to both clock and meter stick, because something strange has happened to time and space. But be careful. That "strange" word is another dangerous, loaded, nonrelativistically correct one. Things are perfectly normal in your frame of reference. That's a meter stick. It measures a meter long. That's a perfect clock. And it keeps perfect time. It's just the time it's measuring, and the space the meter stick is measuring, are not the same quantities that my stick and my clock are measuring. They're different quantities. We'll see later on that they are, in fact, aspects of a more unified, four-dimensional quantity, which we all agree on, that has to do with spacetime. We'll get to that soon.

But in fact, measures of space and time are no longer absolute in relativity. And we simply don't notice the fact that measures of space and time are relative to the observer, because we don't have a lot of observers whizzing around at speeds that are a substantial fraction of the speed of light. But in fact that's the case. Measures of space and time are not absolute. And we now know how to find them out in these simple cases, in cases like the time dilation experiment we know the time on the clock moving from here to here, versus the two clocks that are at rest with respect to each other, is smaller by that factor of the square root of $1-v^2$. And, similarly, the distance between two events is smaller by that factor of $1-v^2$ in this particular kind of example.

So, I've convinced you now of something called *length contraction*. The length of an object—if you want to be really precise about it, the spatial interval between events of the one end of the object passing and another end of the object passing, and so on, if we want to be really rigorous about it—but let's just say the length of an object is greatest in a frame of reference where the object is at rest. When I'm at rest, with respect to this meter stick, it's one meter long. When I move it, like this, it's still one meter long, with respect to me. But it's shorter with respect to you by that factor, square root of one minus velocity, when the velocity is measured in terms of speed of light squared. That's called length contraction. It's also called the Lorentz contraction, again, after that Dutch physicist Lorentz—or the Lorentz-Fitzgerald contraction after the Dutch physicist Lorentz and his Irish colleague, Fitzgerald, who, as I described after the Michelson-Morley experiment, came up with the idea that maybe objects compress in the direction of motion due to their motion through the ether. They came up with the right formula. They came up with this phenomenon. They just had no basis for it whatsoever. They thought there really was a real length and the ether somehow compressed things. Well, things do compress in the direction of motion, but it's not because some physical effect of the moving squeezes them together. Don't look for that any more than you look for something tugging back the hands of the clock to slow down the clock. These are effects that are happening to space and time themselves. Not to the objects that are in space and time.

The meter stick is shorter because space itself is different in that frame of reference, not because the meter stick has somehow squished with respect to some absolute space. So don't get yourself talking nonrelativistically, don't say, it "really" is such. Don't look for the atoms to move a little closer together or something. All measures of space are contracted in that different frame of reference relative to what they were in another frame of reference. But neither frame is justified in saying, I'm the right one.

How does this lead to nonsimultaneity? How does it lead to the relativity of simultaneity? Well, that's actually quite easy to see once you accept the fact that length contraction occurs. Let me give you a couple of examples where length contraction is important though before we go on. It doesn't bother us in everyday life, again, because we don't move at speeds that are near that of the speed of light.

But there are a number of situations, in high-energy physics particularly, where we worry about it a lot. Remember those muons that were coming down Mount Washington. Mount Washington is 6,000 feet high. Not to those muons. The muons see Mount Washington length-contracted; it's only 700 feet high. That's why they last so long. They're only going 700 feet. Time dilation, length contraction, they're aspects of the same thing. The different measures of space and time in different frames of reference.

If you've ever been to Stanford University, driving down Interstate 280 in California, you cross a very, very long concrete structure. It's two miles long. You cross it just south of the northern-most exit to Stanford University. It's the Stanford linear accelerator. It's two miles long. It's a giant particle accelerator that accelerates subatomic particles to speeds almost that of light.

Electrons get up to 0.9999995 the speed of light in that thing. In the electron's frame of reference, the Stanford linear accelerator—two miles long in the Earth's reference frame—is only three feet long. Isn't it really two miles long? No. Because the electron's frame of reference is just as good as yours for doing physics, and for asserting the reality of physical entities, and for doing physical measurements and stating the results. It's really three feet long to those electrons. And if the engineers who designed it didn't take that in to account, it wouldn't work at all.

Similarly, your TV tube is shorter by a factor of the square root of $1-0.3^2$, which isn't a big factor but is measurable to the electrons barreling down it to paint the picture than it is to you and me sitting here in the room where the TV tube is. And if the engineers who designed that tube didn't take that into account, the picture would not be in focus. So, there are cases where length contraction is an important phenomenon. And there are cases where objects are moving at speeds approximately that of light, or close to that of light.

Okay, let's go on now and use this fact of length contraction to look at why events that are simultaneous in one frame of reference can't be simultaneous in another frame of reference. Okay, I'm going to imagine a situation in which two identical airplanes pass each other. And I'm going to draw this situation in three different reference frames. One, in a reference frame where both are moving. So, these two airplanes are going to be coming at each other from opposite directions. In this frame of reference, they're going to pass each other. I'm sitting in this frame of reference. The two airplanes are moving with respect to me. They're moving at the same speed, but in opposite directions. And so they're already going to be length-contracted. So, here are the two airplanes coming at each other. And I've drawn them a certain length. And that is their contracted length. That's not the length they would have if I were at rest with respect to them. That's a shorter length.

And I'm going to ask the question about two events. And the events are the coinciding of the tail of the upper airplane and the nose of the lower airplane. And the nose of the upper airplane and the tail of the lower airplane. Those are the two events I'm concerned about. And again, this goes far beyond airplanes. This is about any two events as judged in different frames of reference. In this frame of reference, the two airplanes are the same length, because in this frame of reference they're moving at the same speed. And although they're length-contracted, they're length-contracted by the same amount.

So, what happens when they get opposite each other? Well, you can see that the tail of the upper airplane and the nose of the right airplane are side by side at the same instant that the nose of the upper airplane and the tail of the lower airplane are side by side. The two events, of the tails and noses coinciding, are simultaneous events in this frame of reference, in the frame of reference where the two airplanes are moving at the same speed.

Now, let's look at that in a different frame of reference. Here's going to be a frame of reference where the upper plane is going to be at rest, and the lower plane is going to be moving—therefore, faster than it was in the frame of reference where both were moving. So, what's it look like? We have a great big airplane—long. That's its rest length. That's a length in a frame of reference at rest with respect to us. And the lower airplane is foreshortened even more. And it's moving very fast, whereas the upper airplane, in this frame of reference, is at rest.

What happens? Well, the first of the two events we were worried about is the nose of the upper airplane coincides with the tail of the lower airplane. Obvious. But at the same instant, the tail of the upper airplane and nose of the lower airplane are not coinciding. In fact, it's a while later that that second event occurs. The two events that were simultaneous in the first frame of reference—the frame in which the two airplanes were moving toward each other at the same speed—those events are not simultaneous in this other frame of reference.

And we can go to yet another frame of reference. We can go to a frame of reference in which the lower airplane is at rest. And what's going to happen there? The lower airplane is at rest. It's now its normal "rest length." The length it has in a frame of reference at rest with respect to it. The upper one is foreshortened. Just like the lower one was last time. And what event is first? It's the event of the tail of the upper plane coinciding with the nose of the lower plane. And a while later, the next event—that occurred first in the previous frame of reference—now occurs later. So, events that are simultaneous in the first frame of reference are not simultaneous in the other two frames of reference.

And it's even worse than that. Their time order is, in fact, reversed. In one case, this event occurs later. In the other case, this event occurred first. Not only are the events not simultaneous, but their time order is even relative. Whoa. Does that mean your death could occur before your birth in some frame of reference? Fortunately, it doesn't. There is only a certain class of events for which the reversal of order is possible. They're events that can be simultaneous, in some frame of reference. Not all can. There is no frame of reference in which your birth and death are simultaneous. I guarantee you. And we'll see why soon. In fact, we'll see why in the next lecture. We're going to explore the question of causality. Is it possible for events to have their time order reversed without throwing out the window the whole concept of cause and effect, because we usually think cause occurs before effect. And if the time order can reverse, what's happened to causality? We'll see that in the next lecture.

Faster than Light? Past, Future, and Elsewhere
Lecture 11

> The fact [is] that events that are simultaneous in one frame of reference are not simultaneous in another, and, in fact, that the time order of those events is also relative. Event A may occur before event B in one frame of reference, and event B before event A in another frame of reference. How can these things be?

That the time order of events can be different in different reference frames seems to wreak havoc with cause and effect. Is there some frame in which your death precedes your birth, for example? No, the only events for which the time order can be different are those that occur far enough apart in space that it would be impossible for a light signal to get from one event to the other.

Time between any two events depends on the frame of reference from which the events are measured (as time dilation examples show). To clarify, an *event* is specified by giving both a place and a time. However, the *time order* of events can differ in different reference frames only if the events are far enough apart in space that not even light travels fast enough to get from one event to another. Because nothing can travel faster than light (more on this shortly), the two events cannot be causally related. Thus, there is no problem with causality. For example, when the Mars Rover was active in the late 1990s, Earth and Mars were 11 light-minutes apart. An event on Earth and one occurring 5 minutes later (Earth–Mars time) on Mars cannot be causally related—and there can be observers for whom the Mars event occurs first. (But doesn't the Earth event *really* occur first? That question is not RC! Think about why not.) Such causally unrelated events define a new realm of time, in addition to past, present, and future.

In relativity, the past consists of those events that can influence the present. For example, your birth is in part the cause of your now watching or listening to these lectures. It is in your past, and all observers will agree that it came before your present moment (although they will disagree about the amount of time between your birth and now). But an event simultaneous with your birth (simultaneous in Earth's frame of reference) at the center of our galaxy, 30,000 light-years away, is not in the past because it can't yet affect us.

Similarly, the future consists of events that the present can influence. Events that will happen tomorrow on Earth are in the future, and all observers will agree that events on Earth today come before those on Earth tomorrow. However, events that will happen tomorrow at the galactic center are not in the future of the present moment on Earth, because there is no way we can influence them.

In relativity, the past consists of those events that can influence the present. For example, your birth is in part the cause of your now watching or listening to these lectures.

Those events that are neither in the past nor the future are in the *elsewhere*. They can have no causal relation to the event here and now, and different observers will judge differently whether they occur before, after, or are simultaneous with the here and now. The elsewhere is not some mysterious realm, forever inaccessible; it's just inaccessible to the here and now. Events that are now in your elsewhere will sometime later be in your past and, at some earlier time, they were in your future. But there's a band of time centered on the present—22 minutes for Mars, 60,000 years for the galactic center, 4 million years for Andromeda—during which events are unrelated to the here and now. Again, this gives us a new realm of time to go with our categories of past, present, and future.

All this depends on nothing being able to go faster than light. But why is it impossible to go faster than light? It follows from the principle of relativity that light goes at speed c relative to any uniformly moving reference frame. Therefore, there cannot be an observer who is at rest with respect to light. Because you would need to reach the speed of light to go faster still and because you can't be at rest with respect to light, then you can't go faster.

More technically, a light wave at rest is simply not a solution to Maxwell's equations of electromagnetism; only a light wave moving at c is. Accepting the principle of relativity—that all of physics, including Maxwell's equations—is valid in all reference frames, then it is impossible to be at rest relative to light and, therefore, impossible to go at speed c. (This is the point Einstein puzzled about at age 16.) Attempting to travel faster than light by "leapfrogging" from one rapidly moving reference frame to another fails, because measures of time and space differ in different reference frames. A more precise statement about faster-than-light travel is that no *information* can be transmitted at speeds faster than light.

Is this something special about light? No, it's about *time*; light only provides the extreme case. If a police car going 50 mph clocks you going 30 mph relative to the police car, are you going 80 mph relative to the road? Not quite! The difference is tiny, but it is there and it results from the fact that measures of time and space are different in the different reference frames. As speeds approach c, the effects of this *relativistic velocity addition* become more dramatic and prevent any material objects ever moving at c relative to each other.

There is one important caveat: It is the speed of light *in vacuum* that is the ultimate speed. Light moves slower through transparent materials, such as glass, water, or even air—and there's no problem with objects moving faster than the speed of light in such materials. High-energy subatomic particles moving through water, for example, often do exceed the speed of light in water and, when they do, they produce shock waves analogous to the sonic booms from supersonic aircraft.

A more sophisticated description of c is not so much that it's the speed of light but rather that it's a conversion factor between units of space and time. If we really wanted to be relativistically correct, we would measure time and space in the same units and c would have the value 1. ■

Essential Reading

Hey and Walters, *Einstein's Mirror*, Chapter 4.

Hoffmann, *Relativity and Its Roots*, Chapter 5, pp. 121–127.

Suggested Reading

Moore, *A Traveler's Guide to Spacetime*, Chapter 8, Section 4.

Taylor and Wheeler, *Spacetime Physics*, Chapter 6.

Questions to Consider

1. Right now it's "the present," but is it "the present" everywhere? Explain your answer.

2. Suppose a technological civilization evolved 25,000 years ago on the other side of our Milky Way galaxy, 60,000 light-years from Earth. Could there be observers for whom technological civilization emerged first on Earth? What if the extraterrestrial civilization had evolved 1 million years ago?

3. What's wrong with defining the past as those events that have already happened?

EVENTS ON EARTH AND MARS

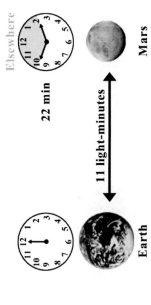

22 min

Elsewhere

11 light-minutes

Earth   **Mars**

This diagram illustrates the idea of the **Elsewhere**. The Elsewhere is a region of spacetime that is neither past nor future. The elsewhere of a given event consists of those events that cannot influence or be influenced by the given event – namely, those events that are far enough in space that not even light can travel between them and the given event.

The distance between Earth and Mars is 11 light-minutes. Events on Mars cannot influence events on earth for a period of 11 minutes. Events on Earth cannot influence events on Mars for a period of 11 minutes. Thus there is a 22 minute interval in which events on Mars cannot influence or be influenced by the present moment on Earth. These Martian events lie in the **elsewhere** of Earth's present.

Faster than Light?

0.8c

0.8c

Speed of small ship
relative to Earth?

~~1.6c~~   0.97c

In this diagram, the large ship is moving at 0.8c relative to Earth. The small ship is moving in the same direction, but at 0.8c relative to the large ship. Classical physics states that their velocities should be added together in order to find out how fast the small ship is moving relative to Earth. This simple addition would give the answer of 1.6c which is faster than the speed of light. Relativity shows that velocities do not combine in a simple addition, but in a more complicated way that does not allow a velocity combination to exceed the speed of light. In this case, the two velocities combine to produce a velocity for the small spaceship of .97c relative to Earth.

Faster than Light? Past, Future, and Elsewhere
Lecture 11—Transcript

Welcome to Lecture Eleven. My title "Faster than Light? Past, Future, and Elsewhere," draws on two of the problems that arise from the kinds of things we've been discussing. The nonrelative it, the relative it, relativity of simultaneity, for example. The fact that events that are simultaneous in one frame of reference are not simultaneous in another, and, in fact, that the time order of those events is also relative. Event A may occur before event B in one frame of reference. And event B before event A in another frame of reference.

How can these things be? They're closely tied with our notions of the past and the future and what those concepts really mean. They're also closely tied with a question that arises frequently when people study relativity. And that is why is it that nothing can go faster than light? I haven't asserted that fact yet, but it's been in the back of my mind in a lot of the things I've been saying, and now we need to bring it to the forefront.

So today's lecture deals with these two questions. What do we really mean by the past and the future? What about causality in a world where the time order of events can be different? And what about this business of faster than light? And how is that coupled with questions of causality?

I want to begin by clarifying a definition I gave you several lectures ago. And that's the definition of an event. In relativity, events are the key things. Events really happen. You were born. That happened. Different observers may disagree about the time interval between your birth and your sitting here watching this course, or listening to this course, for example. But nobody disagrees about the event of your birth having happened. An event is specified by giving the place it happened and the time it happened, and that is complete and enough to specify the event. We'll be talking, again, a lot about events.

Now, we saw several lectures ago that the time interval between two events is a relative quantity. Time intervals are simply not absolute. The time interval between your being born and your sitting watching this course might be, say,

45 years—that is, in a frame of reference at rest with respect to the Earth. But to somebody whizzing by Earth in a high speed spaceship going in uniform motion, straight-line motion at constant speed, that interval may be a very different amount of time. Time intervals are relative. That's something we came to a while ago, and I hope you're a least a little bit comfortable with that by now, even though, again, it is never going to make intuitive sense, because intuitive sense is common sense, and common sense, again, is built on that limited experience we talked about before.

But in the last lecture we came up with an idea that's perhaps even more radical. And that's the notion that the time order of events may be different in different frames of reference. And I left us last time with a question: How is that possibly consistent with our notions of cause and effect? How can the time order of two events be different? How can observers disagree about whether event A occurred before event B, or event B occurred before event A, when we expect of cause and effect that the cause comes before the effect? Surely your being born was a cause of your now watching this course. If you hadn't been born, you certainly would not now be watching this course. Don't observers all have to agree that your birth came before your watching this course? Well, the answer to that question is yes. There are only some kinds of events, only some events, only some pairs of events, for which the time order can be judged to be different by different observers.

I want to tell you a little bit more about what those kinds of events are. The events for whom the time order can be relative are a special class of events that are so far apart in space, in some frame of reference—and this then turns out to be true in all frames of reference—if they are so far apart in space that not even light, emitted at the time of this event, can get to the second event. If events are that far apart, then it turns out that there are different observers who will see the two events as event A occurring before event B, another observer will see event B occurring before event A, and there'll be other observers who see events A and B as being simultaneous. It's that class of events—events that are so far apart, that are widely spaced in space, that not even light can get from one event to another—those events are close enough in time that for some observers they occur at the same instant. They're simultaneous. And for some observers, event A occurs before event B, and for some observers event B occurs before event A. Now, we're beginning to get out of the realm where everything I say about relativity is going to follow absolute logic. So I'm not going to prove that statement to you. I'm simply going to assert it. It takes a relatively small amount of algebra to demonstrate it, but I'm not going to do it. So, you'll have to take this somewhat on faith—the statement that the events for whom the time order can be different for different observers are the set of events for which the distance between the events is so far, so great, that not even light could travel from one event to the other.

Let's go back to the example of your being born. And then you're sitting here, looking at this course, or listening to this course. How far apart are those events? Well, in space, those two events both take place on Earth. So, they're at most, thousands of miles apart. Could light, traveling at 186,000 miles a second, have traveled from the event of your birth to the event of your sitting here now? Well, it could have done so in a tiny fraction of a second. And, yet, maybe 45 years have elapsed. So light has had plenty of time to get from the event of your birth to the event of your sitting here now. Those are events for which the time order cannot be different for different observers. All observers will say, for those events, those events are close together in space and far apart in time. They're 45 years apart in time, judged by an observer on Earth. They're a few thousand miles apart in space, which is a very short distance in terms of how far light can travel in 45 years. Those events are, again, close together in space, far apart in time. No observer will think or measure that your sitting here watching the course occurred before your being born. All observers will agree that you were born before you watched this course.

Not all observers will agree that the interval between your birth and your sitting here now is 45 years, however. For some, that interval will be shorter. It depends on their relative motion, relative to Earth. But, all will agree that the event of your birth comes before the event of your sitting here now.

On the other hand, consider another case of some events. Consider an event that's happening right here, now. You're sitting watching this video and an event that happened at the time of your birth, but in the Andromeda Galaxy, which is 2 million light years a way. Now you were born 45 years ago, say. Well, light leaving the event of your birth could travel a maximum distance of 45 light years. Why? Because a light year is the distance light travels in one year. In 45 years, light travels 45 light years. Remember, a light year is a unit of distance, it's a unit of how far things are separated, and it's measured by how far light goes in one year.

So, in 45 years, light that left the event of your birth could travel 45 light years. Well, that's nowhere near as far as Andromeda, which is 2 million light years away. And, therefore, the event that occurred 45 years ago at the time of your birth, on Andromeda, and the event of your sitting here now watching this course, those two events are events that there could be observers disagreeing about when they occurred—which one occurred first. They won't only disagree about the time interval between them, they will disagree about which one came first.

There are observers for whom the event that occurred 45 years ago on Andromeda actually occurred first. And you might say, "Well, didn't it really occur 45 years ago." Well, if you start using that dangerous word "really," you're saying, oh, there's something special about the frame of reference of Earth and Andromeda, and observers moving relative to them are in weird, special frames that don't really count, and so, things aren't right for them. You can't do that. Relativity is the statement that the laws of nature, the laws of physics, are the same for all observers in uniform motion. So, everybody has to agree about that. So, everybody has equal perspective on deciding which event came first. And people disagree about that. Observers disagree about that. In fact, there will be some observers, moving in just the right way relative to Earth and Andromeda, for whom the event of your birth and the event of your sitting here now, and the event that occurred on Andromeda 45 years ago, as far as the Earth frame is concerned, that those events are simultaneous for some observers. They occur at the same time. And, again, for some observers, your sitting here now comes first. For other observers, that event comes first.

There's no contradiction of causality. Why not? Because of that fact that I'm going to discuss at great length in the second half of this lecture, the fact that nothing can go faster than the speed of light. In particular, no information can go faster than the speed of light. And because of that, for events that are so widely separated in space that not even light can get from one event to the other, there is no way for the events to communicate; there is no way for them to be causally related. And, therefore, there is no contradiction with some observers seeing event A before event B, and some observers seeing event B before event A. The events simply cannot be causally related.

I want to explore that situation in a little bit more detail with a specific example. In the late 1990s, we sent a wonderful space mission to Mars, and we landed a little Rover on the planet Mars. And you may remember the news at the time. The Rover was rolling around planet Mars, backing up to rocks, sticking its spectrometer into the rocks and analyzing them, and so on, and having a great time rolling around the surface of Mars and taking pictures.

At the time that happened, Mars and Earth were 11 light minutes apart. What's a light minute? Well, again, it's a unit of distance. It's the distance light travels in a minute. You can figure out what that is in miles by multiplying 186,000 miles a second by 60 seconds, and whatever that comes out, that's how many miles in a light minute. But it's just easier to talk about light minutes. Mars was 11 light minutes from Earth at the time that the Mars Rover was running around on the surface. Because Mars and Earth are both in orbit around the sun, they don't stay 11 light minutes apart all the time. But that was the distance, at that time.

What does that mean? It means it took 11 minutes for radio signals from Earth to get to Mars to tell the Mars Rover what to do. And it took 11 minutes for images being sent back by radio from the Mars Rover to get to Earth, or warnings that we got a problem—our computer is shutting down, as it did several times, or whatever—to be sent back from Mars. Eleven minutes is the communication time using light or other electromagnetic waves between Earth and the Mars Rover.

Well, I have a picture here that shows Earth and Mars and shows that they're 11 light minutes apart. And I'm showing, in addition to this picture, a couple of clocks—one on Earth and one on Mars. And we're going to assume that Earth and Mars are essentially at rest with respect to each other. That's approximately true. Not exactly. Their relative speeds are some tens of miles a second. And that's so small compared to the speed of light that for all practical purposes, as far as relativity is concerned, they're essentially at rest with respect to each other. They define the Earth-Mars reference frame then. And so these two clocks I've drawn are synchronized in that frame of reference. They both read the same time at the same instant. We talked about that in some detail last time. And we know how to synchronize the clocks.

Now, here's the issue. Suppose something bad is going to happen on Mars in 5 minutes. Here we are, we're sitting in the command center for the Mars Rover, and suppose a giant boulder is going to come rolling down and crush the Mars Rover, or one of those aliens we're looking for is going to emerge from a crack in the Martian surface and reach out and grab the Mars Rover and crush it. Is there anything we can do to stop that? Is there any way we can affect that event? Is there any way we can influence it? That event is going to occur 5 minutes from now. It hasn't happened yet, at least in our frame of reference—in the Earth-Mars frame of reference. It's going to happen 5 minutes from now. Is there any way we can influence that event? Well, if we know about that event, right now, it's the first time we know about it. There is no way we can affect it. There is no way we can stop it. We can send a message to the Mars Rover and say, hey, boulder is going to fall on you. But the Mars Rover will not get that message for 11 minutes. And the boulder is going to fall and crush it 5 minutes from now. Or, watch out, alien coming out of the cracks, going to get you in 5 minutes. That message is not going to get there in time. It's going to be 6 minutes late, because the shortest time a message can get from Earth to Mars is in 11 minutes—if you believe that nothing can go faster than light. And keep that question in the back of your mind, because I'll have more to say about that shortly.

So, my example would still be the case if it were 10 minutes before the alien was going to reach out, or the boulder was going to fall and crush the Mars Rover—even if we're 10.999 minutes. But, if it were 12 minutes from now, then we have the possibility of sending a message which will get there in 11 minutes, and the Rover has time to react and get out of danger. So that 11 minutes is an important break point. Events that are more than 11 minutes from now, in the future, on Mars, are events that we, here on Earth, can influence, because we can send a light signal, or a radio wave, or some other form of electromagnetic waves—the fastest things we have available to us or that any observer or experimenter or being anywhere in the universe has available to them. We can send a light signal or an electromagnetic wave of any kind, and we can warn the Mars Rover. And if the event is going to occur more than 11 minutes from now, the Mars Rover will get our warning and it can react—at least in principle.

What that means is this: There is a time span from now until 11 minutes from now—and I'm going to show that on the Mars clock—that we cannot influence, we can't affect events that are going to occur on Mars in the next 11 minutes. And again, I'm talking about during the era when the Mars Rover is on the surface, and Earth and Mars are 11 light minutes apart. There is a band of time, a band of events, that, as far as we are concerned, we, in the Earth-Mars frame of reference are concerned, have not yet happened. And yet, we can't influence them. Events beyond that time interval? Yep, we can influence them. But events in that 11-minute band, from now until 11 minutes in the future, we can't influence.

Now, what if we're sitting here in Mission Control for the Mars Rover and, in fact, that boulder is not going to crush the Mars Rover 5 minutes from now, but it crushed it 5 minutes ago? Are we unhappy? No, because we don't know that yet. The Mars Rover signal—we're still getting signals from the Mars Rover, they're signals that left 11 minutes ago and 11 minutes ago was before the Mars Rover got crushed by the boulder. It will be a while before we'll learn the bad news. It takes time for information to travel, because the fastest rate it can travel in the universe is c, the speed of light. So, an event that occurred 5 minutes ago on Mars, similarly, cannot influence the here and now on Earth. It can't influence us in the present. Later on we'll know about that event—the event of the Mars Rover getting crushed 5 minutes ago is not lost forever to us. We'll know about it sometime, and then we'll be very unhappy. Or the computer fails, or whatever else goes wrong—something bad happens to Mars Rover 5 minutes ago, we do not know about it now. We can't do anything about it. Not only that, we don't even know about it. It can't affect us in any way. It doesn't make us unhappy. It doesn't make Congress cut NASA's budget. It doesn't make anything happen right now. It may in the future. But right now an event that occurred 5 minutes ago on Mars has absolutely no influence on us. And, in fact, if you think about it, an event that occurred 9 minutes ago on Mars, or 10 minutes ago on Mars, or

10.99 minutes ago on Mars, those events cannot influence the present event here on Earth.

An event that occurred 12 minutes ago on Mars or 11.1 minutes ago on Mars, we can know about those. If the Rover was crushed 11.1 minutes ago, the signal has gone dead already. And we know that. So, events that occurred beyond 11 minutes ago, they can influence us. But there is a band from now to 11 minutes in the past that cannot have an influence on us. And, consequently, there is a band that extends 22 minutes in extent—from 11 minutes ago to 11 minutes into the future of events on Mars—that can have no causal relationship to events on Earth. Why? Because if they're in the past, the light or other information coming from them to tell us about them has not had time to reach Earth. If they're in the future, signals from Earth have not had time to reach Mars and affect those events. And so, those events are, in a sense, out of touch with the present event on Earth. There is a band at Mars, which is 22 minutes wide, which has no causal relationship to events on Earth. Those are the events that are far enough away in space—11 light minutes in this case—and close enough in time—within plus or minus 11 minutes of time from the present event here on Earth—that that causal influence can't exist. Because to have a causal influence, we need to send information telling the Mars Rover to do something, or it needs to send us information telling us it's got troubles or whatever. And in that band—22 minutes wide, plus or minus 11 minutes from now—that causal influence cannot happen, because no influence can travel faster than the speed of light.

So, that's the situation with the Mars Rover. And the events, event A—the here and now on Earth, in the control room of the Mars Rover—and event B—any event in that 22-minute window, centered plus or minus 11 minutes from right now on Mars, that's event B, any event in that window—those are the kinds of events that are far enough apart in space and close enough in time that some observers will see their time order differently. That event 5 minutes from now when the Mars Rover is going to get crushed, there could be observers moving uniformly through the solar system who would say that event occurs before the event of me sitting here right now in Mission Control for the Mars Rover. And there's no contradiction, because events cannot be causally related.

Similarly, there will be observers who say those two events are simultaneous. And there is no question of causality problems, because, again, they cannot be causally related. Events outside that 22-minute interval, it's a different story. They can be causally related.

Okay, well, let's expand this away from just Mars itself. Let's just talk a second about what we mean by the past and the future, in this context. If you ask a historian what the past is, the historian will tell you, it's all the things that have already happened. And what's the future? Well, it's all the things that haven't yet happened. But now we have an ambiguity, because we've got this range of time. On Mars it's 22 minutes wide. On the sun, which is 8 light minutes away, it's 16 minutes wide. At the center of our galaxy, which is 30,000 light years a ways, that band is 60,000 years wide, in time—30,000 years before the present time, 30,000 years from the present time. In that band, events at the center of the galaxy are uninfluencable by events now on Earth, and events now on Earth can't be influenced by them.

In Andromeda, which is 2 million light years away, that band of time is 4 million years wide. That's the band in which events on Andromeda cannot be influenced by the present event on Earth, and they can't influence the present event on Earth. In a sense, those events are not in the past, because they can't influence the present here on Earth. And they're not in the future, because we can't influence them. They're in the *elsewhere*, as it's called in relativity. There's a band of time now, instead it's not just past and future anymore, there's not an abrupt demarcation. Instead, there's a new zone of time called the elsewhere. For Mars, when it's 11 light minutes from Earth, the elsewhere are those events that occur in a 22-minute-wide band, plus or minus 11 minutes from now. In a real sense, those events are not in the past or in the future. They're in this mysterious elsewhere—meaning they can have no causal relation to the here and now on Earth.

Now, don't get carried away by that strange word "elsewhere." It doesn't mean these events are lost forever. They are not connected to the present moment here on Earth. But if I wait a while on Earth, events that are in the elsewhere will become in my past. And a while ago, events that are now in the elsewhere were in my future. It's the present moment, the place and time of being here and now on Earth, that those events in that 22-minute band are in the elsewhere. They're not either in the past or future.

In fact, we really need to redefine what we mean by past or future. In relativity, the past becomes not the events that have already occurred because different observers disagree about that, at least for events that are far enough apart, and the farther away you go, as in the Andromeda case, the wider the span of time involved—4 million years for Andromeda, 22 minutes for Mars. We have to redefine the past as those events that can influence the present moment. And again, that's relative to the present moment and its place. The events in the past on Mars relative to the here and now, the present moments on Earth are those events that occurred more than 11 minutes ago on Mars.

What's the future? It's those events we can influence. On Mars, the future consists of all those events that are going to occur more than 11 minutes from the here and now on Earth. All observers will agree that the here and now on Earth occurs before those events. They may disagree about how much time has elapsed between the here and now on Earth and an event that we think occurs 12 minutes from now on Mars, but they will all agree that that event occurs later. They won't disagree about the time order of events that can be causally related, similarly, for the events in the past.

So, we have now a more complex picture of time. We have the past, the future, and the elsewhere. And they are all relative to that present moment.

All this depends on nothing moving faster than light. And that is one of the big questions that keeps coming up in relativity. Why can't anything move faster than light? Let me give you a few explanations for that. They all boil down, though, to the same thing. The principle of relativity says the laws of physics are the same for all observers in uniform motion. So, none of them goes beyond that. And every one of them is, therefore, logically as good as any other. And the first one is, in a sense, the least satisfying one. But it's also, perhaps, the most simple and profound.

The laws of physics are the same in all uniformly moving reference frames. One consequence of the laws of physics is that there should be light waves, and they should move at speed c. Therefore, all observers in uniform motion should see light moving at speed c. Therefore, there can't be an observer who's at rest with respect to light. Therefore, nobody can go at the speed of light. Now, that sounds a little tautological and trivial, but that's as good an explanation as we need. It isn't a very satisfying explanation, however.

And let me give you some more satisfying explanations, then. A little more technically—and this is the one Einstein worried about when he was 16 years old remember, Einstein worried what would happen if you ran along side a light wave at the speed of light. What would it look like? Well, he'd see a stationary wave. A wavy pattern that wasn't moving. But, a wavy pattern that isn't moving is not a solution to the equations of electromagnetism, to Maxwell's equations, it just isn't. That does not solve the equations. You plug the mathematical description of a nonmoving wave in, and it doesn't work. The only thing that works is a moving wave moving at speed c.

So, a stationary light wave is not a solution to Maxwell's equations. And yet, all observers in uniform motion have to have Maxwell's equations valid, because they have to have all the laws of physics valid. And therefore, there can't be an observer who sees a stationary light wave. Therefore, nobody can move along side a light beam. Nobody can move at the speed of light.

That's also not very satisfying, but it's at least an explanation. And it's an explanation based, ultimately, in principle of relativity that says all observers see the same laws of physics, including the consequence that the speed of light should be c.

Well, you might start thinking of cleverer ways to do this. One way you might think of, you know, those people movers in airports. Well, what if you put on top of one people mover another one, and another one, and another one. Couldn't you get up to speed c? Let me give you a picture that shows the same idea. Here we have a rocket ship going past Earth at 80 percent of the speed of light. Pretty fast. We've been using that hypothetical rocket ship before. Suppose—that's a big rocket ship, made as big as the planet Earth—suppose inside that rocket I build a smaller rocket. And I launch that smaller rocket at $0.8c$, relative to the big rocket. So I've got a small rocket that is barreling down the big rocket at $0.8c$ relative to the big rocket. Well, how fast is that small rocket going, relative to Earth then? Think about that a minute. Sure, it's going 1.6c, right? It's going faster than light relative to Earth.

Sounds logical. But it's not true. In that particular case, the small ship is actually going about 0.97c relative to Earth. Why? Well, I'm not going to go through the mathematics of it, but just think a little bit about what's going on here. Measures of space and time in the big rocket ship that's going past Earth at 80 percent of the speed of light are different than they are on Earth. So, when I say the smaller rocket ship is going 80 percent of the speed of light relative to the big rocket ship, that does not mean it's going 1.6 times the speed of light relative to Earth. And, in fact, the equations that describe how velocities combine in relativity will never ever, ever let you go at speeds faster than light. In fact, if I replace the smaller rocket ship with a light beam going at speed c relative to the big rocket, those same equations will then tell me that light is not going at 1.8 relative to Earth, but it's going at exactly c relative to Earth as indeed we know it must, because the laws of physics, with their implication that the speed of light is c for all observers, must hold in all uniformly moving frames of reference.

So we can't get to speeds faster than light by leap-frogging ahead like this with multiple things going at faster and faster speeds relative to each other. Let me remind you here, although I'm talking about things like spaceships, and observers, and people, the real key in relativity is not that no object can go faster than light, but that no information can be transmitted faster than light. The maximum speed that information can be transmitted at is the speed of light. And it can be transmitted by the things that go at the speed of light, which may be only electromagnetic waves; there may be a few other kinds of entities in the universe—we're not sure—but whatever they are, they are massless entities, and they can go at the speed of light. But that's the fastest anything can go. And light is the only thing we really know for sure about that does that. And the speed of light is the upper limit to the speed at which you can transmit information.

So the key here is information. You cannot transmit information at speeds greater than c. And we're going to get back to that when we talk about a very peculiar quandary in quantum mechanics in one of the last lectures.

Now, is there something very special about light that is going on here? Be careful not to get too hung up on light. Talking about light gives us a lot of insights into relativity. But relativity isn't really about light. The constancy of the speed of light, or the sameness of the speed of light for everyone, is just a minor part of relativity. A minor consequence. What relativity is really about is space and time. In this case, in particular, about time. This business of trying to leapfrog the rocket ship is only one example. There are more mundane cases.

Suppose, for example, a police car is going down the road at 50 miles an hour. And I go past that police car. And the police car clocks me going 30 miles an hour relative to the police car. Can I get ticketed for going 80 miles an hour? Well, in practical terms, yes. But, actually, I'm not going 80 miles an hour relative to the road. I'm going 79.99 and a lot of nines and something, because velocities don't quite add the way you would expect them to. Why? Because the measures of time and space in the different frames of reference from which you are trying to add the velocities are not the same. The effect is negligible when we're talking about 50 and 80 miles an hour. But it becomes more and more dramatic as we talk about speeds close to c, and that's why in my spaceship case one spaceship going at $0.8c$ relative to another spaceship that was going at $0.8c$ relative to Earth, doesn't give me $1.6c$ relative to Earth. There's a law called the Relativistic Velocity Addition Law, and that's what, in fact, again, following from the principle of relativity, makes this all work out.

Now, let me add one important caveat before I finish here. People are often asking me, wait a minute, I just read an article. There actually was an article in recent years in one of the main physics journals. They showed a picture of a bicycle, and on the front it said, "Bicycling at the Speed of Light," because scientists had managed to get light to go as slow as 38 miles an hour. And they think they're going to have it down in a few years to one or two miles an hour in certain kinds of substances. There's nothing wrong with that. There's nothing wrong with light going slower in material media. When light goes through material substances, there's a complicated, electromagnetic interaction with the material and the light is slowed down. That's how your eye glasses work. That's how the lens of your eyes work. That's how my contact lenses are working. The light is slowed down. In the process, the light waves bend, and that's what allows us to focus light with lenses, for example. Light moves slower than c in material media. And there can be particles that move faster than light in those material media. In fact, when that happens, something analogous to the sonic boom of a supersonic airplane occurs, and it's called Cerenkov radiation. And we get a cone of bright light emitted by high energy particles that are moving, say, through water at speeds faster than the speed of light in water.

That's not a contradiction. The issue for us is the speed of light in vacuum. And that's the only real issue. In fact, a more sophisticated look at the speed of light, and if we had understood relativity intuitively when we started doing physics, this would be more obvious to us than what we have now. A more sophisticated way of looking at the speed of light, it's really a conversion factor between space and time.

We didn't really understand that space and time were as intimately related as relativity makes them. And it's c, it's the speed of light, that is the conversion factor between space and time, ultimately. And in the next lecture, we're going to look a lot more about that conversion factor between space and time. And the relation between the two. And we'll also give you a more physically satisfying explanation for why nothing can go faster than light. But, like all the other explanations, it's grounded in the principle of relativity—that the laws of physics are the same for all observers in uniform motion.

What about $E=mc^2$, and Is Everything Relative?
Lecture 12

You'll notice everything I've said so far about space, and time, and causality, and past, and present, and elsewhere, and future, and so on, is without any mention of $E=mc^2$.

Shortly after publishing his 1905 paper on special relativity, Einstein realized that his theory required a fundamental equivalence between mass and energy. This equivalence is expressed in the famous equation $E=mc^2$. What this means is that an object or system of objects with mass m contains an amount of energy given by the product m multiplied by the speed of light squared. Because c is large, this is an impressive amount of energy. The energy contained in a single raisin could power a large city for a whole day.

This equation is mistakenly assumed by many to be the essence of relativity. Although the idea behind the equation is seminal, $E=mc^2$ came as an afterthought to special relativity, and it was not until 1907 that Einstein published a comprehensive paper on the subject. $E=mc^2$ asserts an equivalence between mass and energy.

$E=mc^2$ is commonly associated with nuclear energy, and Einstein's work is, therefore, mistakenly considered responsible for nuclear weapons. A historical aside: Although $E=mc^2$ and Einstein are not responsible for nuclear weapons, Einstein did have a minor role to play in the development of these weapons. Dr. Leo Szilard, a physicist who first conceived of a nuclear chain reaction, prepared a letter to President Franklin D. Roosevelt at the start of World War II urging a U.S. nuclear weapons program to counter German nuclear efforts. Szilard convinced Einstein to sign the letter. This was Einstein's only involvement with nuclear weapons. Incidentally, Szilard later founded the Council for a Livable World, which has worked for decades to oppose nuclear weapons.

Actually, rather than applying solely to phenomena involving nuclear energy, $E=mc^2$ applies to *all* energy transformations, including the chemical reactions involved in burning coal or gasoline or metabolizing your food. Even a stretched rubber band weighs more than an unstretched one, because of the energy put into stretching it.

As mentioned above, $E=mc^2$ does also apply to the nuclear reactions that power the Sun or our nuclear reactors and weapons. In the Sun, nuclear fusion reactions convert some 4 million tons of mass to energy every second as a result of nuclear fusion in the solar interior. Even with nuclear reactions, though, only a small fraction of the total mass is converted to energy.

The most dramatic (and efficient) example of mass-energy equivalence comes from *pair creation*, the creation of a particle of matter and its antimatter opposite out of pure energy. The opposite process, annihilation, occurs when a particle and its antiparticle meet and disappear in a burst of gamma ray energy.

$E=mc^2$ provides another way of understanding why nothing can go faster than light. Because of mass–energy equivalence, energy, like mass, manifests itself as inertia, making an object harder to accelerate. As an object is accelerated to high speed, its energy increases and, therefore, so does its inertia. The object becomes harder to accelerate. Inertia increases without limit as an object's speed approaches c. It would, therefore, take infinite force and infinite energy to accelerate a material object to the speed of light—and that is impossible.

Is everything relative, dependent on one's frame of reference? Believing that Einstein's work declares everything relative has sometimes been used to assert relativity in aesthetics, morality, and other humanistic areas. But even if everything in physics were relative, why should this carry implications for morality, for example?

Clearly, everything isn't relative. The principle of relativity declares one absolute: the laws of physics. They are the same for everyone (at least, at this point, for everyone in uniform motion). A corollary of the laws of physics being absolute is that the speed of light is the same for all observers. So the speed of light is not relative.

There are other so-called *relativistic invariants* that don't depend on an observer's frame of reference. An important case is the *spacetime interval* between two events. Although different observers get different values for the time between two events and different distances between the events, all agree on this interval, which is a kind of four-dimensional "distance" incorporating both space and time.

> **"Henceforth space by itself, and time by itself, are doomed to fade away into mere shadows, and only a kind of union of the two will preserve an independent reality."**
> **—Hermann Minkowski**

An analogy—adapted from *Spacetime Physics*, by Taylor and Wheeler—is the measurement of distance between two points on Earth. To get from point A to point B, you might say "go 3 miles east, then 4 miles north." But suppose you made your map without correcting for the difference between magnetic and true north. Then, your map grid would be tilted, and you would describe the path from A to B differently, but the actual straight-line distance from A to B would be exactly the same.

Relativity is like that: Space is analogous to one direction, say east–west, and time, to the other (north–south). Different observers in relative motion are like the different mapmakers; each imposes a different "grid" on an underlying, objective reality. How that reality divides into space and time depends on the "grid"—that is, on the observer's state of motion; just as in the map analogy, the amount of eastward and northward motion differs with which map one uses.

There is an objective, absolute reality behind the quantities that are relative. In the map analogy, this is the distance from A to B, about which users of either map agree. In relativity, it is the spacetime interval between events. Spacetime is the name for the four-dimensional framework in which physical events occur—a framework that transcends individual observers and their different reference frames and gives the lie to the notion that "everything is relative." In the words of Hermann Minkowski, who had been one of Einstein's mathematics professors and later worked on the mathematical aspects of relativity:

"Henceforth space by itself, and time by itself, are doomed to fade away into mere shadows, and only a kind of union of the two will preserve an independent reality" (Hermann Minkowski, 1908, as printed in Lorentz, et al., *The Principle of Relativity*, p. 75). ■

Essential Reading

Hey and Walters, *Einstein's Mirror*, Chapters 5–6.

Mook and Vargish, *Inside Relativity*, Chapter 3, Section 11; Chapter 4, Section 5.

Suggested Reading

Davies, Paul, *About Time: Einstein's Unfinished Revolution*, Chapter 3.

Moore, *A Traveler's Guide to Spacetime*, Chapters 9–10 (requires a lot of math!).

Taylor and Wheeler, *Spacetime Physics*, Chapter 1.

Questions to Consider

1. You throw a bunch of subatomic particles into a closed box, the walls of which block the passage of matter but not energy. Must the number of particles in the box remain the same? Explain.

2. A coal-burning power plant and a nuclear plant each put out exactly the same amount of energy each second. How do the amounts of mass that they convert to energy in the same time compare?

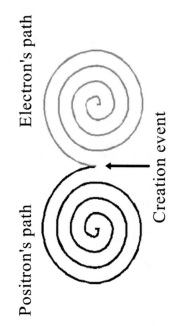

A Pair Creation Event

Positron's path

Electron's path

Creation event

Electron Inertia Increases With Speed

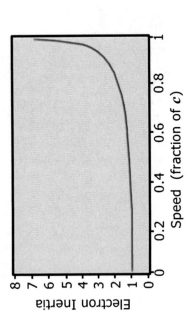

Getting from A to B: The Invariance of Distance

In this first map, a mapmaker has chosen true north as his reference direction. His description of the distance from A to B is "4 units east and 3 units north"

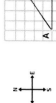

In this second map, another mapmaker has chosen magnetic north as her reference direction. Her description of the distance from A to B is different than the description the first mapmaker gives. The second mapmaker's directions say "4.5 units east and 1.75 units north"

When we superimpose the two maps, we see that the actual distance between points A and B is the same in both maps. The directions differ from map to map but the underlying physical relationship between A and B remains unchanged.

What about $E=mc^2$, and Is Everything Relative?

Lecture 12—Transcript

Welcome to Lecture Twelve, whose title, again, consists of questions. First question, what about $E=mc^2$? And second question, is everything relative? Probably, there's no more famous equation than $E=mc^2$. And in the popular mind, $E=mc^2$ is the essence of relativity. Even in The Teaching Company's advertisements for this course, the headline is often $E=mc^2$. Actually, although $E=mc^2$ contains a fundamental and crucial idea, $E=mc^2$ was not part of Einstein's original 1905 paper on relativity. In a sense, it's an afterthought. Later in 1905, Einstein wrote a short paper that contained the essence of the idea about $E=mc^2$, but it wasn't until 1907 that he elaborated on that idea at some length. So, in some sense, the equation $E=mc^2$ and what it means are an afterthought to relativity.

You'll notice everything I've said so far about space, and time, and causality, and past, and present, and elsewhere, and future, and so on, is without any mention of $E=mc^2$. But when you begin to talk about relativity in terms of dynamics, in terms about the behavior of objects, and their energies, and so on, and forces that act on them, then you begin to worry about things that lead to the equation $E=mc^2$. So, I'd like to describe what that equation means. And I'd like to dispel a lot of popular myths about the equation as well.

What does $E=mc^2$ really say? Well, it asserts a fundamental equivalence between the two major things that make up the universe. What are the things that make up the universe? Well, there's matter, the solid stuff of that tabletop, and of the Earth, and of my body, and the more ethereal stuff of the air, which nevertheless consists of ultimately atoms of "solid matter." We know about matter. It's tangible. It's real. We can grab hold of it.

The other major stuff that makes up the universe is energy. And energy is a little bit less tangible to us. What is energy? Well, when I'm moving along, I've got energy. Energy is motion in a sense, or motion contains energy, or motion is a manifestation of energy. If I lift a heavy object up against Earth's gravity, like this, I've given it energy. There's energy there. If I stretch this piece of bungee cord, I've given it energy. If I have a light wave traveling from the sun to the Earth, the electric and magnetic fields that make up that light wave contain energy. And the light wave is carrying energy to Earth. In fact, it's the energy that keeps us alive here on Earth. And it's coming at a big rate—about 1,000 watts on every square meter of Earth's surface.

Energy is an important part of the universe. If there were no energy, everything would stop. Energy is important. And energy's the other stuff of the universe. And before $E=mc^2$, energy and matter were two separate things. There were separate physical laws describing how they behaved. Both of them are quantities that seemed to exist and never change in amount. If you have a certain amount of matter, you always have that much matter. It may change to different forms. You may burn hydrogen and oxygen, for example, and make water, but you've still got the same amount of matter, so it seemed.

If you have energy, it may change form. It may be electrical energy. You may put it into an electric stove and turn it into heat. You may put it into a microwave oven and turn it into microwaves, and the microwaves may go into a cup of tea and turn into heat. Energy transfers from one form to another. You may have it as the stored energy, ultimately electrical energy, in the bonds that make up the molecules of gasoline in the gas tank of your car, and then you step on the gas peddle and the gasoline burns. It's converted to a different form. The energy now is the energy of motion of your car, and of the heat coming out the exhaust pipe, and the heat going out the radiator, and the heat of friction, and all the other ways the energy in your car is, in fact, wasted. But, it's still there as energy.

So we had energy and matter. And we had separate rules describing their behavior. And one of the rules is that both of those things are basically conserved. If you have a certain amount of energy, you always have that amount. If you have a certain amount of matter, you always have that amount. Even though they may change forms.

What does $E=mc^2$ do for us? It asserts an equivalent between matter and energy. Between mass—if you will, weight, the measure of the stuff we have in matter—and energy. And what it says is this: It says that if I have a hunk of matter, this tennis ball say, and I take its mass, its weight basically, and I multiply it by the square of the speed of light—which is a great big number, because the speed of light is 300 million meters per second; 186,000 miles a second—I multiply the mass of this thing by the speed of light squared, I get a great big number. That's the energy equivalent of this tennis ball. If I could convert this tennis ball entirely to energy, I would have a huge amount of energy. And $E=mc^2$ says there is an equivalence and at least hints at the possibility of converting this tennis ball entirely to energy.

$E=mc^2$ says matter and energy are no longer distinct, separate substances that make up the universe, but that they are interchangeable. And $E=mc^2$ describes, if you will, the exchange rate between matter and energy. Just like you have the exchange rate between two different national currencies, now we have two different currencies, if you will, that carry the substance of which the universe is made—one substance, mass-energy, if you will—and there's a conversion factor between them. And the conversion factor is speed of light squared.

Again, I hinted last time that we should think of the speed of light, not so much as a speed, but as kind of a conversion factor between, ultimately, space and time. And coupled with that is speed of light squared becomes a conversion factor between mass and energy.

And as I pointed out, the speed of light is huge. The speed of light squared is even huger. And, consequently, the energy contained in ordinary matter is enormous. If I could somehow extract from an ordinary raisin all the energy it contains—annihilate the raisin completely and replace it with pure energy—I could power a great big city like New York for a day. So, if I wanted to build a raisin power plant, all I'd have to do is develop the mechanism that makes that conversion. And I'd go to the grocery store and I'd buy a box of raisins, which probably contains a few thousand raisins, which is probably enough for several years. And each morning I'd drop one raisin into my apparatus, and that's all it takes to fuel New York City if I could convert mass to energy with 100 percent efficiency. $E=mc^2$ tells us there is a tremendous amount of energy resident in all the matter around us.

Now, in the popular mind, $E=mc^2$ is associated with nuclear energy. And Einstein and his equation, the theory of relativity, are often held accountable for the existence of nuclear weapons. I remember watching a movie in eighth grade in which there was a nuclear explosion. And then in the aftermath of the nuclear explosion, Einstein's face came on the screen to imply: This is the man responsible for all of this. Well, that's nonsense. First of all, $E=mc^2$ is not only about nuclear interactions. It's not only about nuclear energy. It's about all energy. All energy and all mass are equivalent. Any time we do any process that results in a release of energy, we have decreased the mass, the amount of matter, in that system that released energy. Any time we do something that increases the energy of a system, that stores energy, we have increased the mass, increased the amount of matter, in that system.

Let me give you some examples. Many examples. I take a discharged battery, and I put it on a battery charger, and I charge it up. The battery weighs more after it's been charged. Not because I put any physical substance into it in the form of matter, but simply because it has more energy. You'll never notice that difference weighing it, because a battery is a very inefficient energy storage device and a very small amount of energy goes in there.

If I take material from a plant; if I take the carbon dioxide and the water that go into a plant, and with the light coming from the sun through photosynthesis then make more complicated molecules like sugars, and I weigh all the stuff that goes in—the carbon dioxide, the water—and I weigh the sugar molecule that's produced by photosynthesis, it weighs a tiny, tiny, tiny bit more. Insignificantly tiny bit for chemical reactions like that, because chemical reactions—no offense to the chemists—are rather weak and puny. But they convert matter to energy and vice versa.

When I burn gasoline in a car, the gasoline is converting some, a very small fraction, of its mass into energy. It's not just nuclear reactions. When I blow up a nuclear bomb or run a nuclear power reactor, I'm also converting matter to energy. And the only difference with nuclear reactions is I'm converting a much larger fraction—about 10 million times more mass gets converted to energy in a nuclear reaction than it does in a chemical reaction. It's still a very small and insignificant amount—less than a percent. Nuclear reactions are not very efficient converters of mass to energy. Even though they are about 10 million times more efficient than chemical reactions.

In fact, if you do something as simple as burn a candle. And you weigh the candle before you light it. And you weigh all the oxygen that combines with the wax from the candle as it burns. And you weigh all the soot and all the carbon dioxide that's produced as a result of that burning. You will find the products of that combustion weigh less than what went into it by a very small amount. But that small amount, when multiplied by c^2, tells you the energy that was released from the candle in the form of light and heat, for example.

Similarly, if you take two electric power plants that produce the same amount of energy, the same amount of power—1,000 million watts is a typical large-size electric power plant. Take two—a 1,000 megawatt, a 1,000 million watt electric power plants. One burns coal or other dirty fuel (by the way, that contributes dramatically to global warming) and the other fissions uranium, it's a nuclear power plant. And for the coal burning power plant, you weigh all the coal that goes into it and all the oxygen that goes in to combine with that coal when it burns. And you weigh all the stuff that comes out the smokestack, all the soot that's carried away from the electrostatic precipitators that remove the particulate stuff from the air. And you weigh all the pollutants—everything that comes out the smokestack—all the carbon dioxide, what have you. You will find there is a slight difference in the weight. In over one year you can measure that difference between what went in and what came out. And if you multiply that difference by c^2, that will be the energy the power plant produced. Most of it, by the way, as waste heat that was dumped into a river, probably, or another body of water—some of it into the atmosphere—and about a third of it went out as electricity.

If you do exactly the same thing for the nuclear power plant, you weigh all the uranium that came in—there's no oxygen to weigh because there's no burning going on—you weigh all the uranium that came in and you weigh everything that comes out of the plant, which isn't very much. And you weigh the waste fuel that's left in the reactor after you're done. After a year, you will find that the weight or mass of that system has also gone down. And if these two power plants produce exactly the same amount of energy in a year, the difference will be exactly the same. They'll have lost exactly the same amount of mass.

That's the point I'm trying to make, that $E=mc^2$ is not just about nuclear physics, it's about all interactions that occur. And the nuclear power plant and the coal burning power plant lose the same amount of mass in a year— convert the same amount of mass to energy if they produce the same amount of energy. The difference is much, much more mass went into the coal burning power plant. A 1,000 million watt coal burning power plant typically burns about 10 to 15, 100-car trainloads of coal every week. The nuclear power plant, typically, uses a couple of truck loads of fuel every couple of years. That's the difference. That's that factor of 10 million in the efficiency of nuclear reactions and chemical reactions in converting mass to energy. But the process is, in principle, the same.

There is nothing intrinsically nuclear about $E=mc^2$. And I spent a lot of time on that, because it's a great misconception that $E=mc^2$ is about nuclear energy. If it took $E=mc^2$ to discover nuclear fission, for example, it would have taken $E=mc^2$ to discover fire. It is not necessary to understand relativity to make nuclear bombs unfortunately, or fortunately as the case may be. But it's also not necessary to understand relativity to burn a candle, or to metabolize food in your body, or to burn gasoline in your car. Yet all those processes involve the conversion of mass to energy, as described by the equation $E=mc^2$.

A couple of other examples. Our sun, which is really a large nuclear fusion reactor, has very inefficient nuclear reactions going on in its core. They're the reactions that, ultimately, produce the sunlight that I've described takes that eight minutes to get from the sun to the Earth and keeps us alive. The sun is converting about 4 million tons of its matter to energy every second. The sun is on a big weight loss diet, if you will. It's losing energy and is, therefore, losing mass because $E=mc^2$ and the mass loss rate is about 4 million tons a second. And if you multiply that by c^2, that's a big amount of energy coming out every second. It's about a one with 24 zeros after a 100-watt light bulb's worth, to give you an example.

So the sun is losing a lot of matter—four millions tons of mass every second. Even in situations where we're not burning something or undergoing nuclear fission or nuclear fusion, when we change the energy of a system, we alter its mass by, again, $E=mc^2$ or $m=E/c^2$. For example, I lift up this heavy ball, as I did before. Now the entire system—consisting of Earth and the ball and their gravitational field—has more mass than it did before. It has more energy because I lifted up the ball. It has more mass also. If I let the ball back down again, I get that mass back as energy.

We don't see the details of the energy conversion, but it's going on there. The mass is a very, very tiny fraction of the total mass of the system. It's a very inefficient process in that sense. But it's really happening.

Similarly, if I take this bungee cord, which I stretched for you before and stretch it again. Let's look at this bungee cord stretching in the context of $E=mc^2$. What happens when I stretch the bungee cord? Well, energy conversion processes in my body result in the metabolizing of my food. That actually lowers my mass somewhat, because I'm converting mass to energy. That energy manifests itself as motion as I pull against the bungee cord, you stretch it. If I weighed the stretched the bungee cord now, I would find it actually weighs a little bit more than it did before. How much more? Well, $E=mc^2$. So, $m=E/c^2$. I can tell you how much energy I put in by multiplying the force I pulled with by how far I pulled it. That's how we would do that. And if I take that number E, and I say $E=mc^2$, I divide E by c^2. I get a tiny, tiny number. That's the amount by which the mass of the bungee cord system changed.

Don't look for additional particles to have gone in there. I haven't added physical particles to it. What I've done is added mass in the form, ultimately, of the energy associated with the stretching of the molecules when I stretched the bungee cord. It's ultimately stored actually in electric fields, primarily because electric fields are what hold molecules together. So, even as simple an act as stretching this bungee cord, or stretching this rubber band, anything that stores energy in a system, also increases the mass of that system by an amount you can calculate from the equation $E=mc^2$.

Now, none of these processes I've described are very dramatically efficient at converting mass to energy. Even nuclear reactions convert, typically, less than a percent of the total mass present into energy. And for chemical processes, like the burning of gasoline or the explosion of a chemical explosive, those numbers are, again, down by a factor of 10 million, even lower. And then processes like lifting that ball or stretching the bungee cord are even less efficient in terms of the amount of the fraction of the total mass of the system that gets converted one way or another.

However, there are possible situations where we can convert matter to energy with 100 percent efficiency and back again. We see those situations in subatomic physics cases. And I'm going to show you an example now, which is a simulated picture of a reaction that occurs, actually, quite commonly in large particle accelerators where we accelerate high-energy particles to very high speeds.

So, let's look at this picture. This is a simulation of what's called a pair creation event. I want to show you what we have here. These two spirals, you see, are the tracks of two subatomic particles. One of them is the very familiar electron, which is the particle that circulates in the outer parts of atoms. We'll get to that more in subsequent lectures. And the other is a particle called a positron, which is exactly like an electron, except it carries a positive charge. It's the electron's *anti-particle*. There exists something called *anti-matter*, and every subatomic particle we know about has an anti-particle, which is its anti-matter opposite. It's just like it, except it's opposite in crucial properties like electric charge.

Now, it's a remarkable fact that we can actually see the tracks of individual subatomic particles. The way we do that, they come out in these experiments with very high energies, and they move through a medium like liquid hydrogen, or they move through an array of wires that can detect the passage of their electric charge. And from that, either they disturb atoms along the way or they disturb these wire detectors. And we can piece together what the paths of these subatomic particles look like. And what we're seeing here is a situation where there is a magnetic field. Remember, magnetism and electric charge are intimately related. There's a magnetic field that's actually coming perpendicular to the picture, if you're seeing this in your book, or perpendicular to the screen. And that magnetic field causes the positron and the electron to go in spiral paths. And they spiral, in fact, in opposite directions, because they have the opposite electric charge.

So, there is the electron's path. And there is the positron's path. Now, what happens at the middle, where those two paths join? Well, that's actually a creation event. What has happened at the middle is that that electron and positron have been created out of, well, I want to say nothing. But it's not quite nothing. They've been created out of pure energy—pure energy in the form here of a burst of very high energy electromagnetic waves called a gamma ray—suddenly cease to exist, and in its place appeared a pair of particles, a pair of matter particles—an electron and its anti-matter opposite, a positron.

We haven't created anything new. In the old days, before we knew about $E=mc^2$, we would say, "Oh, we've created matter out of nothing." Well, we haven't created it out of nothing. What we've shown here is there's an equivalence between matter and energy. And we have created matter out of energy. We haven't created electric charge, which really is conserved and can't be changed because we've created the negative charge of the electron and the positive charge of the positron. And the total charge is still zero, which it was before. So that's okay, but we have created mass—matter where there wasn't any before—out of pure energy.

So that's a pair creation event. And that's a 100 percent conversion of energy to matter. We can go the other way too. If an electron and a positron come together, they will annihilate completely and they will send out two bursts in opposite directions of gamma rays, of high-energy electromagnetic waves. And those waves will have a characteristic energy. The energy will be exactly the energy associated with $E=mc^2$ operating on an electron's mass. And, in fact, there are places in the universe where dramatically high energies exist, and where this annihilation process is occurring all the time. And we can identify it by looking at the gamma radiation, the gamma rays, with that particular energy. So this is something that does occur in the universe, and it's a 100 percent conversion of matter to energy.

In order to make my raisin-powered power plant, what I'd have to do is actually go to the grocery store and buy a box of raisins. And then I'd have to go to an anti-grocery store in some anti-matter world, if such a thing exists. We don't think it does. But if it did, I'd have to buy a box of anti-raisins. I'd have to very carefully keep them from touching any matter, and then, somehow, I'd have to drop a raisin and an anti-raisin into my power plant. And poof, they would annihilate in a burst of energy. Or my tennis ball, which I argued had some energy; in fact, a big amount, by $E=mc^2$. If I brought it together with an anti-tennis ball, then I'd have a 100 percent conversion of matter to energy. That's not a process we operate on Earth at any but the smallest scales in our atomic particle accelerators. But it is a process that, in principle, can occur, and it does occur in some places in the universe.

In fact, in the very early universe, the early fraction of a second of the existence of the universe, the temperatures were so high—and temperature corresponds to the motion of particles—that the particles moving around were moving so fast, that when they collided with each other, the collisions often involved energies bigger than that needed to create new particles. And so, in the very early universe, the number of particles, which we think of as sort of fixed here on Earth except for rare events in our particle accelerators or in cosmic ray interactions, we don't see this annihilation or this creation of pairs of particles, but in the very early universe that was very commonplace. And the number of particles simply wasn't fixed as there was an interchange between matter and energy. Now the universe has cooled to the point where that's a relatively rare thing; either here or even out in astrophysical situations. But it can occur, and it causes a 100 percent conversion of matter to energy—or the opposite.

Let me give you just a brief historical aside on this question of nuclear weapons and $E=mc^2$. Although Einstein was not responsible for nuclear weapons, and $E=mc^2$ has nothing to do with them, Einstein played a minor role in the development of nuclear weapons. Leo Szilard, who was a physicist who first conceived the idea of nuclear fission, actually suggested to Einstein that he write a letter to Roosevelt, President Roosevelt, saying that the U.S. should get to work on a nuclear bomb, because the Germans were, in fact, working on one at the start of World War II. And so Einstein agreed to sign this letter, and that helped get the U.S. Manhattan Project going. So there is a slight involvement of Einstein there, but it has really nothing to do with $E=mc^2$. Later, incidentally, Szilard founded the Council for a Livable World, which is an organization that has worked to oppose nuclear weapons.

One more aspect of $E=mc^2$ before we go on. Because energy and mass are, in some sense, equivalent, if I have energy, it exhibits an important property of mass—which is a property we talked about in the context of Galileo—and that is *inertia*. It's hard to accelerate energy as well as mass. And so if an object starts getting accelerated; it's now going faster; it has more energy; it has more inertia. And, therefore, it gets harder and harder to accelerate. And if we measure the inertia of, say, an electron, it looks something like this. I'm going to show you a graph with the electron's inertia as a function of its speed. An electron starts out with its normal inertia at slow speeds. But as the speed increases, the electron's inertia, and anything else's inertia, increases enormously. That makes it harder to change its motion, and harder and harder to accelerate it. So, if you want a more physical argument for why nothing can go faster than light, the argument is, as things get closer and closer to the speed of light, relative to you, they have enormous amounts of energy. And that enormous amount of energy corresponds to an enormous amount of inertia. And it makes it harder and harder to accelerate them. And it would take an infinite force, and an infinite amount of energy, to get anything up to the speed of light.

I think that's a more satisfying answer as to why you can't get to the speed of light. But it really is based in nothing more than anything else we've said, namely, the principle of relativity.

Finally, in the remainder of this lecture, the last one on special relativity, let me move on to one last question. Is everything relative? People have said, "Yeah, relativity says that everything is relative," and they even used that to justify moral relativism, and relativism in philosophy, and relativism in aesthetics. I think that's all nonsense. For two reasons, one is I don't think the physics of relativity has much to do with those other disciplines, at least in terms of making value judgments for them. And second, everything is not relative. Clearly it's not. The essence of relativity is not that everything is relative, but that some things aren't relative. What isn't relative? The laws of physics. They're the same for all observers in uniform motion. What else isn't relative? The speed of light is a consequence of the laws of physics being all the same. And there are other quantities that aren't relative as well. And I want to describe one of them in some detail.

I'm going to use an analogy here, which comes from, quite literally, from a wonderful book called, *Spacetime Physics,* which is in your bibliography. It's a delightfully written book by Taylor and Wheeler, who are experts on relativity. It's at a little bit higher level than this course, but it's worth looking at. And Taylor and Wheeler use an analogy of mapmakers. And these map makers are going to ask about the distance between two objects. So, let's look at a picture that shows that.

Here's point A and B, these are just points on the surface of the Earth. And I want to get from A to B. And one mapmaker lays out a map, a grid of north and south lines. It's based on true north. And in this mapmaker's system, we would say, "How do we get from A to B?" Well, we go one, two, three, four miles east, and one, two, three miles north. And that's how to get from A to B. And if we wanted to calculate the straight-line distance from A to B, it's the diagonal and it's exactly five miles. It turns out you can apply the Pythagorean theorem and get that.

Now, there's another mapmaker. He says, "I'm going to use magnetic north, which is not quite the same as true north, for my map." So the other mapmaker has a system that looks something like this. Here we have the same grid. One mile squares. North, east, south, west directions, but they're slightly different directions. Now how do you get from A to B? Well, a description would be go one, two, three, four, and about two-thirds miles east, and go about, oh, one-and-a-half miles north. And you got from A to B. That's a different description of how to get from A to B. Why is it different? Because the second map maker uses a different definition of north-south versus east-west. But we're still talking about the same underlying physical reality. And if this second mapmaker calculates the actual straight-line distance from A to B, it's still exactly what it was before. It's exactly five miles. And we can see that by imposing both mapmakers' systems on the same picture. The descriptions of how you get from A to B, in terms of how much eastward and how much northward you go, are different, but the description of the actual distance from A to B is one and the same.

Now, what's going on here? In a sense, the second mapmaker's map, it's east-west direction mixes in some of the north-south of the first mapmaker's map. East-west, north-south. Those are different dimensions, but they get kind of mixed up. Each mapmaker's north-south and east-west get mixed together to make the other map maker's east-west, north-south. That's why they have different descriptions of the same underlying physical reality. But the important point is it's the same underlying physical reality. The distance from A to B is invariant. It's the same no matter which mapmaker you use.

Now, what's this got to do with relativity? Well, the analogy is this: Space is analogous to one of these directions. Like, say, east-west. Time is analogous to the other direction. Say, north-south. Different observers in relative motion are like the two different mapmakers. They have different perspectives on what's east-west and what's north-south, or what are spatial intervals and what are temporal intervals. That's why different observers come up with different times between two events, and different spatial distances between them. Just like our two map makers come up with different amounts of east-west to go from point A to B and different amounts of north-south. But there's something they all agree on. And that something is a combination of, in the case of the mapmakers, east-west and north-south. They combine in a way to make an invariant—a true physical quantity that is objectively real and that they all agree on, and that's the distance between points A and B.

In relativity, it's a little bit different. In relativity, there's an objective, absolute "distance." But it's not a distance in space. It's a distance in space and time taken together. And if you combine those time intervals that we've been talking about between events that are different for different observers, and you combine that with the spatial interval between the two events, that's also different for different observers, every observer can calculate an invariant quantity. And it's called the *spacetime interval* between those two events. And they all agree on that spacetime interval. So, between two events, there is some reality in their separation. The reality is not a separation in space and a separation in time, like we thought it was before relativity. But it's a separation in a combined entity, which we're now going to call *spacetime*.

There really is a four-dimensional fabric to our universe. Four dimensions. The three dimensions of space—this way, this way, this way; three mutually perpendicular directions—combined with a fourth dimension, which I can't draw for you or indicate its direction, because I'm working in three dimensional space. But the fourth dimension is time. And the way spacetime breaks out into separate spaces and times differs with different states of motion. Just like the way physical reality of space for our mapmakers breaks out into north-south and east-west depends on their choice of compass directions. Similarly, what's space and what's time differs for different observers in different states of uniform motion. But they all agree on something underlying, which is the invariant spacetime interval. Now, you calculate it in a slightly funny way. Not the Pythagorean theorem, and I'm not going to get into that. But they all agree on that.

And there are, in fact, other quantities they all agree on. There are quantities that arise in electricity, for example, that involve the electric charge and the electric current that look different in different frames of reference. But they come together to form a common factor that everyone agrees on, no matter what their frame of reference. So there really is an underlying physical reality. And this underlying reality was summed up by Hermann Minkowski, who was Einstein's mathematics teacher, and later, as I said in an earlier lecture, became one of the people who gave the best mathematical interpretation of special relativity. And what Minkowski said was this, "Henceforth, space and time, by itself—space by itself, and time by itself—are doomed to fade away into mere shadows. And only a kind of union of the two will preserve an independent reality. And that union of the two is spacetime."

And we've just introduced it, now, as an abstract concept. But it's going to become a major player. It's going to play a major and active role in the development of general relativity, which is the next topic we'll be covering.

A Problem of Gravity
Lecture 13

The general theory of relativity is, remarkably, a theory of gravity. It doesn't seem, at first, that it should be. And in this lecture, we're going to explore why that is.

The special theory of relativity is *special* because it is restricted to observers in *uniform motion*. All such observers have equal claim on the validity of the laws of physics. A *general theory of relativity* would remove the restriction to uniform motion, making the laws of physics equivalent to all observers, whatever their states of motion. We need to proceed with caution at this point: This won't be as logical as the development of special relativity; the ideas are more abstract and the mathematics best left to experts! At first, the idea of a general theory of relativity seems absurd, because we *feel* when we're in accelerated motion (a car rounding a curve, an airplane accelerating down the runway for takeoff, a ship sailing in stormy seas). Historically, general relativity arose from Einstein's attempt to reconcile gravity with the principle of relativity. The incorporation of gravity into the theory solves both the problem of accelerated motion and inconsistencies between Newtonian gravitation and the principle of relativity.

This is the problem of gravity: Newton describes gravity as a force between distant objects; e.g., between Earth and moon, Sun and Earth, or Earth (all the matter of which it consists) and you. In Newton's theory, the force of gravity somehow reaches *instantaneously* across empty space to hold the moon, for example, in its orbit. But this cannot be consistent with relativity for several reasons. Relativity precludes any information moving at speeds greater than the speed of light. Newton's instantaneous action-at-a-distance gravitational force violates the "cosmic speed limit."

Newton's gravitational force depends on the distance between two objects, a distance that often varies with time. But special relativity shows that distance is relative, as is simultaneity. So Newton's law of gravity gives different results in different reference frames and, thus, is inconsistent with the principle of relativity.

Galileo recognized (purportedly by dropping objects off the Tower of Pisa) that all objects fall with the same acceleration. A more massive object needs a greater force to achieve the same acceleration in direct proportion to its mass (Newton's second law, or $F = ma$). Therefore, the property that determines the gravitational force on an object (its "gravitational mass") is the same as the property that determines how hard it is to accelerate ("inertial mass"). Gravity and acceleration are related.

Therefore, the property that determines the gravitational force on an object (its "gravitational mass") is the same as the property that determines how hard it is to accelerate ("inertial mass"). Gravity and acceleration are related.

The effects of Newtonian gravity and acceleration are indistinguishable. This is what makes general relativity a theory of gravity; any attempt to deal with accelerated reference frames brings in the indistinguishable effects of gravity. In 1907, Einstein asserted that in a small freely falling reference frame, gravity is not evident. Imagine you're in an elevator with its cable broken; if you take a ball out of your pocket and release it, it falls with the same acceleration you do—and thus appears weightless. It is impossible to distinguish free fall in the presence of gravity from the complete absence of gravity. The elevator occupant is unfortunate because free fall will soon stop, but an astronaut in a space shuttle is in exactly the same situation—free fall—and, therefore, doesn't feel gravity.

Objects in a small, freely falling reference frame (for example, inside an orbiting spacecraft) behave just as they would if they were in a uniformly moving frame far from any source of gravity. Thus, special relativity applies in freely falling reference frames. In fact, such frames are the closest we can come to the ideal uniformly moving frames of special relativity. Accelerated motion in the absence of gravity is indistinguishable from unaccelerated motion in the presence of gravity.

Gravity in the general theory of relativity. In the general theory, laws of physics should be the same in *all* reference frames. Therefore, something that is present in one reference frame but not in another can't be "real." Gravity can be "transformed away" by going into a freely falling reference frame. Therefore, what we usually think of as "gravity" or "the gravitational force" can't be what gravity really is.

What can't be transformed away are so-called "tidal forces," which in Newton's theory, result from *differences* in gravity from place to place. In a small reference frame, we won't notice the effects. In a large enough reference frame, even in free fall, these differences will be evident. Tidal forces are the "true" manifestation of gravity; however, because there is no underlying Newtonian gravity of which tidal forces are the differences, there must be some other explanation for gravity.

In 1912, Einstein argued that spacetime is *curved*, and gravity (e.g., what a Newtonian would call tidal force) is synonymous with the curvature of spacetime. Einstein's law of motion states that absent any force, an object moves in the straightest possible path in curved spacetime. *Locally*, that path is always a straight line at uniform speed, but on larger scales it reflects the geometry of spacetime—which is different from the Euclidean geometry studied in tenth grade. For example, parallel lines intersect in a spacetime with positive curvature. It is hard to picture spacetime curvature in four dimensions (three of space, one of time).

In summary, gravity is synonymous with the curvature of spacetime. Matter gets its "marching orders" (Taylor and Wheeler) *locally*, responding to the geometry of spacetime in its immediate vicinity. At the scale of a single particle, spacetime always looks locally flat and the particle acts as if it is in a uniformly moving reference frame. Special relativity applies perfectly. For extended objects or several spatially separated particles, the curvature of spacetime manifests itself in the subtle effects that used to be called "tidal forces." Gone completely is Newton's view of gravity as a force exerted between distant objects; in fact, gravity isn't a force at all, and free fall becomes the natural state of motion. The next logical question (to be answered in the next lecture) is: What makes spacetime curved? ∎

Essential Reading

Chaisson, *Relatively Speaking*, Chapter 6.

Hey and Walters, *Einstein's Mirror*, Chapter 8.

Hoffman, *Relativity and Its Roots*, Chapter 6.

Mook and Vargish, *Inside Relativity*, Chapter 5, Sections 1–8.

Thorne, *Black Holes and Time Warps: Einstein's Outrageous Legacy*, Chapter 5, from p. 93.

Suggested Reading

Taylor and Wheeler, *Spacetime Physics*, Chapter 9.

Questions to Consider

1. You drop a large rock and a small rock. Because of its larger mass, the gravitational force on the larger rock is greater. Why doesn't the larger rock fall with greater acceleration?

2. An airplane flying from San Francisco to Tokyo first heads north toward the coast of Alaska. Why? How is this analogous to what happens in general relativity's description of gravity?

The Principle of Equivalence

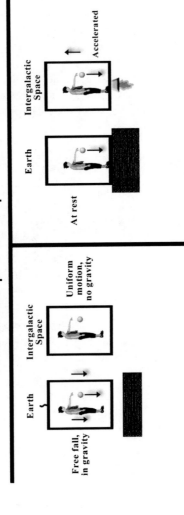

Free fall, in gravity — Earth

Uniform motion, no gravity — Intergalactic Space

The Principle of Equivalence states that there is no difference between being in a freefall toward Earth under the influence of gravity and being in uniform motion in intergalactic space far from any gravitational influence. If you were in a closed box with no windows, there is no way you could tell the difference between the two states.

At rest — Earth

Accelerated — Intergalactic Space

The Principle of Equivalence states that there is no difference between being in a state of rest on Earth and accelerating at 1 g in intergalactic space. If you were in a closed box without any windows, there is no way you could tell the difference between the two states.

Tidal Forces

In this diagram, we see the effects of "tidal forces" upon objects. In this case, a laboratory that is a substantial fraction of the size of the Earth is descending toward the planet. Inside this hypothetical laboratory is a very large person and two balls that are a good distance away from each other. Because the force of the Earth's gravity points toward the center of the Earth, the balls will move together as they fall closer to Earth. Notice that the person and the laboratory are also affected by the non-uniform gravity. They are stretched lengthwise and is squashed horizontally.

A Problem of Gravity

Lecture 13—Transcript

We're now halfway through the course. And at this point, you have a pretty solid understanding of Einstein's special relativity—its grounding in the principle of relativity, that the laws of physics are the same for all observers in uniform motion—and many of the consequences of that statement. And you understand them with a logical grounding, which is pretty solid. In the rest of the course, we're going to be a bit more descriptive, a bit more brief. And I'm not going to be able to ground everything as much in that logic.

In the next three lectures in particular, we're going to look at Einstein's *general theory of relativity*—the extension of special relativity to the more general case, where we don't have to worry about things just being in uniform motion. The general theory of relativity is, remarkably, a theory of gravity. It doesn't seem, at first, that it should be. And in this lecture, we're going to explore why that is.

What would a general theory of relativity be? Well, in essence, a general theory of relativity would state: The laws of physics are the same for all observers, period. The laws of physics are the same in all frames of reference, period. It doesn't matter what their state of motion is. No state of motion is special.

But that seems absurd on the face of it. After all, if I'm in that airplane moving through smooth air, I can eat my dinner. But as soon as I hit turbulence, my French fries are flying all over the place. The peanuts are jumping off the tray table and going all over the airplane. It doesn't seem that the laws of physics work there the same way they did in a uniformly moving reference frame.

Making general relativity is much more difficult. It took reformulating the laws of physics in such a way that, indeed, they are independent of your point of view, that the laws of physics are basically the same for all observers. And that was a difficult thing to do.

Again, the airplane, a ship in a stormy sea, that twin who turned around when she got to the distant star—they feel these forces on themselves. When you're in a car rounding a curve, you feel forces on yourself that aren't there when you're going in a straight line. Something is different. How do we reconcile that? How do we make a theory of relativity that says all those states of motion are perfectly good states for doing physics? It seems difficult. It seems, perhaps, impossible.

But historically, general relativity arose as a theory about gravity. And it did so because as soon as one tries to ask about motion that is not uniform, one is dealing with acceleration. And as soon as one starts dealing with acceleration, one finds there's a very close link to gravity. And in this lecture, we're going to explore that link.

Let's talk a little bit about Newton's theory of gravity that I introduced way back in the beginning of the course. Newton describes gravity as a force. And it's a force that's exerted between distant objects. The Earth, here, reaches out and pulls on the moon, and the moon, as a result, goes in its circular orbit, falling, if you will, away from its straight-line path at a rate that's set by the force that the Earth is exerting on it. Well, that's an interesting description. And it's one that serves us very well for sending spacecraft to Jupiter, or putting satellites in Earth's orbit, or calculating the positions of the planets or the position of the moon. So we can see when there's going to be an eclipse of the sun, or so we can land a spacecraft there, or whatever.

But nevertheless, it's a description of gravity that is philosophically unsatisfying, in a way I described quite a while ago, when I introduced the field concept. And now that we know special relativity, it's unsatisfying for two important reasons. First of all, this action-at-a-distance idea can't be right. It can't be consistent with relativity. Because if something happens to the Earth, over here, we know that the moon cannot respond instantaneously. Information about something happening to the Earth—it disappears, or whatever—cannot reach the moon in any time shorter than about a second and a half, which is about the travel time, a second and a quarter, the travel time, for light from the Earth to reach the moon. Because no influence, no information can travel faster than light. That's the whole point I was making in the last few lectures. So, the instantaneous action at a distance built into Newton's theory of gravity is simply inconsistent with special relativity.

The other problem is a little bit more subtle. In Newton's theory, the force of gravity depends on the distance between the objects. Remember, that's how he reasoned that there was a $1/distance^2$ fall-off in the strength of the gravitational force. Newton actually calculated the acceleration of the moon, compared the distance the moon was from the center of the Earth with the distance the falling apple was from the center of the Earth. Squared things. And, lo and behold, it worked out right.

But, there's a problem with distances. Because we know that distances depend on your point of view. Distance is not an absolute. Distance depends on the observer. And so there's a great big ambiguity in Newton's law. How do we apply Newton's law in a situation where distance is not a well-defined quantity, where distance is relative, where distance depends on the point of view, on the observer, on the observer's state of motion?

Something is wrong with Newton's law of gravity. It cannot be consistent with special relativity. Einstein knew this. So he knew that if he was going to develop a theory of gravity, he was going to have to somehow move beyond Newton. And at the same time, by incorporating gravity, it turns out the problem of making a theory that deals with non-uniform motion also gets solved. The mathematics of general relativity and the conceptual basis of general relativity is much more difficult than that of special relativity.

And I can tell you right now, you won't come out of my three lectures on general relativity with anything like the depth of understanding you have of special relativity. But I hope to give you a glimpse of this theory in all its richness. And I hope to spend some time on some of the contemporary applications of general relativity in astrophysics, particularly to black holes.

So let's move on and look at really what is the fundamental basis of the theory of general relativity. The fundamental basis, in a way, goes back to Galileo. Remember Galileo's famous experiment that may or may not have occurred? Galileo is purported to have dropped two objects of different mass off the Leaning Tower of Pisa and discovered that they fell with the same—

not speed—but the same acceleration. Objects experience the force of gravity in Newton's description. And they fall toward Earth with a constant force acting on them, at least near the surface of the Earth, where gravity doesn't change much with height. They fall with an acceleration, and their speed increases at a steady rate. And the point is that all objects do that in the same way.

So I'm going to drop this massive basketball and this much less massive tennis ball, and they basically fall with exactly the same acceleration. They hit the ground at the same time, despite the fact that this one is much more massive. Now, why is that? Well this one is, indeed, much more massive.

Many lectures ago, I put these two balls on the table. And I hit each one with the same force. And you found the massive one accelerated much less than the tennis ball. Why? Because same force, bigger mass, less acceleration. Newton's law: $F = ma$. So why do these two things fall with the same acceleration? Well, it must be that there's a bigger force acting on the big basketball than there is on the little tennis ball. And big and little here mean massiveness, not size. This is more massive. It weighs more. So, when I drop these two objects, they both fall with the same acceleration. Apparently, because the force acting on each one—the force of gravity—is proportional to its mass. That's a profound point. That's a point that, for many years, was regarded as simply some kind of strange coincidence.

Let me tell you a little bit more about that, that point. When I accelerate this basketball by pushing it on the table, I can actually determine how massive it is by applying a known force and seeing how much it accelerates and using Newton's law, $F = ma$, to calculate what its mass is. If I apply the same force to the tennis ball, it accelerates more, and I can determine that it is less massive, and I can calculate that quantitatively. Those experiments and the property, the mass of those objects that I'm thereby measuring, have nothing at all to do with gravity. I never mention gravity. I'm moving those things horizontally on a tabletop. I could do the same experiment in interstellar space, or intergalactic space, far from any gravitating object. It has nothing to do with gravity.

On the other hand, I do another experiment. I weigh these two objects by hefting them against gravity. And I find their weights, which is basically a measure of the gravitational force Earth exerts on them, is in direct proportion to their mass, as I determine it by trying to push on them. Now that's a strange coincidence. This property of these objects that makes them hard to accelerate, a property that has nothing to do with gravity, seems to be proportional to the same property that determines how much force Earth exerts on them. And it's because of that proportionality, or might as well call it an equivalence, that these two objects fall with the same acceleration. The tennis ball is less massive. And it, therefore, takes less force to give it a given acceleration. But on the other hand, the force of gravity is less in exactly the same proportion.

Sometimes those two properties, the property that determines how hard it is to accelerate something—that property is called *inertial mass* because it's about the inertia of the object—and the other property that determines how much force the Earth's gravity, or any other object's gravity, exerts on the object—that's called the *gravitational mass*.

And the remarkable fact is that the gravitational mass and the inertial mass seem to be the same thing, even though they deal with very different aspects of reality and the properties of these objects—one with how hard it is to push it, with nothing to do with gravity, and the other with what force gravity exerts on it. And that has nothing to do with pushing this. Yet, they seem to be the same property. That equivalence of gravitational mass and inertial mass was, for centuries, an unusual coincidence.

Einstein made it the cornerstone of general relativity. Einstein declared that is the basic principle. Inertial mass and gravitational mass are exactly the same thing. That principle, that they are the same thing, is basically called the *principle of equivalence*.

Let me now describe the principle of equivalence and exactly what it implies for acceleration and gravity. Because the principle of equivalence, the principle that the inertial properties (how hard it is to accelerate an object) and the gravitational properties (how much force it feels as a result of the gravitational attraction of Earth or some object), the fact that those two properties are the same leads to an equivalence between the effects of gravity and the effects of acceleration that makes the two indistinguishable. And ultimately, if you want to know why general relativity becomes a theory of gravity, it's because of that indistinguishability.

If you'd like, logically, to generalize special relativity, remember that was specialized to the case of uniform motion. If you want to generalize it to non-uniform motion, you're then talking about accelerated motion; motion that changes; motion that isn't straight line at constant speed. And I've just told you that acceleration, the effects of acceleration, and the effects of gravity are indistinguishable because of the principle of equivalence. And therefore, the minute you start talking about accelerated motions, you're also talking about gravity. And that's why general relativity becomes a theory of gravity.

Let's take a look, in some detail, at the principle of equivalence. I've got a picture here, which shows an unfortunate human being who's in an elevator in which the cable has been cut. And this poor individual is in so-called free fall. That is, the individual, the elevator, and all the contents of that elevator are falling freely toward the Earth. There are no other forces acting on it. We're going to neglect air resistance. It's falling freely. It's not sitting on the ground, where the ground would be pushing up on it and supporting it.

What if I take a ball out of my pocket and I drop it in this freely falling elevator? Well, what happens? The elevator is falling downward with a certain acceleration. (It happens to be 32 feet per second of speed every second you fall, or 9.8 meters per second every second you fall. But the number isn't important.) The important point is, as Galileo showed us, it's the same for all objects. So me, the ball, and the elevator are all accelerating downward with the same acceleration. And what that means is if I take the ball out of my pocket and put it there, it just sits there. It seems to me to be weightless. It isn't really weightless, because gravity is really pulling down on it and the force of gravity is its weight. But it seems to be weightless.

Now this is a bad situation to be weightless in because pretty soon that falling elevator is going to crash into the ground. And then forces that are not the force of gravity, but rather the forces of the Earth pushing up, which are ultimately the electric forces of the crushing of the atoms in the Earth, and so on, those are going to act with disastrous consequences for the elevator and its passenger. But there are situations where one can be in "free fall" without running into this danger. And the most obvious one is being in an orbiting spacecraft.

As I described when I talked about Newton's vision of orbits, a spacecraft in a circular or even elliptical orbit is simply falling. It's falling around the Earth. It's deviating toward the Earth from the straight-line path it would follow if there were no force acting. It is just as much in free fall as this elevator. In fact, it's more so, because there's no air resistance up there.

But weightlessness in a space ship has nothing to do with being in outer space. The unfortunate number of books, especially children's books, on the subject of science that say, "Oh, people are weightless in space, because there's no gravity in space." Hogwash! If there were no gravity in space, the space shuttle would go off in a straight line at constant speed, obeying Newton's first law—the law of inertia—and we'd never see it again. And, oh, if it's going in a circular path in Newtonian terms, that's because there's a force acting on it.

Why are astronauts weightless? Because they and their spaceship and everything in the spaceship are in free fall. They all experience the same acceleration. And that's ultimately because their inertial masses—how hard it is to accelerate them—and their gravitational masses are the same thing. That's why astronauts are weightless. So the astronauts are, in fact, in this situation, but more safely so, because their path does not happen to intersect the Earth.

Another example. If you've seen the movie *Apollo Thirteen*, Tom Hanks floats weightlessly around the cabin of an airplane, which is made to go in a free-fall trajectory. It's basically in free fall during the time the astronauts are weightless. At the time *Apollo Thirteen* came out, the newspapers were full of wonderful reviews saying how cleverly they had simulated the weightlessness of outer space. Nonsense. The weightlessness that went on in that airplane was exactly the same as the weightlessness that occurs in an orbiting spacecraft. It's the weightlessness that occurs for any object or any system that is falling freely, that is moving under the influence of gravity. I don't care whether it's falling down or falling around. It's moving under the influence of gravity and that's enough.

Okay. Now what's this all got to do with the principle of equivalence? Here we are; we're weightless in this freely falling elevator. There's another situation in which we would be completely weightless—actually be completely weightless. And that is if we were in intergalactic space, far from any gravitating object, so far from any galaxy that gravity is completely negligible. And if I were in that elevator in that situation and I take a ball out of my pocket, it just sits there. There's no force acting on it. Those two situations, according to the principle of equivalence, are, in fact, indistinguishable.

If I'm in that freely falling elevator and I don't have any windows or anything, and you don't have any way of knowing what's going on outside, and you start pulling balls out of your pockets or any other physics apparatus and you do experiments, there is no way to distinguish the situation of being in free fall near the surface of the Earth from the situation of being in intergalactic space, where there is no gravity, until you, unfortunately, smash into Earth and other forces act. But those two situations are indistinguishable. That's what the principle of equivalence is saying.

Another thing it's saying then is that we can actually take away gravity. We can eliminate gravity. We aren't really eliminating it, in some sense. But we can make it seem to go away by going into free fall. We can transform it away. And I'll have a lot more to say about that in just a minute. So being in free fall, you don't notice, ironically, the very gravity that is, in principle, making you fall.

By the way, there's another important consequence of this. In that situation where I'm in uniform motion in intergalactic space and there's no gravity involved, if I push on that ball, what will it do? Well, it will do what the first law says. It will accelerate briefly while I push on it. And then it will move in a straight line at constant speed, obeying the law of inertia—the first law of motion.

This is a situation that acts exactly like that. This is a uniformly moving situation—exactly the kind of situation in which special relativity is valid. But the situation of being in the freely falling elevator or being in an orbiting spacecraft is equivalent. And therefore, we have an answer to that question I posed back with the beginning of special relativity.

How do we find a frame of reference that is truly in uniform motion? Well, we don't. But the next best thing we have is a freely falling frame of reference. It's a little bit hard to believe because that freely falling frame of reference is accelerating with respect to Earth. But what this says is that freely falling frame of reference is the closest we can come to a uniformly moving frame. And if it's very small—and I'm going to say what that means in a few minutes—it essentially becomes like a uniformly moving frame of reference. The closest we can come to the frames of reference in which special relativity is valid are freely falling frames that are small. And I'll show you why small in just a minute.

Now, there's another equivalence. And it's the equivalence between this situation, where I'm sitting at rest on planet Earth in my elevator, and I take a ball out of my pocket and I drop it. And the ball falls with an acceleration of 32 feet per second, per second to the ground. That situation is equivalent to another situation in which I'm in intergalactic space. And I'm in a space ship, which is, in fact, accelerating due to a rocket motor, at exactly that same acceleration. And if I take a ball out of my pocket, it obeys the law of inertia. It basically just sits where I left it, having whatever motion my hand had when I let go of it. But the whole rocket or rocket-powered elevator is accelerating, in this case, upward, in the direction upward on your screen, and so the bottom of the elevator pretty soon bumps into the ball. And the ball appears, from the point of view of the elevator, to accelerate downward.

So those two situations are equivalent. And we can't tell again, in any small, localized laboratory, which of those situations we're in. Are we at rest, near a gravitating body? Or are we off in interstellar space, far from any gravity, and we're being accelerated? Acceleration and gravity effects mimic each other. In fact, in big space stations that we plan to build, space stations will be set rotating to simulate the effect of gravity, because of the accelerated motion. Acceleration and gravity have indistinguishable effects, at least in the small scale.

What does this have to do with general relativity? Well, here's where Einstein's genius kicks in. Einstein said, "Look, if we're going to make a general theory of relativity, the laws of physics have to be the same in all reference frames, including these freely falling frames and so on."

If there's a quantity that exists in one frame of reference, but not in another, that quantity doesn't have objective reality. It's like in special relativity, measures of time and space differ in different frames of reference. In some frames of reference, two events are simultaneous and there's zero time between them. And in another frame of reference, they're not simultaneous and there's some time between them. That time does not have an objective reality, because it depends on your point of view.

And Einstein says the same thing here. Quantities that are there in one frame of reference and not in another can't be objectively real. So, Einstein makes this enormous leap. Because I can transform gravity away, because I can make it not seem to be there by jumping into a freely falling frame of reference, gravity, as we know it, is not a real thing.

That's a remarkable statement. But that's what Einstein says. This big, heavy force I feel pulling me toward Earth—that's not real—because there's a frame of reference I can get into, by going into free fall, by taking myself away from contact with the floor, and for a moment, I'm in a frame of reference in which I don't feel gravity. And so gravity can't be real if that's to be as legitimate a frame of reference as any other frame of reference. So gravity can be transformed away.

"What, then, is gravity?" says Einstein. We've got to find something that can't be transformed away. And there is something that can't be transformed away. And it's something subtle. Something that was known about in Newton's time. And it's called, in Newton's theory of gravity, the *tidal forces*.

These are forces that are due to slight deviations in gravity from one place to another. For example, if I put this tennis ball here, it feels a gravitational force toward the center of the Earth. If I put it over here, the center of the Earth—because the Earth is round—is a slightly different direction. If I put it here, it feels a stronger gravitational force than here, because the gravitational force weakens with distance from the Earth.

The fact is, gravity varies slightly from place to place. And those deviations are responsible for the tides. Gravity of the sun and moon pull differently on one side of the Earth than on the other, and that's what causes the tides. That's why those are called tidal forces. Let's take a look at those tidal forces.

Here we are again in a freely falling laboratory. And I'm going to take two balls out of my pocket, and I'm going to let them fall. And now I've drawn the laboratory big enough that I notice the curvature of the Earth. Down below is the Earth.

What do those balls do? They fall toward the center of the Earth. But you'll notice, as I just described, toward the center of the Earth isn't quite the same direction for the two different balls. So what happens? As they fall toward the center of the Earth, on paths that take them toward the center, they actually come closer together.

Now that's something all observers agree about. The observer in the freely falling frame of reference notices, doesn't know about falling, but notices these two balls drifting slightly together. An observer on the Earth says, "Oh, I see what's happening. They're falling on these converging paths."

They disagree about the interpretation of what's going on. But the phenomenon is real for both of them. Those two balls appear to drift together. That's something real that cannot be transformed away. Notice, it's not something that you could experiment with, with a single point of particle. A single tennis ball wouldn't tell you that. Well, something as big as a tennis ball would, ultimately. But a single point, like an electron or something, you couldn't learn that from one point. You've got to have an extended object or a couple of objects to tell you that.

So what happens as we fall? Well, two interesting things happen. The balls, which are now this far apart, come closer together. Something else has happened here. If you look carefully at this picture, you'll see it. What's happened to me? Well, I've become squeezed a little bit this way, because the particles in me are also falling toward the center of Earth, coming closer together. Furthermore, I've become stretched in the vertical direction. Why? Because gravity is slightly weaker at my head than it is at my feet and so I'm somewhat stretched. Now for Earth, these effects are something we never, ever feel because Earth's gravity is so weak.

If we were as big as this picture has us, if my height were a substantial fraction of the size of the Earth, then I would feel these effects, and they would be very noticeable. But if I'm in a small enough frame of reference, a small enough freely falling laboratory, I don't notice these effects. If I'm in a laboratory the size of a real elevator, I don't notice them. But if I'm either in a large laboratory—large compared to the size of the Earth, comparable to the size of the Earth—or, if I'm in a place where gravity is extremely strong (and we'll be talking about those places in the next couple of lectures), then these effects will become dramatic.

And the point is these effects cannot be transformed away by going to a different frame of reference, so they must be real. So for Einstein, the reality of gravity is what for Newton was a very subtle residual in the differences in gravity, for place to place. Einstein takes away Newtonian gravity itself. It simply doesn't exist. And it's replaced with this very subtle effect. And Einstein doesn't think of it as a force. It's not that those two balls are attracting each other. That's a different story. We're making them small enough that their own gravitational attraction for each other is truly negligible. But what we're saying instead is there is something that's causing them to come together. What is that something? Well, Einstein identifies that something with the geometry of spacetime. And this is probably the hardest idea to grasp in general relativity.

In the last lecture, I introduced the notion of spacetime. It was sort of a framework, a stage, in which physical events took place. And I showed you how different observers broke spacetime into space and time in different ways. But spacetime was still pretty much like a combination of ordinary space that we understand and ordinary time that we understand.

And now in general relativity, Einstein says spacetime has properties. And what are the properties? It has geometry and its geometry is not necessarily the geometry that you learned in 10th grade—the geometry of Euclid. The geometry that says, for example, if you have a line and another line, there's one line parallel to it and they never intersect. Or a triangle has 180 degrees if you add up the angles. Those are statements of Euclidean geometry. They seem to make intuitive sense. But Einstein says the geometry of the universe is not Euclidean geometry, not the geometry of parallel lines that never intersect, not the geometry that you learned in 10th grade. But nevertheless, it's a four-dimensional geometry. It is, in some way, curved. And the curvature, the geometry of spacetime, is what gravity is all about. And this is a very, very difficult thing to picture.

But here's what Einstein says. I'm going to try to do this by an analogy. I have here a large ball. It's a chalkboard ball. I can write on it with chalk. And I'm going to pretend that the two-dimensional surface of this ball, (It's two dimensional because I can move in one of two mutually perpendicular directions on it.) I'm going to pretend that the two-dimensional surface of that ball is an analogy—an imperfect one, but an analogy, with the four-dimensional spacetime that is curved—because the ball is, in fact, curved. If spacetime were not curved, I wouldn't be showing you this ball, but I'd be showing you a flat blackboard. And that would be like Euclidean spacetime. That would be a two-dimensional analog of four-dimensional Euclidean spacetime that we would have in the absence of gravity.

But Einstein says gravity is the curvature of spacetime. And then he formulates a very simple law of motion. Remember we raised the question: Is there a natural state of motion? And by the time we got to Galileo and Newton and even into special relativity, the answer was yes—it's uniform motion in a straight line.

In general relativity, the natural state of motion is to be in free fall, to be moving under the influence of nothing but gravity. But in general relativity, gravity isn't some force that influences you. It's the geometry of spacetime itself. And the law of motion becomes very simple. It says an object that's not under the influence of any forces—and you might want me to say, "except gravity," but I'm not going to because, for Einstein, gravity is not a force—it's the geometrical structure of spacetime. For an object that's not moving under the influence of any forces, electric forces, pushes and pulls from people, whatever, then the object moves in the straightest possible line in curved spacetime. The straightest possible line in curved spacetime.

But, that line is not necessarily a Euclidean straight line like you learned about in 10th grade. Let's try to draw a straight line in curved spacetime. You've all actually done this exercise. At some point, I'm sure, you've studied how an airplane gets from San Francisco to Tokyo. It doesn't go west across the Pacific Ocean. It goes north. Why? And then south. Because that's the great circle. That's the shortest route on the curved globe. You fly from New York to Europe and you go a lot further north than either New York or Europe. Why? Again, because the shortest path on a curved surface is the so-called great circle, part of a diameter. And that's the best way to get from point A to point B. That's the shortest, straightest possible path.

So, here's point A and here's point B, and I want to get from one to the other. I go on the straightest possible path in curved spacetime. And it's a path like that. That path is straighter than other paths I can draw. That's a straight line in curved spacetime. And if I drew the whole straight line, it would go all the way around the globe and connect back on itself—something no straight line in Euclidean geometry ever did.

Now there are other straight lines. For example, here's another straight line that intersects that one. But a while later, those two lines are parallel. They're running completely parallel to each other. So those are parallel straight lines. And yet, the parallel straight lines intersect.

This is not Euclidean geometry. It's a different geometry. And it's that intersecting of the parallel straight lines that tells us something unusual is going on here, that we have a curved geometry of spacetime. Now again, this is an imperfect, two-dimensional analogy for something that's happening in a four-dimensional, curved spacetime. Hard enough to think of and hard enough to picture.

But go back to my example of the two objects falling near the surface of the Earth and coming together. They were falling on essentially straight-line paths. But those straight-line paths are in a curved spacetime and they begin to converge. And that convergence is analogous to the convergence of these two parallel lines, in this curved spacetime, in this imperfect analogy, for the four-dimensional curved spacetime of general relativity.

It's very hard to picture that. It's especially hard to picture that because you want to say, "Oh, but there's a third dimension in which this is curved." But, in Einstein's theory, there isn't a fifth dimension in which the four-dimensional spacetime is curved. The four-dimensional spacetime is all there is.

It's as if I said the surface of that globe is all that exists. And if that were true, you could nevertheless do experiments with geometry, and drawing lines and things and you would have to conclude that there was something funny about your surface—that it had this property called curvature. And for Einstein, that property of curvature is gravity.

So, let me summarize what Einstein says about gravity. Gravity is the curvature of spacetime. And how does matter move? It moves very simply. In the words of Taylor and Wheeler, it gets its marching orders locally. It simply samples the local geometry of spacetime and moves in the straightest possible path. What could be simpler? There's an unanswered question in all that. And the unanswered question is, why is spacetime curved? What curves spacetime? That's the other half of the general theory of relativity. And that's the part we'll deal with in the next lecture.

Curved Spacetime
Lecture 14

It took Einstein nine years after the special theory of relativity to develop the general theory. It took him about six weeks of actually working to develop the special theory. So this is quite a change. And the reason it took so long is, as you can see, from the idea of curved spacetime, general relativity is a conceptually very difficult idea, compared with special relativity in many ways.

Gravity is synonymous with spacetime curvature, but what causes spacetime to curve? Einstein gave the answer: Matter and energy curve spacetime in their vicinity. In 1914, Einstein published his complete and fully quantitative theory. This is one of the crowning achievements of the human mind, because there was no experimental confirmation at the time that he developed the theory. The essence of general relativity consists of two simple statements:

- Matter and energy cause spacetime to curve.

- In the absence of forces, objects move in the straightest possible paths (*geodesics*) in curved spacetime.

We can use a simple analogy to demonstrate spacetime: Stretch a sheet of clear plastic and roll a small ball across it. With the sheet stretched flat (no curvature), the ball rolls in a straight line. A larger ball resting on the sheet distorts it; now the small ball's path is no longer straight because of the curvature of the sheet.

General relativity makes definite predictions that can be verified through observations. Where gravity is relatively weak, as in our solar system, the predictions of general relativity differ only slightly from those of Newtonian gravitational theory. But modern astrophysics offers examples in which general relativistic effects are dramatic. Elliptical orbits should not remain fixed in space but should rotate slowly about the gravitating body.

In our solar system, the planet Mercury shows the greatest effect, because it is closest to the Sun, but even here the effect is only about 1/100 of a degree of angle every century. Einstein knew of this subtle deviation from Newtonian gravitation and was delighted when his new general relativity could account for it. Today we know of collapsed stars in such close orbits that this *precession* effect is much more obvious. A famous case is the binary pulsar discovered in the 1970s and studied ever since by Joseph Taylor and Russell Hulse, who won the 1993 Nobel Prize for this work. This system includes a *neutron star*—an object with the mass of an entire star compressed into the size of a city—in orbit around another collapsed star. The neutron star spins rapidly and in the process emits regularly spaced radio signals, like a ticking clock. Studying these signals reveals orbital details, including the precession effect.

Time should run more slowly in regions where gravity (i.e., spacetime curvature) is stronger. A simplified explanation is that light loses energy "climbing" away from a gravitating mass. Light can't slow down, but the frequency of the light waves is reduced. To an observer looking toward a region of strong gravity, the effect is to see time running slower in that region. This is called *gravitational time dilation*. In a very sensitive experiment at Harvard in 1960, physicists used nuclear radiation to verify gravitational time dilation, effectively measuring differences in the rate of time over a distance of a mere 74 vertical feet.

Gravitational time dilation is also verified by sensitive measurements of the frequency of radiation emitted by the Sun. The effect is much more obvious in collapsed stars, in which the dense concentration of matter results in much greater curving of spacetime. Such stars include *white dwarfs*, which have the mass of the Sun crammed into the size of the Earth, and the neutron stars described above. Gravitational time dilation is also important in the round-the-world atomic clock experiment described in Lecture 9. Even though curvature of spacetime in Earth's vicinity is slight, the Global Positioning System (GPS)

> **Time should run slower in regions where gravity (i.e., spacetime curvature) is stronger. A simplified explanation is that light loses energy "climbing" away from a gravitating mass.**

is so precise that its position determinations would be off by a significant fraction of a mile if gravitational time dilation were not taken into account.

Light travels in the straightest possible path, but in curved spacetime that path is not a straight line. General relativity predicts that light should be bent by gravity. The equivalence principle shows why this must be so. When starlight passes

Gravitational lens.

W.N. Colley and E. Turner (Princeton University), J.A. Tyson (Bell Labs, Lucent Technologies) and NASA.

by the Sun, its path is bent slightly, making the apparent positions of the stars change relative to their positions when the Sun is not near the light path. Observations of this can be made only during an eclipse of the Sun. This effect was first observed on May 29, 1919, by Sir Arthur Eddington. A Quaker, Eddington had been granted an exemption from service in World War I so he could undertake a test of Einstein's theory as soon as possible after the war ended. By happy coincidence, the first available eclipse was May 29, 1919—a date when there happened to be many bright stars near the Sun. Confirmation of Einstein's prediction catapulted Einstein to world fame.

Today, astronomers routinely observe distant objects whose light is bent significantly by massive galaxies. Called *gravitational lensing*, this effect can produce multiple images of a single object. Gravitational lensing is also used to search for dark, massive objects that might constitute the "missing mass" in the universe. When such an object passes in front of a star, its gravity momentarily focuses the star's light, producing a bright flash. This effect is called *microlensing*.

General relativity predicts the existence of *gravitational waves*—"ripples" in the fabric of spacetime that travel at the speed of light. Gravitational waves should be produced in certain high-energy astrophysical situations, such as with dense objects in close orbits or the merging of black holes. Early attempts to detect gravitational waves involved huge aluminum bars that would vibrate in response to the waves. Gravity wave detectors that are currently in the design phase include space-based devices similar to the Michelson-Morley experiment, some with arms thousands of miles long. The binary pulsar, discussed above, should lose energy by radiating gravitational waves. The waves haven't been detected directly, but changes in the orbit agree with general relativity's prediction for the energy loss.

Finally, general relativity predicts the existence of *black holes*—a topic worthy of an entire lecture. Stay tuned! ∎

Essential Reading

Chaisson, *Relatively Speaking*, Chapter 7.

Hey and Walters, *Einstein's Mirror*, Chapter 9.

Will, *Was Einstein Right? Putting General Relativity to the Test*, Chapters 1–7.

Suggested Reading

Wheeler, *A Journey into Gravity and Spacetime*.

Questions to Consider

1. In special relativity, we stressed that time dilation is reciprocal: When we're moving relative to each other, I see your clock running slow, and you see mine running slow. Now we have gravitational time dilation in general relativity: If you're closer to Earth or another gravitating body than I am, I see your clock running slow. Do you expect this effect to be reciprocal too, or will you see my clock running fast?

2. Gravity seems a pretty formidable force if you're trying to lift a heavy object or scale a cliff. In what sense, though, is gravity on Earth (and indeed throughout our solar system) weak?

Orbital Precession

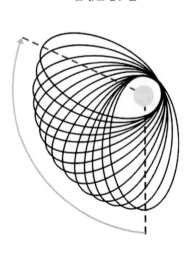

Newtonian physics says that a planet's orbit should repeat exactly the same path forever. But general relativity predicts a gradual rotation, or **precession**, of the orbital axis. Within our solar system the effect is very small, but for the planet Mercury it *is* measurable.

Bending of Light

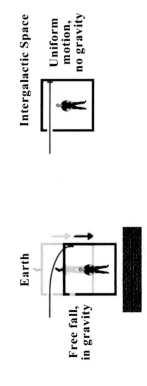

Earth

Free fall,
in gravity

Intergalactic Space

Uniform
motion,
no gravity

The right hand frame shows a small laboratory in intergalactic space, far from any gravitational influences. Futhermore, it's not accelerating. A beam of light enters the lab through a hole in one wall and strikes the wall directly opposite the hole.

The left hand frame shows a similar small laboratory, now falling freely under the influence of Earth's gravity. The Principle of Equivalence asserts that the two situations are indistinguishable as far as the laws of physics are concerned. So a light beam that enters the hole in the side of the laboratory must strike the wall opposite where it entered. But the laboratory accelerates downward as the light moves across its interior, and therefore the light path must curve in order for the light to strike the wall opposite the hole. Thus the Principle of Equivalence shows that gravity must bend the path of a light beam.

Bending of Light

Light from a distant star (left) would normally take a straight-line path (faded line) to Earth. But in the case shown, the sun lies between the star and the Earth, and the Sun's gravity curves spacetime in its vicinity. The light follows the straightest possible path in this curved space-time, and is thus bent around the Sun. An observer on Earth would have to look in different directions to see the star, depending on whether or not the Sun were between the Earth and the Star. Thus the apparent position of the star in the sky would differ in the two situations. With the Sun's relatively weak gravity, the change in observed star positions is very small, much smaller that the full bending around the Sun shown here.

Bending of Light: A Gravitational Lens

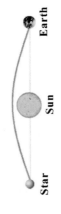

Much more dramatic effects occur when a massive galaxy lies between the Earth and a distant quasar. The galaxy's curvature of Spacetime can create multiple paths for light from the quasar. This may result in multiple images of the same object, or in distortions where spacetime curvature has smeared the quasar's image into a series of arcs or even a complete ring. This phenomenon is known as **gravitational lensing**.

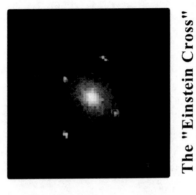

A Gravitational
Lens as seen with the
Hubble Telescope

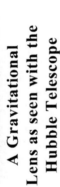

The "Einstein Cross"
as seen with the Hubble
Telescope

Curved Spacetime
Lecture 14—Transcript

In the previous lecture, I introduced the general theory of relativity and with it Einstein's description of gravity not as a force, but as the geometrical structure, a four-dimensional spacetime itself, as the curvature of spacetime. And I introduced Einstein's notion that the law of motion becomes the simple statement that objects move in the straightest possible lines. But they're not Euclidean straight lines any more in this curved spacetime.

I left dangling though the question: What causes spacetime to curve? And Einstein gave an answer to that question, too. And it's a very simple answer. Spacetime is curved by the presence of matter and the presence of energy—the two basically being the equivalence that we discussed when we talked about $E=mc^2$. So, it's matter and energy that curve or warp spacetime. And matter, in turn, responds to curved spacetime by moving in the straightest possible paths where spacetime is curved. Those are not straight lines.

The end result is Einstein's general theory of relativity, developed in 1914, or published in 1914. It took Einstein nine years after the special theory of relativity to develop the general theory. It took him about six weeks of actually working to develop the special theory. So this is quite a change. And the reason it took so long is, as you can see, from the idea of curved spacetime, general relativity is a conceptually very difficult idea, compared with special relativity in many ways.

Although, again, the essence of it is simple: matter and energy curve spacetime. And matter responds to curved spacetime by moving in the straightest possible lines. Another reason general relativity is difficult, and it's a subject that's difficult enough most undergraduate physics majors don't study it, is that the mathematics used to describe four-dimensional curved spacetime is very, very complex.

Einstein basically had to learn that mathematics. It was a whole branch of geometry called non-Euclidean geometry. Not a very commonly used subject, because the geometry of the everyday world we deal with seems to be, essentially, Euclidean. So Einstein had to learn non-Euclidean geometry and all the complicated mathematics for dealing with that and for describing laws of physics in curved spacetime. But he did so. And in 1914, he published the general theory of relativity.

And historically, I think, the general theory has to rank as one of the really crowning intellectual achievements of humankind. Why? Partly because Einstein developed this theory entirely from his own mind. It came out of his saying, "This is the way I think the world has to be. How can I make gravity consistent with what I already know, namely, the principle of relativity? How can I make gravity consistent with special relativity? How can I generalize special relativity so your state of motion, whatever it is, doesn't matter? How can we formulate a physics that works for all observers?"

In the process, gravity and acceleration got mixed up, as we saw in the last lectures, through the principle of equivalence. And general relativity became a theory of gravity. But this all came out of Einstein's mind. He had absolutely no experimental evidence. No observational evidence telling him Newtonian gravity is wrong. There was one minor, minor, unexplained deviation from Newton's gravity that existed in solar system observations, and I'll get to that later in this lecture. But that wasn't Einstein's motivation at all.

Einstein's motivation was simply gravity, as Newton described it, cannot be consistent with the principle of relativity. How can I make it consistent? In the process, how can I generalize special relativity to get rid of that restriction to uniform motion only? So Einstein thought, cogitated, learned mathematics, and developed a theory that emerged full blown in 1914, with, at that point, essentially no experimental confirmation and no obvious way to verify it. And Einstein nevertheless said, "This is correct." And later came verifications, which I'll also be discussing in this lecture.

But let me give you the essence of general relativity, because here it is, and it consists of two simple statements. The general theory of relativity states, first of all, that matter and energy curve spacetime. They shape the geometry of spacetime. And the second statement, the very simple one I introduced last time, objects move in the straightest possible paths through this curved spacetime. But again, because spacetime is curved, those straightest possible paths are not straight lines.

It's very difficult to picture this. My globe is an imperfect analogy. I'm going to give you another analogy in a minute, which is also imperfect. The reason they're imperfect is because it's not just space that's curved, it's spacetime, and spacetime is a four-dimensional entity. And motion through both space and time is involved in determining these straight-line paths or straightest line paths in curved spacetime.

Now, I want to do a very simple demonstration that gives you the flavor of what it means to have matter curve spacetime and to have objects move in curved spacetime. It's, again, an imperfect analogy. And what I'm going to use is this circular embroidery hoop. And I have stretched across it a piece of shrink-wrap plastic. That shrink-wrap plastic is right now completely flat. And that represents a flat Euclidean spacetime, and I'm now going to set up a demonstration, which will show you what it means to warp spacetime. And what it means, then, to move in the straightest possible path in curved spacetime.

Here we are, with a two-dimensional analogy for general relativity with its curvature of four-dimensional spacetime, the curvature being caused by the presence of matter and energy. And matter itself, moving in response to that curvature in the straightest possible lines. I have here a small steel ball. That's going to represent my piece of matter that moves through curved spacetime. I've taken the embroidery hoop, with the sheet of plastic that I showed you, and I've laid it on a horizontal surface. And I have a camera pointing directly down at it. And right now, there is no curvature to my spacetime. It's stretched taut and flat.

And I'm going to take my small steel ball, and I'm going to roll it across the curved spacetime, I'm going to give it a push. And let's watch what happens. So here it goes. It moved in a perfect straight line. It did exactly what Newton's law would say, and in fact, what Einstein's law also says. It moved in the straightest possible line in curved spacetime. In this case, because spacetime is not curved, or it's got zero curvature, or it's flat, that straightest possible path is a true Euclidean straight line.

Let's do it again in a different direction. Here it goes. Straight-line path, constant speed, obeying the law of inertia. Again, spacetime is not curved. And the ball rolls straight across it in a straight-line path at constant speed. No curvature. That's the second part of Einstein's general relativity—that matter responds to the curvature of spacetime by moving in the straightest possible path. But in this case, spacetime was not curved.

Now, I'm going to warp my two-dimensional analog for spacetime. And here's where the analogy becomes a bit imperfect. I'm going to take this large steel ball. This might represent a massive object, like, for instance, the sun, or the Earth. And my small ball will then represent a satellite, or the Earth going around the sun, or the moon going around the Earth, or whatever. And I'm going to place the large steel ball in the center of my sheet. Now, you can't see the curvature any more than you can see the curvature of four-dimensional spacetime in which you live. But that large ball has distorted the sheet, the plastic sheet. It's done so through the agency of gravity, but don't get that confused. That's only to distort the sheet. We're not here demonstrating Earthly gravity. We've simply distorted that sheet, in some way or other.

And this small ball, which is constrained to move on that sheet, is now going to move in the straightest possible path it can describe, in what is now a curved, two-dimensional analog to a curved, four-dimensional spacetime.

So what happens? If I roll it past the big ball, you see the path curves as it goes by the big ball. It's as if it were attracted to the big ball. It's not attracted to it. What it's doing is getting its marching orders, as Taylor and Wheeler tell us, from the local geometry of the spacetime through which it's passing.

That's the second part of general relativity. Objects move in the straightest possible paths in curved spacetime. But those straightest possible paths are, in this case, not the straight lines of Euclidean geometry but curved paths that reflect the curvature of spacetime. If I'm very careful, I can get it actually to go into an orbit, for a little while, before friction slows it down. There it goes. It's in an elliptical-like orbit now around that massive object. Again, it's not responding to any pull from the massive object. It's responding to the curvature of spacetime in its immediate vicinity.

You can't see the difference, because you can't see the warping of that plastic sheet, but it's there. That's the curved geometry of spacetime. And this small ball simply senses the curvature of spacetime that it moves through. And it moves in a curved path. That's an imperfect, geometrical analogy, or two-dimensional analogy, for the geometry of four-dimensional spacetime. There is no attraction between these two balls. One of them simply distorts the curve, distorts the geometry of spacetime—in this case, that two-dimensional sheet—and the second one then describes a curved path instead of the straight line it would describe otherwise in that curved, two-dimensional spacetime.

So that's our analogy. An imperfect one, but nevertheless an analogy that lets you see both parts of what general relativity has to say—that the presence of matter or energy curves spacetime and that other matter then responds by moving in the straightest possible paths in curved spacetime.

And those straightest possible paths may be anything but straight. They can be things like the orbit of the Earth around the sun, the moon around the Earth, or my small steel ball around the large steel ball. So that's an analogy for the entire content of general relativity.

So you now know, in essence, what general relativity has to say. It talks about how matter and energy curve spacetime and how matter then responds to the curvature of spacetime.

But what predictions does it make? What does it say about how the world should behave that will allow us to tell whether this theory is correct? What predictions does general relativity make? In fact, general relativity makes a number of predictions about the behavior of matter—predictions that, in places where gravity is weak, agree almost, but not quite, with the predictions of Newton's law of gravity.

That's not surprising. Newton's law does a very successful job of predicting the motions of the planets, or allowing us to launch spacecraft to Jupiter, or put satellites in orbit around the Earth. Newton's theory of gravity is pretty close to being correct. In places where gravity is weak—and by the end of today's lecture, and particularly in the next lecture, I'm going to define exactly what I mean by weak and strong gravity—but in places where gravity is weak, the predictions of Einstein's theory differ only slightly from the predictions of Newton's theory of gravity.

For most of the twentieth century, we had access only to places where gravity was relatively weak. And therefore, tests of whether general relativity was correct or not were very sensitive tests, which hinged on measuring very minuscule differences between the predictions of Newton's theory and the predictions of Einstein's theory. And general relativity, for that reason, has not, at least until recently, been verified nearly as strongly as was special relativity.

But in recent years, with the discovery of extreme situations in astrophysics, particularly situations where gravity does become very strong, we're much, much more confident now that general relativity is correct. And general relativity is really an essential theory now to understanding very extreme objects in astrophysics, and indeed, to understanding the overall structure of the whole universe. We're still not as convinced of its correctness in every detail as we are of special relativity. But we're becoming very convinced.

There are a few competing theories, most of which have been ruled out. And general relativity, really, is beginning to reign as the correct theory of gravity. So we're quite confident of it. Although, through most of the twentieth century, we certainly didn't have the kind of dramatic confirmation that, for example, the muon experiment provided for special relativity.

Now in the case of special relativity, I was able to lead you from the basic principles to understanding rigorously the predictions of time dilation, for example, or non-relativity of simultaneity and things like that. In general relativity, I can't do that. I can't take you from those simple statements about curvature of spacetime and matter moving in the straightest possible paths. I can't take you from that to a rigorous understanding of the predictions. So I'm simply going to tell you what they are and run you through them and give you a sense of what observations and experiments have, indeed, helped us to confirm that general relativity is right.

Well, the first thing, or one obvious thing that general relativity will talk about is the orbits that planets follow. In Newton's case, the prediction is, as we saw, and as Kepler had first told us, that the orbits of the planets should be perfect ellipses. And in the case of Kepler and Newton, those ellipses should simply repeat again, and again, and again, and again—exactly the same ellipse.

General relativity predicts a slight deviation from that prediction. General relativity predicts that the ellipse should swing around and be in a slightly different orientation with each subsequent ellipse. I've got a picture of that over here. It's called *orbital procession*. So an elliptical orbit starts out, initially, with its long dimension this way. And with each successive orbit, the long dimension swings slightly, until, toward the end of this operation, the long dimension is pointing that way. There's been a procession from one orientation of the long dimension, of the orbit, to another. That's called orbital procession. That does happen in the solar system.

It happens, most dramatically, although hardly dramatically at all, as you'll see, for the planet Mercury. Why? Because Mercury is closest to the sun, the most massive object in the solar system, and the one with the greatest gravitational force. So Mercury exhibits this procession. But in Mercury's case, it amounts to about 100th of a degree of angle. And a degree is a pretty tiny angle—about 100th of a degree of angle every century.

Nevertheless, years and years of accumulated, observational evidence from astronomical observations is enough to see that effect. And at the end of the nineteenth century, it was known that the orbital behavior of Mercury deviated slightly from the predictions of Newton.

This was not a crisis in physics, because there possibly were other explanations. Maybe there were unknown planets somewhere that were exerting subtle effects, or something like that. Or maybe the sun wasn't perfectly spherical. Or other effects could explain that. It wasn't a crisis for physics. But Einstein did know about it. And one of the first things he did with his general theory of relativity was to figure out what the orbit of Mercury should be. And, lo and behold, he was delighted when he found he could explain this procession of the orbit of Mercury precisely with his theory of general relativity. So this very, very sensitive test, based on this

100th of a degree of angle per century, helped confirm Einstein's theory. But it's hardly a very dramatic effect.

Today, however, we know of systems, astronomical systems, where gravity is really strong. They consist of collapsed stars. A collapsed star is a star. A thing with, roughly, the mass of the sun—the sun is about a million miles in diameter—that's collapsed into the size of, oh, the Earth in some cases, or in more extreme cases, the size of, say, a city, to an object a few miles across. Such an object is called a *neutron star*.

These things tend to spin very rapidly, as much as a few hundred to a thousand times a second. And if they have hot spots on them, or magnetic regions, they emit bursts of radiation, or they emit a beam of radiation. As that beam comes by Earth, we hear bursts, boop, boop, boop, boop, boop, like that. These things are called *pulsars*. And they give us a very regular timing signal, from which we can determine a lot about the system in question.

Well, in the 1970s, a very famous system was discovered called the *binary pulsar*. It consists of one of these very densely collapsed stars in a very tight orbit around another collapsed object. And because these two objects are very, very dense and very collapsed, a lot of mass crammed into a small space, they have very strong gravity. They're in a very close orbit. And these effects are multiplied many, many, many fold.

Studies of this binary pulsar, which have taken place since it was discovered in the 1970s, and its discoverers, or, the people who worked on it, Joseph Taylor and Russell Hulse, who also (Taylor discovered it) won the 1993 Nobel prize for this work. They'd been painstakingly studying this system for many, many years. And I'll have much more to say about that a little bit later in this lecture.

But the regularly spaced emission of radio signals from this pulsar allows us to calculate just what it's doing in its orbit, as it comes toward us, or moves away from it. The frequency of those signals changes slightly and we can calculate out what the orbit is doing. And we can verify directly that this procession is occurring. And it's occurring at a much greater rate than it does for Mercury. And it confirms, again, general relativity.

There's another prediction of general relativity and that's a prediction that time should run slower in regions where gravity is stronger. Another funny effect on time. Not surprising, because after all, general relativity is about the distortion of the geometry of spacetime by mass. In regions where gravity is strong, or to put it in general relativistic terms, where spacetime is very sharply curved, time runs slower.

This is a *gravitational time dilation*. And what it means is this: It means that if I take two clocks and I put them at different heights in the Earth's gravity, they'll run at different rates. This clock runs slower than this one. And if I hold them in this position for a very long time, this clock will be seen to be lagging behind this one. Now over this distance of a few feet, that's an absurdly difficult thing to measure with clocks like this. But, nevertheless, it's a real effect.

And the prediction of general relativity is that time should run slower as you get into regions of more intense spacetime curvature. That is, regions that, in Newtonian terms, we would call regions of strong gravity.

Can we measure that? Well, remarkably, in 1960, scientists at Harvard, using a 70-foot tower, basically, 74-foot tower and, well, it's a stairwell, basically, in one of the Harvard University buildings, measured the gravitational time dilation using a very, very sensitive timing device, which consisted essentially of the frequency of decay of a radioactive atom. And they were able to measure the shift in the frequency from the bottom to the top of that 74-foot height. Not a very big height, compared to the radius of the Earth. A very sensitive test. And they proved that gravitational time dilation occurred.

Gravitational time dilation is now verified much more dramatically by looking at emission of signals from the surfaces of these compact stars, where gravity is very strong and we see atomic processes, for example, slowed down to an extent that is commensurate with general relativity. We can even see this for processes on the sun. Atomic processes, vibrations that give rise to light waves, happen at a slightly slower rate. Again, the sun is not a source of very strong gravity. So these effects are very subtle in the solar system. But we can, nevertheless, see them, even for the sun. We can see them much more dramatically for these compact objects.

Gravitational time dilation is also important in situations here on Earth, for example, in the around-the-world atomic clocks experiment that I described to you and that's listed in your bibliography. I used that as an example of the twin paradox, showing us that the twin situation really happened. We flew clocks around the world. They came back 300 nanoseconds slower than the clocks that were left behind.

But I pointed out that there were effects, both of special and general relativity. The special relativistic effect was due to the motion of the clock relative to the clock that remained behind on Earth. And it was non-uniform motion, because it went around in a circle. That clock came back younger.

But there was another effect. And that is the clocks that went in the airplane went up higher. They went up to regions where gravity was weaker. And therefore, they actually ran faster for that reason. And both those effects had to be taken into account.

Finally, although I've emphasized that spacetime curvature, or equivalently the strength of gravity in the solar system, is in some absolute sense rather weak in the sense that the gravitation predictions of Einstein's theory don't differ significantly from those of Newton, nevertheless, they do differ slightly. And in very, very precise measurement applications, those differences can be important.

And a case in point is the global positioning system. People think, "What on Earth does general relativity have to do with me? What's its relevance to the everyday human being?" Well, as more and more of us use the global positioning system to navigate our cars, or in the airplanes we fly in, or whatever, or to find ourselves when we're lost in the woods, we're relying on a positioning system that involves timing—a precise timing of signals from a whole constellation of satellites above the Earth. And the programs that calculate where you are based on the timing of those signals have to take into account, not only the motion of the satellites, which affect their internal timekeeping because of special relativistic effects, but they also have to take into account the gravitational time dilation associated with the altitudes of the satellites.

If those general relativistic effects were not taken into account, by about one day, the global positioning system would accumulate errors of a good fraction of a mile. And this is a system that's supposed to localize you within a matter of meters. So general relativity really is relevant. It's right there in the global positioning system. If we didn't take general relativity into account, the global positioning system would fail utterly, at least at localizing us to any kind of reasonable precision. So general relativity, although it's a very abstract theory about the curvature of spacetime is, nevertheless, important to us here on Earth.

There are some other examples of general relativity's predictions that are important. General relativity predicts that light should be bent in a gravitational field. Light should be bent by gravity. Why is that?

Let me show you. Here's a situation where I'm going to use the equivalence principle to convince you of something that must be the case. Here I am again, in my laboratory, falling elevator, whatever you want to call it. I'm in intergalactic space. I'm moving uniformly. There's no gravity anywhere near me. And I'm going to imagine a light beam enters my spaceship, or elevator, or laboratory, whatever it is, traveling horizontally. And what does it do? It comes in through a hole in the spaceship and it hits the wall opposite itself. No problem. But because we believe the equivalence principle, we believe this situation is equivalent in every respect to this laboratory, elevator, spaceship, whatever, freely falling in Earth's gravity.

If the thing is freely falling in Earth's gravity, the light enters the left side. A while later, the light has to hit the right side. If these two situations are equivalent, it has to hit the right hand wall exactly opposite where it entered. The problem is, in the left hand situation, when we're freely falling near Earth, we've fallen and the right hand wall is somewhere else. The light still hits at the same place on that wall, because these two situations are equivalent, and therefore, it had to have curved. Light must be bent by gravity.

And the most dramatic confirmation of general relativity came in May of 1929. And there's an interesting historical, 1919 rather, May 29th of 1919, the experiment was carried out, or the observations were carried out by the

British astronomer Sir Arthur Eddington. Eddington had been a conscientious objector in World War I. He was a Quaker. And in return for not serving in the military, he agreed to carry out an expedition to observe an eclipse as soon as the World War I hostilities were over. Well, it turned out that the first occasion for that after World War I was May 29, 1919.

And it also turned out that at that time, the greatest number of bright stars were in the sky in the immediate vicinity of the sun. And you'll see in just a minute why that was important. It's important for this reason. Let's take a look at the situation.

Here's the Earth and some star. And we want to observe the light from this star. And normally, the light from that star comes straight to Earth. And we point our telescopes straight at the star and we see it. And what I'm about to show you is highly exaggerated. This is a very subtle effect for the sun. What if there's some time of year when the sun, whose motion relative to the Earth and the sky is different from the stars—the sun comes between the Earth and the star, or something like that. Well, we can't observe the star by looking directly through the sun. But what happens? Light that leaves the star, if the prediction of general relativity is correct, will be bent by gravity and may come to the Earth, nevertheless. And we look in a rather different direction than we would if the sun had not been in that position in order to see that star.

Now, again, this is highly exaggerated. We don't really see stars that are behind the sun in this situation. What we do see are slight changes in the positions of stars that appear to be very close to the sun already. So this is highly exaggerated. But it shows you what we see, or the idea of what we see. So this is the gravitational bending of light.

And this gravitational bending of light was observed by very carefully taking photographs of the positions of stars at the time the sun was between the Earth and the star. And you might say, "Well, how on Earth can you see stars, when the sun is right there in the sky very near them?"

Well, the answer is, the only time you can do that is during a total eclipse of the sun, when the moon moves between the sun and the Earth and blocks out the sun. And that's what happened on May 29, 1919. And that's when Eddington was able to make these measurements. And it was a very happy coincidence that the first available eclipse coincided to a time when there were so many bright stars near the sun.

Now today, we routinely make similar measurements, except instead of using the bending of sunlight by, or bending of starlight by the sun, we look at very distant objects—distant galaxies, or quasars, which are actually, sort of, infant galaxies, so far away from us that we're looking at them at a time before they've evolved into galaxies. I'll say more about those later.

And here's the typical situation. We see a quasar. In between the quasar and Earth, there's some galaxy. And light from the quasar reaches us by either that path, or that path and a remarkable thing we see when we look at the quasar, then, is we see, actually, we see it looking in one direction. We see it looking in a slightly different direction. Or equivalently, if we take a picture of the quasar or of the sky where the quasar is, we actually may see multiple images of the quasar.

If the situation were perfectly symmetric, as I've drawn it, we might actually see the quasar stretched out as a ring. If it's not perfectly symmetric, we will see multiple images of it. And some of the images will be distorted. We call this event, we call this situation, *gravitational lensing*. The gravitational field of the intervening galaxy, or in relativistic terms, the warping of spacetime, acts as a lens that can focus or distort or actually create multiple images. And I have a couple of examples of that to show you.

This is a dramatic image taken with the Hubble space telescope. Near the center of it, you see a bright cluster of galaxies. Not terribly distant galaxies. A bright cluster of galaxies. And in the regions around there you can see some in here, you can see a blue structure that's sort of elongated. Out here, you can see almost a ring of blue, distorted-looking structures. Those are multiple images of the same galaxy.

The galaxy is actually very distant. It's behind that nearby, or nearer clump of galaxies, cluster of galaxies. And the intense gravitation, the intense warping of spacetime by this cluster of galaxies is distorting the images. Distorting the paths of light coming from way behind it, from these distant galaxies and making this ring-like structure of distorted-looking objects. That's an example of gravitational lensing. And the Hubble telescope, as well as other telescopes now, has given us dramatic examples of gravitational lensing, by typically galaxies or more distant galaxies or quasars.

Another example of a gravitational lensing is this object. What we're seeing in the yellow object in the middle is a galaxy, some hundreds of millions of light years away. And that is imaging a distant quasar, which is billions of light years away. And the quasar appears here, not as a single image, but as four dots, four images of the quasar, because of the particular geometry of the lensing here. This particular object is called the "Einstein Cross" because of its cross-like shape. And obviously, named after Einstein, who gave us the notion that gravitational fields should bend light; that the bending of spacetime itself should cause light itself to travel in the straightest possible paths. But they aren't necessarily straight lines.

Today, gravitational lensing has become not just a way of confirming that general relativity is right, it's become a way of searching for objects. One of the problems we have in cosmology and understanding the origin of the universe is that we don't know where most of the mass of the universe is. More on this in lecture twenty-three.

But one way we can search for dark objects that we can't otherwise see is if a dark object passes in front of a distant star. There may be a temporary increase in the brightness of the light from that star as the gravitational field, or if you will, the warping of spacetime by this massive but dark object, causes a lensing effect on the light from the distant stars. These are called *micro-lensing events*. And they're now used in a search for kinds of objects that hypothetically may make up some of the mass of the universe. That's called *micro-lensing*.

Finally, general relativity predicts the existence of gravitational waves. If some dramatic event happens to a massive object, and it starts wiggling, or two black holes collide, or something like that, that will start waves, literally waves, in the fabric of spacetime. And those waves propagate outward with guess what speed? The maximum speed they can, the speed of light, c, and they carry information about that event to the distant universe.

For several decades, astronomers have been trying to detect gravitational waves. They first did it by building enormous aluminum cylinders with delicate sensors at the end that would sense any vibrations set up in the cylinder by the passage of a gravitational wave.

We're moving toward a much more sensitive form of gravity wave detector now. We're building Michelson-Morley–like experiments, in which we send laser beams, not a few yards, but thousands of miles in space-based Michelson interferometer-type situations. And these should be in orbit in the next few years. And we should be able to look at gravitational waves produced by all kinds of dramatic astrophysical events in the distant universe. And that will open for us literally a whole new window on the universe.

The binary pulsar that I mentioned earlier ought to emit gravitational waves. And, in fact, we haven't detected those waves. But as it emits those waves it ought to lose energy. And the orbits ought to change. And we have detected those changes in orbit. That's part of what got Taylor and Hulse the Nobel Prize. By studying that event, they saw the energy loss, and the energy loss is exactly as predicted by general relativity, if the energy is carried away by gravitational waves. Finally, general relativity predicts the existence of an entirely new and dramatic kind of entity—the black hole. And the black hole deserves a whole lecture in itself. And we'll be talking about them next time.

Black Holes
Lecture 15

Throw the ball up, down it comes. Up, down it comes. Throw it a little faster. It goes up higher, stays there a little longer, but down it comes. Throw it faster still, it's up higher, stays longer, but back it comes. What goes up must come down, right? No.

The speed an object must have to escape forever from Earth or any other gravitating body is the *escape speed*. Escape speed for Earth is 7 miles per second, but it can be much higher for objects that are both massive and small, such as the white dwarfs and neutron stars that form at the ends of some stars' lifetimes. Escape speed provides a measure of how much the predictions of general relativity diverge from those of Newton's gravitation. In regions where escape speed is small compared with the speed of light (i.e., weak gravity), the two theories are in close agreement, and general relativistic effects are subtle. This is the case everywhere in our solar system. Where escape speed approaches the speed of light (i.e., strong gravity), only general relativity provides an accurate description of gravitational phenomena.

General relativity predicts the existence of *black holes*, objects whose escape speed exceeds that of light. Black holes require extreme concentrations of matter. To form a black hole from Earth, the planet would have to be compressed to a sphere about one inch in diameter. For the Sun to become a black hole, it would have to be squeezed from its current million-mile diameter to a diameter of about 4 miles.

Contrary to popular opinion, a black hole does not "suck in" everything in its vicinity.

Let's take a further look at black holes. Nothing that falls into a black hole can escape. The boundary of the region of no return is the hole's *event horizon*, where escape speed becomes c. Contrary to popular opinion, a black hole does not "suck in" everything in its vicinity. At significant distances from the hole, gravity behaves just as it would around any other gravitating object.

Gravitational time dilation becomes infinite at the event horizon—meaning an outside observer would never see an object actually cross the horizon. To a small-size observer falling into the hole, however, everything would seem perfectly normal. (Remember that free fall is the "natural state of motion" in general relativity.) However, the falling observer would experience destructive tidal forces either before or after reaching the horizon, depending on the size of the observer and the hole.

Do black holes exist? How can they be formed? Black holes may be formed in the intense *supernova* explosions that end the lifetimes of massive stars. These explosions leave a collapsed remnant that may be a neutron star or, if more massive than about three times the Sun's mass, must become a black hole. Such stellar-mass black holes may form in binary star systems, in which case they can be detected by their effects on the companion star. Typically, gas flows from the companion to form a disk of gas orbiting the hole. The matter heats up through friction as it spirals toward the event horizon, emitting copious x-rays.

Supermassive black holes—with the mass of millions or billions of Suns— seem to lurk at the centers of most galaxies, including our Milky Way. The intense radiation emitted by matter falling into the hole early in a galaxy's life may account for *quasars*, distant objects with colossal energy output. Galactic holes grow gradually as stars fall into them.

It is speculated that rotating black holes may be able to form *wormholes*, tunnels connecting remote parts of spacetime. ■

Essential Reading

Chaisson, *Relatively Speaking*, Part IV.

Suggested Reading

Kaku, *Hyperspace*, Chapters 10–11.

Lasota, "Unmasking Black Holes," *Scientific American*, vol. 280, no. 5, pp. 40 (May 1999).

Thorne, *Black Holes and Time Warps: Einstein's Outrageous Legacy*, Chapters 7–10.

Questions to Consider

1. If the Earth suddenly shrank to become a black hole, with no change in mass, what would happen to the moon in its circular orbit?

2. If you were falling into a black hole and looked at your watch, would you notice time "slowing down"? Justify your answer using basic principles of relativity.

3. For another fascinating look at black holes, wormholes, and spacetime, we recommend The Teaching Company course *Understanding the Universe: An Introduction to Astronomy, 2nd Edition* by Professor Alex Filippenko of the University of California at Berkeley.

Black Hole in a Binary Star System

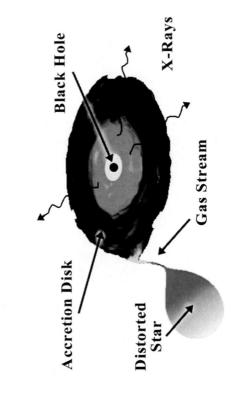

Black Hole

X-Rays

Accretion Disk

Gas Stream

Distorted
Star

Black Holes

Lecture 15—Transcript

Here we look at spacetime curvature and the extreme predictions of general relativity in their absolute, utmost extreme.

Let me begin though with a very simple question. What goes up must come down, right? Throw the ball up, down it comes. Up, down it comes. Throw it a little faster. It goes up higher, stays there a little longer, but down it comes. Throw it faster still, it's up higher, stays longer, but back it comes. What goes up must come down, right? No.

And the reason what goes up must not necessarily come down has to do with the fact that gravity, either as we describe it in Newton's theory or in general relativity, is a force—if you look at the Newtonian picture, or a distortion of spacetime if you look at the Einsteinian picture—that gradually decreases with distance from the source.

When I put that massive steel ball on my plastic sheet, it distorted the plastic strongly in its immediate vicinity. But far away from the big steel ball, the plastic was hardly distorted. It was almost flat. If I had rolled the small ball very far from the big one, I would have hardly noticed the deviation from a straight-line path because the distortion weakened with distance from the big, massive ball.

Similarly, in Newtonian theory, gravity falls off as the inverse square of the distance from the gravitating center, like the sun, or the Earth, or whatever we're talking about. Well, it falls off so rapidly—again, as $1/distance^2$—that once you begin to get some distance away from a gravitating object, it gets very easy to get further and further away.

And what that means is it takes only a finite amount of energy—not an infinite amount of energy, but only a finite amount of energy—to get infinitely far from a gravitating object. And the way that manifests itself is with a number we call the *escape speed*.

If I'm standing here at the surface of the Earth, and I throw my tennis ball up in the air—I don't throw it very fast—down it comes. Again, if I throw it faster, it goes farther. If I throw it fast enough, and for Earth's surface that speed is about seven miles a second, it will leave the Earth. It will get farther and farther away from the Earth. As it gets farther and farther away, Earth's gravity will weaken. And at a speed of seven miles a second, it will have enough energy that it will never be brought all the way to a stop—never turn around and come back to the Earth's surface. It will escape to infinitely far from the Earth. And by the time it gets very, very far from the Earth, it will hardly be moving. But it will have just barely enough energy to get infinitely far from Earth.

If I throw it a little faster than seven miles a second, then it can easily get infinitely far from the Earth and have energy left over to spare. So there's a special speed, called the escape speed, which is the speed it takes an object thrown, if you will, or given an initial impulse from the surface of a planet, or the surface of the sun, or any point in space for that matter, to escape the gravitational influences around it.

For the Earth, escape speed is about seven miles a second. For the sun, escape speed is about 400 miles a second. For the moon, escape speed is less. We have sent objects into space with speeds that exceed escape speed from Earth, and even with speeds that exceed escape speed from the sun's influence, at least when you're already out as far as the Earth's orbit.

The Pioneer and Voyager spacecraft, for example, are traveling beyond the outermost planets already. And they have escape speed relative to the sun. They will escape the gravitational pull of the solar system, which is dominated by the sun, the most massive object in it, and they will travel forever through the galaxy. Not escaping the galaxy, but escaping the solar system, because they left Earth and left the solar system with a speed greater than the escape speed necessary to do that.

So there's this concept of escape speed. And that's an entirely Newtonian concept. We've known about that for hundreds of years. We knew what escape speed was. We know how to calculate it. It's a simple idea. It's shocking to you, perhaps, because you think what goes up must come down.

Earth's gravity does extend infinitely far through space. The distortion of spacetime caused by the Earth does, in principle, extend forever. But, in fact, it weakens so radically with distance that you can throw something with a finite amount of speed—seven miles a second for Earth—and it will escape forever.

What's this got to do with general relativity and the subject of this lecture, namely, black holes? Well, it's got the following to do with it. What determines escape speed? Well, one obvious thing that determines it is the mass of an object. If I doubled the mass of the Earth, that would make escape speed increase. The other thing that affects it is the concentration of that mass—how close one is to the center. If I took the Earth without changing its mass and compressed it smaller, I would make escape speed from its surface bigger. If I compressed it still further, escape speed would get bigger and bigger and bigger and bigger and bigger.

Well, we know from special relativity that there's an ultimate upper limit to the speeds we're allowed to achieve. And the ultimate upper limit is the speed of light. What if I were to compress the Earth more and more and more and more and more until its escape speed approached the speed of light? If I compressed it to that extent, I would have an object where escape speed was the speed of light. And I simply could not throw the tennis ball up or any other object up, because I can't achieve speeds greater than that of light.

I was talking in the last lecture about how gravity in the solar system is weak, how the effects of general relativity are subtle in comparison with the effects of Newtonian gravity. There the distinctions are very small and difficult to measure. I said I'd tell you now what I mean by weak gravity. And now I'm in a position to do that.

Weak gravity corresponds to situations where the escape speed is much less than the speed of light. Escape speed from Earth—seven miles a second— sounds pretty fast. But it's nothing compared to 186,000 miles a second. Escape speed from the sun, 400 miles a second. Impressive number. But it's still nothing compared to 186,000 miles a second for the speed of light. On the other hand, if I compress the Earth to the point where the entire mass of Earth—you, me, the Pacific Ocean, Mount Everest, the molten core of the

Earth, all the things that make up the Earth—were compressed to the size of a basketball, well, escape speed would still be considerably smaller than the speed of light. But if I compressed Earth to the size of about a ping pong ball, an inch or so in diameter, at that point the escape speed would approach the speed of light.

Strong gravity means being in a situation where the escape speed is some reasonable fraction of the speed of light—10 percent, now you're getting there—50 percent, 90 percent—that's extreme gravity. General relativity tells us that we can achieve situations where escape speed is a significant fraction of the speed of light.

It takes major compression of an object to do that. It would take squeezing the Earth to a diameter of one inch. It would take squeezing the sun to a diameter of about four miles before escape speed of the sun approached the speed of light. It wouldn't take as much to squeeze the galaxy. And the entire universe may already almost be at the density where escape speed from it is the speed of light, whatever that might mean. And we'll get back to that in lecture twenty-three.

So in principle, it would be possible to squeeze the Earth to the point where its escape speed is the speed of light. In principle, it's possible to squeeze the sun to that point, but it would take extreme squeezing to do either of those things. Well, the possibility of escape speed equal to or greater than the speed of light is what defines the object of my lecture today, a *black hole*.

Black holes are objects whose escape speed exceeds the speed of light. That's the definition of a black hole. Why is it a black hole? It's a black hole because nothing can escape from an object whose escape speed exceeds light. Nothing. Not even light. So black holes, by definition, are black. No light can escape from them. They're completely invisible in that sense; although there are some subtle reasons why they're not, in quantum mechanical terms. And we'll talk about those briefly in another lecture.

But black holes are essentially, in this picture we're describing them with, completely black, because nothing, not even light can escape from inside the black hole. A black hole is simply an object that has been compressed to the point where its concentration of mass is so high that the escape speed from its surface is greater than the speed of light.

So what is a black hole like? Well, right now I'm talking about them as theoretical objects that general relativity tells us can exist. At the end of this lecture, I'll describe to you why we think black holes really do exist and how we think we've detected them.

But let me talk a little bit abstractly about a black hole first. Here's this object. Maybe it's the Earth that we've compressed smaller and smaller and smaller and smaller, and finally its escape speed has exceeded the speed of light. And the object compresses still further, perhaps. And now there's a tiny mass inside there. We'll see later, it may actually shrink to zero size. But it's surrounded by a region at which the escape speed is the speed of light. Within that region, the escape speed is greater than light. Outside that region, the escape speed is less than light. So there's a special region in space—it may not have any matter there, it may be completely empty—but there's a region where, once you cross it, you're inside that black hole, in the sense that you're now in a region where the escape speed is greater than the speed of light. And there's no way you can escape from that region. That region is called the *event horizon*. It's called a horizon because, like the horizon you see when you look out across the Earth, it's a sort of point of no return. Once an object crosses that event horizon, it can't get out. It's trapped inside that black hole. It's lost forever to the outside universe.

Now what happens inside a black hole? Well, I have to tell you one other thing about a black hole. Remember gravitational time dilation? Remember when I held up these two clocks and I argued that the clock at the lower level was running actually slightly slower because it was in a region of stronger gravity—a region of greater spacetime curvature than the upper clock? A minuscule effect here, an effect that nevertheless could be detected in that experiment in a 74-foot tower at Harvard University. But there is a gravitational time dilation. Now imagine this smaller clock being near the surface of a black hole. Gravitational time dilation becomes dramatic. In fact, at the event horizon, this point of no return of a black hole, gravitational time dilation becomes effectively infinite. Time essentially slows down.

If you were to fall into a black hole, here's what would happen. Suppose I were watching you. Suppose I were sitting on Earth and I were watching you fall into one of these black holes. Unfortunate for you, but let's let that happen. I would see you falling and falling and falling and let's suppose you communicate with me by sending a radio signal. Or you're on the radio and you're talking to me all the while, or on your cell phone or whatever.

What would happen? I would see you falling toward the black hole. But as you fell toward the black hole, you go into regions of stronger and stronger gravity. That is, greater and greater spacetime curvature. The gravitational time dilation increases and increases and increases. And what that means is things, you see, getting to me seem slower and slower and slower. And as you approach the event horizon of the black hole, I see all passage of time, as far as you're concerned, come to a stop. And you freeze on the edge of the black hole. And I never see you cross the horizon.

In fact, if you have a powerful enough rocket, before you cross the horizon you can blast your way back. And you've found another way to travel into the future. Because as time has slowed down for you near the boundary of the black hole, it's ticking along for me, at the regular, ordinary rate it does back here for me. A lot of time elapses for me. Very little time elapses for you. Before you've crossed the event horizon, you can blast your way out and come back to me. And a lot of time will have elapsed for me and very little for you. And we have the twin situation again. But this time done gravitationally, instead of with a trip to a distant star and turning around and coming back because of our high speeds and special relativity. This is a gravitational kind of twin situation.

So a black hole is very strange in that, right at this event horizon, the gravitational time dilation becomes infinite. In fact, the Russians have a very different name for black holes. They don't concentrate on the blackness of them because no light can escape. They concentrate, instead, on this gravitational time dilation going to infinity at the horizon, the event horizon of the black hole. They call them frozen stars, because many black holes, not all of them, are believed to originate from stars that have collapsed. And as that collapse occurs, we, watching it from outside, will see the collapse happen, first rapidly, and then slower and slower and slower. And then we would see the material tend to freeze at the event horizon. It's called a frozen star. It's frozen in that process of collapse.

In fact, you can debate philosophically whether black holes even exist at all, because when a black hole starts to form, we in the outside universe never see the material cross the event horizon. Because the gravitational time dilation becomes infinite. It's a remarkable thing that black holes do to time.

By the way, there's something that black holes don't do, though, that a lot of people think they do do. People imagine black holes go around sucking up everything in the universe. If there's a black hole, it sucks everything in. Well, nothing could be further from the truth. Black holes are like any other gravitating object. If the Earth, for example, were suddenly compressed to that ping pong ball-sized diameter that would make it a black hole, nothing different would happen to the moon in its orbit.

The moon in its orbit, which senses the total mass of the Earth, not whether it's compressed into a black hole or in a sphere 8,000 miles in diameter— that has nothing to do with how the moon behaves. All the moon notices is the total mass of the Earth. The moon would continue, perfectly happily, in its orbit. In fact, if the Earth shrunk to a ping pong ball-sized black hole, many objects that now hit the Earth, like meteors, and so forth, comets occasionally, would be much less likely to do so, because the Earth would be so much smaller.

The difference with a black hole and the thing that makes a black hole's gravity inexorable is once something has crossed the event horizon, there's no going out. So black holes can only grow. They can gather in more and more matter. And they don't go around sucking up things, but if something happens to fall into a black hole, it doesn't get out. And that's the origin of this idea that black holes go around sucking up things. They don't. But if things do happen to intersect them, if things happen to be attracted by the black hole's gravity and are on such a path that they'll collide with it, then they fall in and they don't come out again.

Now, what happens to you falling into this black hole, from your point of view. Well, that's a very different story from what I see happening.

Remember, I see you falling toward this black hole, going slower and slower and slower—freezing on the surface of the event horizon.

But now we get into the principles of relativity again. Remember what relativity says. It says a freely falling frame of reference is the closest we come to those truly uniformly moving frames of special relativity. If you're in a small enough freely falling frame of reference, small enough that you don't notice those tidal forces due to variations in gravity from place to place, those tidal forces due to the curvature of spacetime and even at the event horizon of a black hole, that curvature is not infinite. So you can get into a small enough frame of reference that you won't notice it. Relativity says if you're in a freely falling frame of reference, this is the simplest place to do physics. Everything seems perfectly normal. Things obey the law of inertia. They go in true, straight lines in your frame of reference. You don't feel any forces. You don't feel anything unusual.

So what happens to you as you fall into the black hole? Why, you simply sail right across the event horizon and in you go. Now, sooner or later, you're going to be in trouble. The reason you're going to be in trouble is sooner or later those tidal forces are going to become so dramatic, the curvature of spacetime, to put it in relativistic terms, is going to be so big, that you'll be stretched, as the picture I showed in the last lecture showed. You'll be stretched from head to foot as you fall in and you'll be compressed the other way. And eventually, you'll be torn apart, unfortunately. But that need not happen before you cross the event horizon. It certainly would happen if you fell into an Earth's-mass black hole that was the size of a ping pong ball, because you're much bigger than that thing. You'd certainly notice strong tidal forces if you fell into a sun-mass black hole four miles in diameter. And even if you didn't notice them at the event horizon, you'd be crushed pretty soon thereafter.

But you could fall into a black hole with the mass of the galaxy and you wouldn't notice anything. And it'd be a while before you felt those tidal forces. But you're drawn inexorably to the center, and that's the end of you, eventually. But the point is, and this is where relativity comes in, relativity says freely falling frame of reference, everything seems perfectly normal.

You cross that event horizon, and you don't notice anything unusual. And it's got to be that way, if you believe what relativity says.

Relativity says the principle of equivalence makes a freely falling frame of reference the simplest possible place for doing physics. But here's the remarkable fact. When you cross the event horizon, because the gravitational time dilation has become infinite, that means you're already in the infinite future of the folks back home. That's one way of looking at why you can't possibly go back. It would be absurd if you did. So you're sitting there. You watch, you jump out of your spaceship and you start falling toward the black hole. Fire a little jet pack to get you going. Start falling toward the black hole. You watch your watch. And it's a finite amount of time. Ten minutes. An hour. Three years. Whatever it takes, depending on the size of the hole and how far away you are from it when you cross the event horizon. But you don't notice anything unusual. But at the moment you cross the event horizon, because of gravitational time dilation, all of eternity has passed back home.

Very strange things black holes do to space and time. Well, that's an abstract view of black holes and what they do and the curvature of spacetime they engender. Do these things exist? Well, in that philosophical sense I just mentioned they may or may not exist. There may be things that are in the process of forming black holes, but these are frozen stars, on the surface of which, on the event horizon of which nothing has yet crossed, as far as the outside world is concerned. But that doesn't make any difference, because if we jumped in with the matter that was falling into those objects, we would see that, indeed, there were black holes. There were event horizons. And we would cross that horizon and be done for.

So practically speaking, it doesn't matter whether the black holes have yet formed or not. If you get out of our frame of reference, far away from the hole, and get into a frame of reference with the matter falling into the hole, the hole is really there. So black holes could exist.

How could they be formed? Some process has to occur that is dramatic enough to compress matter in ordinary situations like the size of a star, say, to the size of a few miles across.

Do such processes exist? Well, we believe they do. There are explosions called *supernovas* that mark the ends of the lifetimes of particularly massive stars—stars of a mass, say, 10, 20 times that of the sun. These stars end their life often in a dramatic explosion that results in an outburst and a spewing out of much of the material of the star into interstellar space.

Some of that material, by the way, goes on to form new stars and planets and life, and that is, in fact, where we came from. The material of which we're made was cooked up in a star that exploded and spewed that material out into space.

We're not concerned with that here. Take The Teaching Company astronomy course to look at that in more detail. We're concerned with what's left after the star has exploded and puffed off its outer part. There's a core, which is sometimes compressed from the size of, say, the sun, a million miles across to sizes of a few miles across. And we have good theoretical reasons to believe that if this core has a mass of more than about three times the size of the sun, once it gets down to these small sizes, it simply can't sustain itself against its own gravity. And it must collapse, inexorably, to a black hole.

General relativity predicts that this must take place. And so, what we see out in the universe are several kinds of situations where we think black holes exist. One situation is where one of these supernova explosions has occurred. And it's occurred in a binary star system. Now, I used the example of a binary star system several lectures ago, when I wanted to convince you that the speed of light didn't depend on the speed of its source. So we had two stars going around each other like this. And we were talking there about light from these two stars, and how we knew that meant the speed of light didn't depend on the speed of the source.

But what if one of these stars undergoes a supernova explosion—say, this yellow one. It blows up and it collapses to form a very dense core. And what if these two stars are very close together? What can happen is matter from the remaining normal star can start flowing through space toward the black hole. Typically, it will form a disk of material, spiraling around the black hole. And that material will rub against itself and several things will happen.

As it rubs against itself, it will lose energy and it will spiral tighter and tighter until it actually falls into the hole. But another thing that will happen is rubbing against itself, it will heat up. And it will emit copious amounts of energy, typically in the form of x-rays.

So our guess, our best evidence for black holes of a stellar sort, that is black holes formed by stars, might come from looking in binary star systems. And I have here an artist's conception of what such a system might look like. On the one side, we have a distorted star. It's some kind of normal star, but it's very close to this black hole. Close enough that it's been distorted by the black hole's gravity.

This is, essentially, a tidal effect. The ocean tides are a modest distortion of the Earth by the gravitational effects of the sun and the moon. Here the effect is much more dramatic because these two objects are in a very close orbit.

Now we can't see the black hole itself because it's black. But you see this disk of materials. Call it an *accretion disk*, because the material is accreting, or falling onto or into the black hole. We see this accretion disk. We see the black hole. We see the gas stream and the material forms this disk and spirals around and into the black hole.

Now, we never can see this system as this artist's conception shows it. These systems are all far too distant for us to actually image this. But what we can do is observe the different parts of the system spectroscopically. We can look at the light that is part of that gas stream. We can look at the light that is part of the original star. We can look at the light that's coming from the gas, that's spiraling around the black hole. And we can infer things about the motions of those gasses without actually being able to see or resolve the individual structures there.

And we've studied a great many of these binary, so-called compact binary star systems. And they come in several classes. In one case, we have, as the compact, collapsed object, not a black hole but a *neutron star*. And those situations are easy to identify, because these neutron stars typically spin around and they emit regular pulsations. So if we see the regular pulsations, that's evidence that this is not a black hole system. If we don't see the regular pulsations, that's some evidence that it may be a black hole system.

Furthermore, we know Newton's laws of gravity, or Einstein's general relativity, although Newton's laws are good enough for the motion of the two stars in this case. We can figure out from the observed motion, and we can observe the motion by looking at the light coming from the two objects— even if we can't see the two separate objects—from the observed motions we can infer the masses of the two objects, because we understand the laws of gravity. And if we determine that the small object, the compact one, the one that we can't actually see at all, is bigger than about three times the mass of the sun, we're pretty justified in inferring that it's a black hole.

In recent years, we've had even better evidence for that. We've seen systems where the energy emitted by the system is much less than we might expect it to be. And the interpretation there is that the black hole is dragging the energy associated with the in-falling material; it is falling across the horizon of the black hole before it has a chance to be released as, say, visible light, or x-rays, or some other form of radiation we might detect back on Earth.

So the energy is actually being dragged into the black hole. And because nothing can come out of the black hole, energy itself can't come out. And so that energy is lost to our universe forever. And these black hole systems, or these systems in which the energy appears to be less than we might have expected coming out, a good evidence that we're actually seeing material that is disappearing or is about to disappear across the event horizon.

So we've been studying black holes in binary star systems for perhaps about, almost 30 years. About 30 years. And these systems are now well classified. In the first couple of decades of their study, there were a lot of skeptics who didn't really believe objects as bizarre as black holes could exist. But for a handful of systems, it's now become quite clear that there is no interpretation known to the laws of physics that makes any sense except that these systems contain black holes.

And so we think that black hole systems are probably fairly common throughout our galaxy. And we've definitely seen and identified a few of them. And this change has occurred in astronomical thinking, I'd say, only over the last 10 years. And particularly in the last five years of the twentieth century we've come to realize that black holes really are a standard part of, if you will, the astrophysical repertoire. They're out there. They're real things. And we have to think about them.

And again, the first place we really saw them was in these binary systems, where the material spiraling into the black hole emits copious amounts of x-rays. And so we can see that material. Again, we're not seeing the hole itself. That's why it's easy to identify the hole in a binary system, because in a binary system, we can see it through its effect on surrounding matter.

If a black hole were out there isolated in space, and there may be some, we would have difficulty seeing it, although it might do gravitational lensing on passing objects, on objects that pass in front of them. There might be other ways to detect it. But not as easy as detecting these x-ray emitting systems of gas spiraling around black holes and binary star systems.

But another place we think there are black holes and sort of outstripped the binary systems as the most common location of black holes in the universe are the centers of galaxies.

We now believe that black holes, very massive black holes, probably lurk at the centers of most galaxies. It's almost certain that our own Milky Way galaxy has a rather modest three million sun-mass black hole at its center. Some of our neighbor galaxies, which are not terribly different from our own, have black holes of many tens of sun masses—40, 50, 100 sun masses. Black holes at the center of the galaxies.

How do we know this? Well, one of the instruments that we've used to find this out is the Hubble space telescope, which can look with remarkable precision and not only with visible light, but also with x-rays, ultraviolet, and a host of other approaches, a host of other regions of the electromagnetic spectrum can look into the cores of galaxies. And it can look through clouds of dust with different wavelengths of electromagnetic radiation.

So, we can probe the inner cores of galaxies. And what we see when we look in there are stars. And then, further in, gas moving at enormously high speeds. And we can calculate those speeds. In some cases, those speeds are close to the speed of light. And we can calculate what kind of object, how big the object would have to be, and how massive it would have to be for gas to be spiraling around it at these enormous speeds. And we come up with a very clear answer: That these objects are so small that they can't be big, extended objects, like clusters of stars. They have to be very small, compact objects. And black holes are the best and in some cases the only candidates.

We see gas spiraling and spiraling around these black holes, or these objects at the center of galaxies, and we're quite convinced now that they are black holes.

We believe that the black holes at centers of galaxies like ours aren't doing much anymore. That's because although some material is still falling into them, the galaxy has evolved enough that much of its matter is in stars. And the stars aren't falling rapidly into the black hole at the center.

But we believe in the early years of galaxies, when the galaxies had a lot more gas and dust in them, those black holes at the center were basically powerhouses of energy. As the material fell into the black hole, it again rubbed against itself, got very hot, and emitted copious amounts of energy.

And we believe now that's the mechanism that powers what, for years, were the mysterious quasars. Distant objects, billions and billions of light years away that seemed as bright, or brighter, than an entire galaxy of stars, and yet, were very small and compact.

We believe we were looking at the black holes that were formed at the centers of then-primordial galaxies. So we think black holes are the engines, if you will, that power quasars by drawing in all this matter, having matter spiral around and fall into them. Heating up and giving off all these x-rays.

So we believe that super-massive black holes, millions or billions of suns, lurk at the center of most galaxies. We believe they even lurk at the center of our own galaxy.

Well, that's real astronomy. That's evidence that black holes exist in our universe. There are some more speculative things we can say about black holes, too. Black holes are actually very simple objects. They've been likened, sometimes, to elementary particles like electrons, because black holes have only mass, electric charge, and they may be rotating. And there's a number that describes how rapidly they're rotating.

They don't have any other properties. Why not? Because anything that falls into the black hole can't come back out again. So we don't care what that stuff was. The black hole contains no information about what went into it, other than total mass, electric charge, and how fast it may be rotating. So black holes are destroyers, if you will, of information. And there they're for very, very simple objects. They have it characterized by only a few basic numbers—mass, electric charge, and how fast they're rotating.

And if you examine rotating black holes—and the mathematics of rotating black holes is much harder than the mathematics of non-rotating black holes and wasn't solved until the second half of the twentieth century—but, if you look at rotating black holes, you find some remarkable interpretations of the mathematics.

For instance, there are interpretations that suggest that rotating black holes can lead to so-called *wormholes*, which are bridges through spacetime that might connect to other parts of the universe. And such wormholes, if they existed, might be at the other end white holes, in which matter that was pouring into the black hole in this region of the universe, might be pouring out in the other region of the universe.

Now, there's absolutely no evidence for such things. Although occasionally this idea has been evoked to explain inexplicable astronomical phenomena. But the equations seem, at least, to permit that kind of interpretation.

If you've seen the movie, *Contact*, Jodi Foster goes rattling along through a wormhole that this advanced civilization has discovered to a distant region of spacetime and gets there much quicker than she could by the normal route, because this is a bridge, a tunnel from one region of spacetime to another, which is possible in a curved spacetime. Well, I think that's very speculative.

But black holes, connecting to remote regions of spacetime? Who knows? It might be possible. I think black holes make an interesting point. When the equations of general relativity first predicted the existence of black holes, people kind of dismissed that. Then they began to take them seriously. And today, most astrophysicists tell you, "Yep. Black holes exist." I think it's an example of a richness that our universe has that we need to appreciate.

Here are these equations of relativity. They predict something that might be possible. And, lo and behold, bizarre as these objects are, we find them. It's as though the universe is trying to make possible all the things that might be. And general relativity certainly predicts some bizarre things. And the black holes are probably the most bizarre of those things.

Into the Heart of Matter
Lecture 16

We want to understand the rules of physics that govern the behavior of the universe at the sub-atomic and atomic scale. And those rules are bizarre, indeed. Those are the rules of quantum physics, or quantum mechanics.

We are headed into quantum physics, the implications of which are even stranger than those of relativity. Relativity asks that we alter our conceptions of space and time but with modified meanings, our common-sense language still applies. In quantum physics, though, our everyday language is completely inadequate to describe physical reality. The next lectures address quantum physics by:

- Examining the nature of matter through early 20th-century understanding of the atom and highlighting problems with atomic models based on classical physics.

- Resolving these problems with the idea of the *quantum*.

- Developing the ideas of *quantum physics*, which governs the behavior of matter and energy at the atomic scale.

- Resuming the descent toward the ultimate heart of matter, looking at elementary particle physics.

- Applying both subatomic physics and relativity to an understanding of the evolution of the universe.

- Describing attempts to merge relativity and quantum physics into a "Theory of Everything."

Democritus (c. 400 B.C.) proposed that matter consists of indivisible particles called *atoms* (the term itself means "indivisible"). John Dalton (early 1800s) organized elements by atomic weight and set forth the seeds of modern atomic theory. Dmitri Mendeleev (1869) developed the periodic table of the elements based on an orderly, repeated arrangement of elements with similar chemical properties.

In the late 1800s, subatomic particles were discovered. Henri Becquerel discovered radioactivity and the Curies (Marie and Pierre) explored the new phenomenon. J. J. Thomson discovered the electron, which led to William Thomson (Baron Kelvin) proposing the "plum-pudding" model of the atom (1900), in which electrons are embedded in a "pudding" of positive charge. American Robert Millikan won the 1923 Nobel Prize in physics for his measurement of the charge of the electron.

Democritus.

In 1909–1911, Ernest Rutherford, Hans Geiger, and Marsden performed experiments in which they shot high-energy alpha particles from a radioactive substance toward a thin gold foil. Most went right through or were deflected slightly, but a few bounced back in the direction from which they had come. Rutherford interpreted this to mean that the atom is mostly empty space, with nearly all its mass concentrated in a tiny, positively charged nucleus. He proposed a "solar system" model for the atom, with electrons held in orbit around the nucleus by the attractive electric force.

These early models of the atom presented some problems. The first problem was that atoms shouldn't exist! Maxwell's equations of electromagnetism predict that accelerating electric charges should emit electromagnetic waves (light). The electrons in Rutherford's atom are accelerating, because they are moving in circles. They should lose energy by radiating electromagnetic waves. Then, like a satellite losing energy because of friction with Earth's upper atmosphere, they should spiral into the nucleus. All this should happen in a split second! Even if Rutherford's atom didn't collapse, the solar system model offers no explanation of atomic spectra—the discrete colors of light emitted by atoms of each different element.

Another problem related to classical physics (Maxwell's electromagnetism and thermodynamics) is the so-called *ultraviolet catastrophe*. Hot, glowing objects should give off electromagnetic waves (light), because of the vibrations of their constituent atoms participating in the microscopic energy we call heat. Indeed, hot objects do glow (picture a hot stove burner or the filament of a light bulb). Classical physics says that they should also give off an infinite amount of electromagnetic radiation, concentrated toward the shorter wavelengths (ultraviolet being the shortest wavelength of electromagnetic waves known at the time). Obviously, this doesn't happen. How can that be explained? ∎

> **These early models of the atom presented some problems. The first problem was that atoms shouldn't exist!**

Essential Reading

Hey and Walters, *The Quantum Universe*, Chapter 4.

Wolf, *Taking the Quantum Leap*, Part I.

Questions to Consider

1. How did the rare occurrence of an alpha particle's being bounced back in the direction it came from imply that the mass of an atom is concentrated in a tiny volume?

2. How did the discovery of subatomic particles alter Democritus's original concept of the atom?

For additional coverage of the topic presented in this lecture, we recommend The Teaching Company course *The Joy of Science* by Professor Robert Hazen of George Mason University and the Carnegie Institution of Washington.

Atomic Models

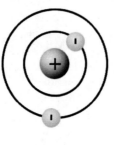

Rutherford (1911) "Solar System"

Based on the results of his experiment, Rutherford devised another model of the structure of the atom in 1911. In this model, a central "nucleus" with a positive charge was surrounded by electrons with negative charges that circled the nucleus in orbits. Due to its parallels to the Sun and the planets, this model was called the "Solar System" model of the atom.

Thompson/Kelvin (1900) "Plum Pudding"

In 1900, Thompson and Kelvin hypothesized that the structure of the atom consisted of a fixed number of electrons, carrying "negative" charge, embedded in a mass that had an equal but opposite "positive" charge. Due to its structure, this model was called the "Plum Pudding" model of the atom.

Rutherford's Experiment

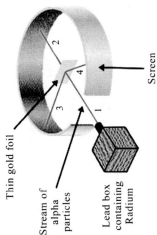

Thin gold foil

Stream of
alpha
particles

Lead box
containing
Radium

Screen

In Ernest Rutherford's classic experiment, a lead box containing the radioactive element radium, was placed in front of an open cylindrical screen. As it decays the radium emits high-energy subatomic particles, called alpha particles (and now known to be helium nuclei.) A small hole was cut into the side of the lead box that faced the screen allowing a stream of alpha particles (1) to shoot out toward the screen. In the center of this open screen, Rutherford suspended a sheet of gold foil which served as the target for the alpha particles.

Rutherford predicted that the alpha particles would shoot directly through the gold foil (2) and hit the opposite side of the screen, uninterrupted by the gold foil. And indeed, most of the alpha particles did just that. But to Rutherford's astonishment, a few (3 and 4) were deflected at large angles. This led Rutherford to postulate the existence of a tiny but massive atomic nucleus, and led to his "Solar System" model of the atom.

Atomic Spectra

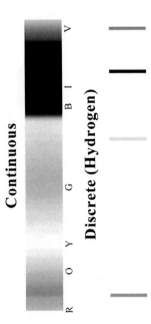

This diagram shows the phenomenon of discrete emisson in atomic spectra. The top band is a continuous spectrum of visible light from red to violet. The bottom band is the spectrum emitted by hydrogen atoms, consisting of discrete colors. Each element has a different pattern of these spectral lines.

Into the Heart of Matter
Lecture 16—Transcript

Here we're going to leave relativity aside—although we'll come back to it in lectures twenty-three and twenty-four as we try to develop a theory that describes everything there is to say about the universe—and we begin a look at the smallest systems we know in nature—atomic and sub-atomic systems. We want to understand the rules of physics that govern the behavior of the universe at the sub-atomic and atomic scale. And those rules are bizarre, indeed. Those are the rules of quantum physics, or quantum mechanics.

We had a problem with relativity. The problem with relativity was I made you stretch your common sense notions of space and time. But the space and time we were talking about in relativity were at least still related to your common sense notions of space and time. You had to stretch them. You had to recognize that your common sense notions were approximations, or limited versions of the richer space and time that a relativistic universe gives us. But at least the language—space, time, causality, matter, energy—that language still worked. The language still described physical reality, even if we had to redefine and stretch and enlarge our definitions of the concepts.

With quantum physics, it isn't going to be that way. With quantum physics, even the language of everyday reality fails us. I've been waving around tennis balls and my big globe and basketballs, and everyday objects to try to illustrate for you some of the meaning of relativity. And those analogies worked. They aren't perfect, but they worked. Those analogies fail, utterly, in the quantum realm. Occasionally, I'll still have to resort to them. I'll pick up a tennis ball and I'll say, "This is an electron." But that's not what an electron is. An electron is not a little hard sphere like a tennis ball.

What is an electron? Well, I don't know. Is it a wave? Is it a hunk of energy? Is a little chunk? What is it? We don't really know. We don't have a good feel for it. Our language is not adequate to describe the world of quantum physics, the world of the atomic and sub-atomic realms. That's going to prove very frustrating. Quantum physics is, in a way, more abstract than relativity. Some people lump relativity with classical physics. It's still understandable in that sense. Quantum physics is going to wreak havoc with our notions of causality. It's going to give us a world which is basically governed, not by cause and effect, but some kind of a statistical association, which philosophers and physicists are still arguing over the meaning of.

Quantum physics is a strange thing. We're going to try to get at that strangeness as best we can. Once again, I can't do it with the logical precision that I used to lead you to special relativity. It's going to be more descriptive, like my analysis of general relativity. But I am going to try to show you historically again, why quantum mechanics had to arise. Why we were forced to some very bizarre conclusions about the way the sub-atomic world works.

Here's the plan for doing all this in the next few lectures. What I want to do in this lecture is look at questions of what matter is made of. How did we begin to get an understanding of the nature of matter through, say the early twentieth century?

People have been asking questions about the nature of matter for thousands and thousands of years. And we've been gradually narrowing down toward a sort of correct answer as to what the nature of matter is.

But just about the turn of the twentieth century, problems arose in that description of matter, problems that made classical physics at odds with what, in fact, was observed in laboratory experiments.

We want to look at the problems that these classical ideas of the nature of matter gave rise to—problems that seemed irreconcilable when you tried to apply classical physics to describe what was going on in the sub-atomic world. Then we want to resolve those problems. And the way we'll resolve them is with the ideas of quantum physics, the idea of the quantum itself—of the discreteness, not only of the matter that makes up the universe, but also of energy.

And then after we've explored quantum physics in several lectures and gotten somewhat used to the bizarre notions of matter that isn't matter, energy that isn't energy, waves that are particles, and particles that are waves—then we'll resume this descent further into the heart of matter and look at the most fundamental particles that make up the universe, as far as we know it. Look at the really elementary things—the quarks, the electrons. And understand how they combine to make the matter of which the everyday world is constructed.

After we've done that, we need to look at how sub-atomic physics (the physics of the very small) and the physics of the very large (general relativity, cosmology, the physics governing the large-scale universe)—how we can put those two together to some extent and understand the evolution of the entire universe.

And indeed, some of the greatest progress in physics and astrophysics over the last few decades has come from merging the studies of the very smallest sub-atomic physics with the studies of the very large-scale universe. And we have now a very clear picture of how the universe evolved from what we think was a beginning a certain number of billion years ago. And we think we know that number pretty well.

We have that understanding because we know the large-scale structure of the universe, determined by the curvature of spacetime, described by general relativity, and we think we know what happens on the small scale. So we'll do that in the next to the last lecture.

Then in the very last lecture, we're going to get rather speculative and ask: Is it possible to do something that most physicists say is not yet near to being done—and that is to develop a theory that combines quantum physics—the physics of the very smallest scales, the physics of the sub-atomic realm— with general relativity—the physics of the large-scale curvature of the universe, of gravity, of the behavior of large-scale objects?

There are deep, profound, physical and philosophical problems to doing that, but we're going to look at some promising leads—string theory, in particular. And we're going to ask: Are we anywhere near a *theory of everything*? Are we anywhere near one equation, or one simple theory that tells us why the universe is the way it is—why it has electrons, why it has atoms, why they have the mass they do, why it has curved spacetime? Are we anywhere close to that? Are we near a theory of everything? And we'll end with that. So that's where we're headed in the remainder of the course.

In this particular lecture, we want to look at the question of the nature of matter and then we want to see how that question gave rise to deep concerns that led, ultimately, to quantum physics.

I raised the question, a few lectures ago: What would happen if I took a glass of water and I began pouring water out? Well, I could pour out half the water and I've still got water; half of that, and I've still got water; half of that, and then I still have water. And at some point in human history, it was an unanswered question whether I could continue to do that forever. Is water continuously subdividable or isn't it?

Well, the Greek philosopher Democritus gave an answer to that question in 400 B.C. I don't think Democritus had a lot of good experimental evidence for his answer. But he declared that ultimately, the universe is made of indivisible particles, which he called atoms, and these atoms were tiny, tiny entities. He didn't describe them in any detail. They came in different varieties. And they, the atoms, were themselves indivisible, and they made up all the rest of the matter in the world.

This is, essentially, the first description of an atomic theory, of a theory that says, ultimately no, you can't subdivide that water infinitely many times. Ultimately, you come down to individual particles. And they are completely intact, indivisible. You can't divide them any further. Now the joke is on Democritus because we, of course, now talk about splitting the atom.

But now we talk about smaller particles—quarks and electrons that are presumed to be indivisible. So there's Democritus in 400 B.C. first proposing atomic theory. And indeed, his word "atoms" means that they are indivisible, although certainly the modern day understanding of atoms does not carry that connotation.

The chemist Dalton, in the early 1800s, looked at the different elements that make up the world—substances like oxygen, carbon, hydrogen, helium, nitrogen, iron, silver, gold. These are the fundamental, elemental substances.

They combine to form things like water, gasoline, rust (iron oxide) and different, more complicated combinations. But by Dalton's time, it was known that there were certain fundamental elements that couldn't be made any smaller, they weren't composed of different substances any more than that. So Dalton looked at the elements and he made a table, arranging the elements in terms of their, what's now called atomic weight, in terms of the mass of the individual atoms, the individual particles that made them up.

And Dalton was trying to explain the properties of chemistry. Dalton was trying to explain how the atoms combined to make different substances, like hydrogen and oxygen atoms combining to make water. And so this was a rudimentary sense of atomic theory. But it turns out the weights or masses of the atoms are not a very good clue to their chemical behavior.

What really is a good clue to their chemical behavior is their electric charge—in the number of electrons these atoms carry. Now, nobody knew about electrons and what they were at this time. But it turns out that that electric charge is what's crucial in determining an atom's chemical properties. And that electric charge—the number of electrons in the complete atom is called the *atomic number*. And as opposed to the *atomic weight*—the atomic weight is the mass of the whole atom—the atomic number is the number of these charged electrons in it, although again, in the 1800s, nobody knew about electrons, at least early in the 1800s.

Well, in 1869, the Russian Mendelayev developed a table of the elements, which was based, ultimately, on this charge, although he didn't know exactly what he was basing it on. And he arranged the elements in this table and found that there were regular repetitions of similar chemical properties. For instance, the elements fluorine and chlorine have similar chemical properties. The elements helium and neon have similar chemical properties, namely, they don't react very much with anything. The elements sodium and potassium have similar chemical properties, and, in fact, their property is that they react violently with lots of things.

And Mendelayev noticed this similarity. And if you've studied any chemistry at all, you've certainly seen the periodic table of the elements. What the periodic table of the elements is is a classification of the elements, ultimately by this number of electrons in the atom. And that number of electrons determines—in fact the outermost of these electrons determine—how the atom interacts electrically, ultimately, with other atoms. And it's that electrical interaction that leads to all of chemistry happening.

So Mendelayev, in 1869, developed the periodic table. And so he's beginning to understand chemistry ultimately in terms of the properties of the atoms, of the individual elements, although the nature of those atoms was not, at that point, understood. They were still atoms. They were still indivisible, as Democritus had proposed in 400 B.C.

But then in the late 1800s came a series of remarkable experiments. Becquerel, in France, did an interesting thing. He accidentally left some uranium-containing compounds in a drawer with some photographic film and to his surprise, when he developed the film, it had indication that light had hit it. Well, it wasn't light that had hit it. It had marks on the film; it was some kind of emanation from the uranium-like material. And Becquerel had discovered nuclear radiation. He had discovered radioactive materials. He had discovered the radioactivity of uranium.

And Becquerel's work was carried on by Marie and Pierre Curie and later by Marie's daughter Irène and her husband Frederic Joliot Curie. And together, the Curies particularly, discovered many, many other radioactive elements. They studied the properties of radiation. They learned a lot about the fact that there were some materials that seemed to emit substances from their atoms. They had discovered sub-atomic particles. They had discovered that the atom was divisible. Democritus was wrong. "Atom" might have meant indivisible, but atoms, in fact, are not indivisible. And through the study of radioactivity, we came upon the notion that atoms are divisable, and that, indeed, some atoms spontaneously spew out other particles.

I'll have more to say about those other particles and some of their uses soon. That was Becquerel's discovery.

Other physicists were experimenting with simple systems in which they would take a glass tube; they would put a couple pieces of metal in the ends of it; they'd connect those metals to the outside world, to a battery; and if they took most of the air out of the tube, they could see a glow occurring in the tube. And they could conclude that some kind of rays were emerging from one piece of metal and heading toward the other piece of metal. And they studied these rays. They were called cathode rays.

And to this day, the TV tube in your TV set, the picture tube, or the tube in your computer monitor, these are called cathode ray tubes, because they involve these cathode rays. They involve metal electrodes in which these rays are made to move. Well, studying these rays, Thompson—J.J. Thompson—discovered in the late 1800s again, he concluded that these rays were not continuous emanations, but they were made up of individual particles. And Thompson had basically discovered the electron. And he was able to measure properties of the electron.

He wasn't able to tell you separately its mass and its electric charge. But he was able to tell you the ratio of those two. He was able to learn something about these particles. And he concluded again that we had individual particles. These are sub-atomic particles. These are parts of atoms. Atoms are no longer indivisible. Thompson has discovered the electron. And on the basis of Thompson's discovery, we have the first really viable model of the atom.

The model was put together by Lord Kelvin, or invented by Lord Kelvin using the results of Thompson's discovery of the electron. And I have a picture of it here. It's called the plum pudding model of the atom, because what it consists of is a number of these newfound electrons. These discreet, individual particles, which Thompson discovered carried negative electric charge.

Remember, it was Ben Franklin who defined positive and negative. And they don't really mean anything is missing or there. They're just names. But there are two kinds of charge. Thompson and Kelvin's plum pudding model said, "Okay, we have a number of these electrons. That's a certain amount of negative charge. Atoms are electrically neutral. They have no overall electric charge and these negative electrons are embedded in a plum pudding-like gelatinous structure, which consists of positive charge."

How much positive charge? Exactly as much positive charge as there is negative charge in the electrons. So, the Thompson-Kelvin model is the plum pudding model of the atom. And it's a model that suggests there are these individual, discrete—very light by the way, not at all massive—electrons embedded in a kind of amorphous pudding of positive charge. And the whole thing makes the atom electrically neutral.

By the way, experiments by Millikan with electrons later verified the charge itself on the electron. And therefore, since we knew the ratio of charge to mass, we also knew the mass of the electrons. Millikan's experiment showed that electric charge is a quantized substance, meaning it comes only in certain discrete hunks. The electron charge seems, or seemed through most of the twentieth century, to be the minimum amount of electric charge one could have.

We'll now see that the real minimum amount is about a third that, or we'll soon see that. But nevertheless, electric charge comes in these discrete little hunks. It's something that's quantized. Electrons themselves, their existence, shows you that matter itself comes in discrete little hunks. We could say it's quantized. It comes in discrete little pieces. It isn't continuously subdividable. And I'm going to use that word quantized again and again and more and more as we go on.

So there's the plum pudding model of the atom. Now in 1909 through 1911, Ernest Rutherford, who was a New Zealander working at that time in England and his associates, Geiger, of Geiger counter fame, and Marsden did a series of experiments in which they probed the structure of matter by shooting high-energy particles at, in their case, a thin gold foil.

And I want to describe this experiment in some detail, because it's kind of a paradigm for the way we still probe the structure of matter. What we do is produce high-energy particles of some sort, subatomic particles. Today, we produce them in high-energy particle accelerators. And the desire to probe deeper and deeper into the structure of matter is what drives particle physicists' desire for ever-bigger and more expensive particle accelerators, by the way.

But this is what we do: We accelerate these particles and then we smash them into matter and we see what comes out. It's a pretty crude way. We can't get in there with tweezers and microscopes and look at the structure of subatomic matter. So what do we do? We accelerate particles to high energy. We bang them into matter and we see what results.

A whole spew of particles comes out. We study those particles. We study their tracks. I gave you an example of some particle tracks earlier, when I showed you the phenomenon of pair creation. We study those particle tracks and we try to figure out what went on, and from that we infer the behavior of the structure of matter.

Well, Rutherford's experiment was one of the first experiments that did that sort of thing. Rutherford didn't have access to a particle accelerator, popularly called an atom smasher, although I don't really like that term. Rutherford, instead, used the newly discovered radioactive particles that emanated from radioactive materials, like the material radium that Marie Curie had discovered.

Radium, in particular, puts out a very high-energy particle called an alpha particle. It was later discovered and known to be a nucleus of a helium atom, but at that point, it was just called an alpha particle. It was a positively charged particle, and it came out at high speeds.

And if you took a source of radioactive radium, and you put it in a lead container so these alpha particles couldn't escape, and you cut a little hole in it, and made a little tube of the lid, you could get a beam of these alpha particles coming out. Rutherford and Marsden used that beam to probe the structure of matter by shooting the alpha particles to this gold foil.

Now, let me give you an analogy with that experiment. Here's my analogy— some toilet paper—pretty thin stuff. This toilet paper represents the gold foil that Rutherford and Geiger and Marsden used. It was a very, very thin sheet of gold.

Now imagine doing something like shooting a BB gun at this. What's going to happen? Well, BBs are very tiny. They're going to go shooting right through. They're going to tear a little hole where they go, but that's it. You would be very surprised if you shot a BB gun at this piece of toilet paper and most of the BBs went through but a few of them bounced right back in your direction. That would be unusual. Remarkable. Very difficult to believe.

That is exactly what happened in Rutherford's experiment. Nearly all the alpha particles went right through the gold foil. They experienced almost no change in direction. They went straight through. They were undeflected. But a very few of them bounced back in the direction they came from.

How could that possibly be? You think this piece of toilet paper is essentially just continuous stuff. But what if the toilet paper had embedded in it very dense, relatively massive, but tiny objects. If, by chance, one of your BBs hit one of those objects, it could bounce back and come back. Most of the time it would go straight through. But once in a while, it would bounce back.

And Rutherford and Geiger and Marsden were forced to conclude from their experiment that the nature of matter must be such that it's basically mostly empty space. Because most of those alpha particles went right through as if they hit nothing. But occasionally they hit something that was very, very small and very, very dense, but very, very massive.

And so they developed a picture of the nature of the atom that was very, very different from Thompson and Kelvin's plum pudding model. They said instead, there must be a very, very tiny object at the center of the atom. They called it the nucleus. It must be extremely small, but it must carry most of the mass of the atom and most of the rest of the atom must be empty space. And that empty space has in it the few, the handful, or whatever number it is, of electrons that are whizzing around that nucleus.

And so they developed a picture of the atom that looked something like this. This is the Rutherford model. It looks like a miniature solar system. This is a very bad picture. It's not to scale and there are a lot of other problems with it. It's not at all consistent with quantum physics. But it's a first guess at what an atom might look like. I'm going to point out, though, that it's not to scale.

That positive charge at the middle—that represents the nucleus. At that point, Rutherford knew nothing more about what the nucleus was composed of. Today we know a lot more about that. That thing is about one ten-thousandths to one one-hundred thousandths of the diameter of the atom. So if the atom were the size of a baseball stadium, or something like that, the nucleus might be a baseball right at the center. And all the rest of it is empty space occupied by a very small population of electrons.

And what are the electrons doing? According to Rutherford, they're in orbits around the nucleus, just like the planets are in orbits around the sun. Except the force that's holding them in orbit is not the gravitational force, but the electrical force of attraction between positive and negative. And so Rutherford worked out this miniature solar system model of the atom.

And by 1911, that became the prevailing model for what an atom ought to look like. And we have a very different sense of matter. No longer is matter this solid stuff. Matter that seemed so solid is really mostly empty space.

That was Rutherford's solar system model of the atom. And it's a model we have in our minds today. Witness the picture of the atom behind me. That is Rutherford's solar system model of the atom, although we're showing several electrons in each orbit. That's actually oxygen up there. It has eight electrons in its orbits.

It's not really a very good picture of the atom, as we'll see when we look further at quantum physics, but again, it's a good first approximation and it's certainly a lot better than the plum pudding idea of an amorphous atom. The atom and, therefore, matter, as we know them, are mostly empty space.

And that's true of almost all the matter in the universe we know of, with the exception of those objects, like neutron stars, that I talked about earlier, which are, in fact, collapsed objects miles in diameter with the whole mass of the star. And they are collapsed to the density of an atomic nucleus or bigger. But all the other matter we know about in the universe is basically mostly empty space with most of the mass concentrated in these tiny, tiny little nuclei that Rutherford discovered by hitting them with the alpha particles that he was bombarding this gold foil with.

Well, that's great. Now we've got a model for the atom. But there are problems with the models for the atom that were developed at that time. And the problems are deep and profound and we need to look at them and you need to appreciate them.

First problem—atoms shouldn't exist. Why not? Well, we've got to go back to electromagnetism again. Let me pick up an electron. If I move an electron, say, jiggle it up and down, I create a changing electric field. Remember what a changing electric field made? It made a magnetic field, which is also changing and that makes a changing electric field and so on. That's the way I make an electromagnetic wave. I make an electromagnetic wave, in fact, any time I change the motion of an electric charge, like an electron. Any time an electric charge accelerates, if I shake it up and down, I generate changing electric fields, make changing magnetic fields, make changing electric fields, and so on. And, in fact, those electromagnetic waves that result are carrying away energy.

Where is the energy coming from? It's coming from me, shaking the electron. If I stop shaking it, the electron will soon stop shaking as the electromagnetic waves carry away the energy. If I make the electron go in a circle, its motion is changing. It will generate changing electric fields that will make changing magnetic fields and it will also generate electromagnetic waves. And those waves will carry away energy. And I'll have to keep supplying energy to make that happen.

Well, what's happening in Rutherford's atomic model? Electrons are going around in circles like this. And if you calculate using Maxwell's electromagnetic theory, which we're quite sure is correct, if you calculate the energy that will be emitted by electrons spiraling around the nucleus, you will find that that energy is lost in a very, very short time. And the electrons will spiral into the nucleus and crash into it, just like a satellite that's in low enough orbit that it has drag from the atmosphere will soon spiral down to Earth and crash to Earth. The same thing would happen to the electrons and the Rutherford atom would go out of existence in a fraction of a second.

That's a pretty profound problem with this model. The thing simply should not be able to exist. And yet atoms do exist.

Even if these atoms don't collapse, there's another problem. When we look at matter and its emission of electromagnetic waves, there are two ways we can look at matter. If we take a dense piece of matter and do something like heat it up, we will find it emits electromagnetic waves in a whole variety of wavelengths, frequencies, vibrational frequencies of the waves, or equivalently, if it's visible light, it's emitting colors. That's what I've shown here—a continuous rainbow spectrum of colors.

On the other hand, if we take individual atoms that aren't crammed together in a solid—and the way we might do this is to take a glass tube and fill it with gas at relatively low pressure and then either heat up the gas or maybe run an electric current through it—that's basically what a fluorescent lamp or a neon lamp is, by the way. We find that instead of getting a continuous band of light, we get a discrete spectrum. We get individual colors of light emitted and mostly darkness in between.

The sodium lights, for example, that light our city streets emit a spectrum of discrete lines, like that. Neon signs emit a spectrum. If you took the light from a neon sign and sent it through a prism that breaks up light into its colors, you would find a few or many discrete colors. That's the spectrum of the neon atoms. The Rutherford model can't explain that spectrum at all. There's no understanding of why an atom with electrons spiraling around in orbits of different heights and diameters, depending on how much energy they have, should always produce a light of single discrete colors, wavelengths, frequencies, you name it.

Why should that be? No explanation whatsoever. And yet, we observe very clearly and had observed for 100 years before Rutherford's model, the existence of these spectra. And we knew how to characterize them mathematically, but nobody knew why they arose. And Rutherford's model can't explain that. So that's a second profound problem.

First problem, atoms shouldn't exist at all. Even if they do exist, they emit light in discrete colors. And there's no explanation, in terms of Rutherford's solar system model, for why that should be.

There's another problem, which is a little bit more subtle, I mentioned. What happens if we take a solid object and heat it up? It will emit electromagnetic waves also. It won't emit discrete frequencies, though, because the atoms are all kind of jostling together and getting their individual atomicness smeared out a little bit. That's a crude explanation for why we don't get these discrete spectra when we have a solid object.

But something else does happen with a solid object, and I'm going to demonstrate that here. My solid object is the filament of a little light bulb here, a high-intensity tube of light. And I'm going to demonstrate what happens. This light happens to equipped with a dimmer, and I'm going to demonstrate what happens when I turn this light on. And for this light, you can picture any object, which I heat up. As I heat up an object, it starts to glow. And at first, it glows a kind of dull red, and this is about the lowest I can get this object. It's glowing a kind of red–orange.

If you want it to glow even cooler, think of an electric stove burner. When you turn it on low, you can't see it glowing. And then you turn it up and it just barely begins to glow a dull red. And maybe on high it's glowing not quite as bright as this filament is.

Things can even glow with infrared light that you can't quite see, like if you put your hand near a hot iron or near a hot woodstove or above a stove burner that's on low, an electric stove burner; you can feel that energy radiating from it. That's infrared light that you can feel, but you can't quite see. Well, by the time it gets hot enough, it emits visible light.

And here's an example. This filament of this light bulb is emitting visible light. What happens if I turn up the electric current through that filament, giving it more energy? Well, not just one thing happens, but two things happen. One thing that's obvious is it gets brighter. But a second thing happens is the color shifts from the red it started with, the reddish orange it started with, it becomes yellowier—or from red to more orange and then it becomes yellower and it becomes brighter. But it also becomes whiter and whiter and the color changes. The color moves from the red end of the spectrum of rainbow colors toward the yellow. And if I could heat it up enough, it would move toward the blue and so on. It would ultimately move toward the ultraviolet and the x-ray, but the filament would long since have burned out before then.

So two things happen as I increase the temperature of a hot, glowing object. One thing is its color changes. Its color tends more and more toward the blue. That's the dominant color. It's emitting a whole range of colors. It's emitting a whole rainbow-like spectrum. But the dominant color is shifting more toward the blue. And the other thing that happens is it overall gets brighter.

But what does classical physics have to say about that? Well, if you apply classical physics, which is basically electromagnetic theory, the theory of how matter emits electromagnetic waves when you jostle around the electric charges that make it up and you apply thermodynamics, which is the study of heat energy—heat and energy as heat, a subject which is ultimately grounded, largely, in Newtonian mechanics—you apply those to a hot, glowing object and you say, "What should it emit? How much energy should it emit at different wavelengths?"

Well, what we would like to say is when it's cool, it emits mostly in the red. When it gets warmer, it emits more in the yellow. When it gets even warmer, it's beginning to emit more toward a greenish and so on. Or it comes out as white, in this case. It's emitting a whole range of colors.

Well, that's not what classical physics says. What classical physics says is it emits a little bit of energy in the red, more in the orange, more in green, more in the blue, more in the violet, and most in the ultraviolet. If you make it hotter, it emits more in all those colors. But it always emits the most in the ultraviolet. And in fact, it emits an infinite amount in the ultraviolet, which is patently absurd. But that's what classical physics predicted. And that phenomenon, that prediction of classical physics, was called the *ultraviolet catastrophe*. It was called the ultraviolet catastrophe because of this prediction that the most energy would be emitted as ultraviolet light.

Actually people didn't know, in those days, about x-rays and gamma rays, which are even higher frequency forms of electromagnetic radiation. And the classical prediction is actually, it emitted infinite amounts of energy at infinite frequencies, which is where the gamma rays are getting up toward. So it could have been called the gamma ray catastrophe, but it was called the ultraviolet catastrophe, because that was the highest energy electromagnetic waves known at the time.

And the ultraviolet catastrophe, like the existence of atoms, is another place where classical physics really confronts, in a very dramatic, contradictory way what reality has to show us. Classical physics says every hot, glowing object—your body is a hot, glowing object, so is the Earth—all these things should emit an infinite amount of energy. It should be emitted in very, very short wavelengths, very, very high frequencies, like ultraviolet. Obviously, it doesn't happen or we'd all be bathed and boiled in infinite amounts of electromagnetic radiation.

So the ultra-violet catastrophe, the existence of atoms, the existence of atomic spectra, these all present profound contradictions between the predictions of classical physics and the reality that we can easily see in experimental situations.

It's going to take quantum mechanics to resolve those situations, and that's where we move to next time.

Enter the Quantum
Lecture 17

Now, we're sitting at about the turn of the 20ᵗʰ century, plus or minus 10 years, depending on which phenomena we're talking about, and we know that there are atoms. Unlike Democritus suggested though, they're not indivisible.

Experiments beginning in the 1880s showed that light shining on a metal surface in vacuum can eject electrons from the metal. This is called the *photoelectric effect* and is the basis of an early kind of "electric eye" used in everything from automatic door openers to sensitive light measuring instruments. Classical physics attempts to explain the photoelectric effect. Electrons are jostled by the alternating electric field of the electromagnetic wave that is the light. They absorb energy from the wave, eventually gaining enough to escape the metal. Because the wave energy is spread over a wide area, it should take a long time for any one electron to be ejected. Therefore, there should be a delay between the light's striking the metal and electrons being ejected. The color of the light should not matter.

The experimental results are at odds with the classical predictions. Electrons are ejected as soon as the light shines on the metal. The color of the light does matter. If the light is too red, no electrons are ejected, no matter how bright the light. If the light is blue enough, electrons are ejected, and their energy increases as the color of the light moves toward violet and ultraviolet.

Let's quickly review the four problems faced by classical physics at the turn of the 20ᵗʰ century, in the order in which I've introduced them. First, there was the problem of the existence of atoms: Atoms shouldn't last. Second, there was the problem of the spectra of light that atoms emit. Atoms should emit light of all colors and not, as had actually been observed, discrete spectral lines unique to each element. The third problem was the ultraviolet catastrophe: Hot objects should glow with an infinite amount of ultraviolet light. And finally, there was the problem of the photoelectric effect: Electron ejection should take a long time and should be independent of color.

Now let's resolve the problems, here in historical order, although the first is the most obscure. Max Planck (1900) showed that the ultraviolet catastrophe could be resolved by assuming that atomic vibrations are *quantized*, occurring only in multiples of a certain basic amount. That required basic amount was given by the formula $E = hf$, where E is the energy of a vibrating atom, f is its frequency (how many vibrations per second), and h is a new constant of nature that became known as *Planck's constant*.

Einstein (1905, same year as special relativity!) explained the photoelectric effect by declaring that the energy in a light wave is not spread uniformly over the wave but is concentrated in particle-like "bundles" called *photons*. The energy of a photon is quantized: For light of frequency f, the energy E of a photon is given by $E = hf$, where again h is Planck's constant. Einstein's proposal explains the photoelectric effect because it takes a certain amount of energy to eject an electron from the metal. If light is too red (too low a frequency f), then the energy of its photons is lower than that required to eject electrons, and none will be ejected. Bluer light can eject electrons and, as the light frequency increases, the photons can impart more energy to the light.

Ejection occurs immediately because the light energy is concentrated in photons, and an electron need be hit by only a single photon to be ejected.

In 1913 Niels Bohr proposed the *Bohr model of the atom*. The Bohr model explained

Atoms radiate electromagnetic waves (light) only when electrons jump among orbits, emitting specific colors of light. This explains the spectra of atoms.

both the existence of atoms and their spectra but had no deeper theoretical basis. Atomic orbits are quantized, with only certain discrete orbits allowed. These orbits correspond to discrete values of the electrons' energy. Bohr's actual quantization condition is that the allowed angular momentum, L, of an orbiting electron (a measure of rotational motion) is given by $L = h/2p$, where p is the electron's momentum and h is again Planck's constant. Electrons in allowed orbits don't radiate electromagnetic waves; they don't crash into the nucleus, but rather stay in orbit. Atoms radiate electromagnetic waves (light) only when electrons jump among orbits, emitting specific colors of light. This explains the spectra of atoms.

Common to all these resolutions is *quantization* involving Planck's constant *h*. Planck's constant is a measure of the fundamental "graininess" of the universe at small scales. Planck's constant is very small (about 10^{-33}, or 1/1,000,000,000,000,000,000,000,000,000,000,000 in the standard meter-kilogram-second system of units). For that reason, the effect of quantization is noticeable only at the atomic scale and smaller. If Planck's constant were truly zero, then the universe would be continuous and classical physics would hold. But *h* is not zero, and that makes all the difference. ■

Essential Reading

Gribben, *In Search of Schrödinger's Cat*, Chapters 3–4.

Han, *The Probable Universe: An Owner's Guide to Quantum Physics*, Chapters 1–3.

Wolf, *Taking the Quantum Leap*, Chapters 3–4.

Suggested Reading

Pagels, *The Cosmic Code*, Part I, Chapter 4.

Spielberg and Anderson, *Seven Ideas that Shook the Universe*, Chapter 7, through Section D4.

Questions to Consider

1. What do Bohr's atomic model, Planck's resolution of the ultraviolet catastrophe, and Einstein's explanation of the photoelectric effect all have in common?

2. If quantization is such a basic feature of the universe, why don't we notice it in our everyday lives?

3. Speculate on what life would be like in a universe in which Planck's constant was much larger—so much larger that the minimum energy for

a photon of visible light was about the same as the energy of a tennis ball just after being served.

The Photoelectric Effect

This experiment consists of an evacuated glass tube (1) containing two metal electrodes (2 and 3). One electrode (2) is connected to the positive terminal of a battery. The other electrode (3) is connected to the negative terminal of the battery, by way of a meter. Because of the gap between the electrodes, no electric current can flow.

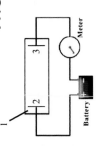

Light incident on the metal electrode may cause electrons to be ejected. The electrons are then attracted to the positive electrode giving rise to an electric circuit. The meter registers this current.

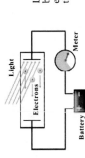

If the frequency of the light is too low (too long a wavelength, or of a color toward the red end of the spectrum), then no electrons are ejected, no matter how intense the light is.

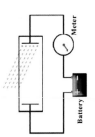

Enter the Quantum
Lecture 17—Transcript

We've come a long way now in our search for the structure of matter, a long way from Democritus in 400 B.C. proposing that the world was ultimately composed of indivisible atoms. Now, we're sitting at about the turn of the twentieth century, plus or minus ten years, depending on which phenomena we're talking about, and we know that there are atoms. Unlike Democritus suggested though, they're not indivisible. We've already seen subatomic particles. We've seen that the atoms have nuclei; they have electrons going around those nuclei. So we know quite a bit about the structure of matter. But classical physics—the physics of Newton, the physics of thermodynamics, the physics of electricity and magnetism that was well known by the end of the nineteenth century—is in big trouble at this point. And we have already three reasons why it's in big trouble.

First of all, we've got these atoms that shouldn't even exist. The electrons spiraling in their orbits around the nuclei of Rutherford's atom should, according to Maxwell's theory of electromagnetism, be emitting electromagnetic radiation. They should lose energy. As a result, they should spiral into the nucleus. That should take place in a fraction of a second, and we shouldn't have atoms at all.

Even if, for some reason, we could have these atoms, we have no explanation for the fact that atoms emit particular, discrete spectral lines, that is, particular colors. They don't emit a whole rainbow of colors. They emit particular spectral lines, unique to each different element. No explanation for why this is.

And finally, we have the ultraviolet catastrophe, which I discussed in the last lecture also, which suggests that all of us, or any hot, glowing object—and we, ourselves, are at least warm glowing objects—should be emitting infinite amounts of ultraviolet energy, and clearly that's absurd and doesn't happen.

I want to add one more insult to classical physics. And I want to go over it in a little bit more detail because its explanation is actually one of the clearest instances of why we must have quantum physics.

Hold on. By the end of this lecture, we'll have resolved all these crises, but we'll have resolved them by introducing an entirely new worldview—the worldview of quantum physics. So, let's take a look at this phenomenon that, again, adds another insult to classical physics.

It's called the *photoelectric effect* and it's a phenomenon that was studied since the late 1800s, since about the 1880s. And in the photoelectric effect, we do a very simple thing. We take a glass tube; put two metal electrodes in opposite ends of it—one of them a flat metal surface at least that we can expose to light. We connect the apparatus up to a battery and a meter to tell us if electric current is flowing. And we shine light through the glass tube onto one of the metal surfaces.

So what happens? We shine in the light. The light hits the metal surface. And, lo and behold, electrons are ejected from the metal surface. And if we have the battery connected up, the electrons can flow through the empty space of the evacuated tube to the other electrode and then through the surrounding wires and through the meter, and they constitute an electric current. And we can measure how much current, by measuring the flow of current with the meter that I've indicated in the picture here. We can also adjust the strength of the battery, that is its voltage, and see what effect that has on the electrons, and that way, we can actually measure the electron's energy.

So this is the photoelectric effect. And it was well known by the turn of the twentieth century. In fact, it's the basis of an early version of the so-called electric eye, basically a light sensor—the kinds of things that are used in automatic door openers. They're used in your garage door openers to sense whether there's an object in the way of the garage door opener. They're used in a whole variety of applications where we want to sense whether light is there or not. This particular kind of device, a glass-evacuated tube with metal electrodes, has been supplanted by solid-state devices, semi-conductor devices that work much more reliably and are much simpler and cheaper to build.

But nevertheless, this constitutes an early technological device of some importance for sensing light electrically. That's beside the point here, though. We want to ask the question: How does classical physics explain the photoelectric effect? Well, classical physics can explain the photoelectric effect. The photoelectric effect is the effect where electrons are ejected from a metal surface when light strikes the surface.

How does classical physics explain that? Well, in classical physics, light is an electromagnetic wave. An electromagnetic wave consists of electric and magnetic fields. An electron is an electrically charged particle. It experiences forces when there's an electric field. So here comes a light wave, an electromagnetic wave, onto this metal surface. Metal is characterized, by the way, by its electrons—at least some of them—being entirely free to move throughout the metal. They're not tied to particular atoms, like they are in insulating materials. That's what makes a metal different.

So, here are these electrons. Here comes the light. The electrons experience the light and they begin oscillating up and down, vibrating up and down in response to the vibrating electric field of the light wave. Remember, a light wave: changing electric fields make changing magnetic fields, and so on. So that's what the electrons do. And they gain energy from the light wave. And eventually, they should gain enough energy that they can break the electrical bonds that hold them to the metal. And they can come flying out. So far, so good.

The problem is this: In a wave, energy is spread out over the entire wave. Think of an ocean wave coming at you on the beach. The energy of that ocean wave is not concentrated in one place. It's spread over the entire wave. And so a single electron, which is a tiny, tiny entity is intersecting only a tiny, tiny fraction of the total energy carried by that wave.

It should take a long time for the electron to build up the amount of energy that would be needed for it to eject itself from the metal. And you can calculate that time for typical conditions. And it's a long time. Minutes, hours sometimes, depending on the strength of the light. So that's what classical physics says. It says electrons should be ejected from the metal surface and they should be ejected after a while—after the time it takes them to build up the right amount of energy. Now if we turn the intensity of the light up, that time should go down a little bit.

Classical physics says something else. It says the color of the light shouldn't matter. Who cares what color the light is? If the light is redder, the electromagnetic wave is oscillating more slowly, but it's still carrying the same energy. And the electrons should take about the same amount of time to build up the energy needed to eject themselves. So the color of the light shouldn't matter. And it should take a long time for the electrons to build up the requisite energy.

So that's how classical physics explains the photoelectric effect. The electrons absorb energy from the light wave. And eventually, they're ejected. And the color of the light, really, is irrelevant to that process.

What really happens? What really happens in the photoelectric effect experiment is somewhat different. You turn on a light and, bingo, electrons come out immediately, even if the light is quite dim. Turn on a light. Bing. You've got electrons coming out immediately. Why?

Furthermore, color does matter. If you have light that's toward the blue end of the rainbow spectrum or into the ultraviolet, you get electrons coming out, and they come out with substantial energy, and the bluer you go, or the more into the ultraviolet you go, the more energy the electrons have coming out. You can measure that by varying the voltage on the battery and seeing how much electric field it takes to stop the electrons.

So you can measure the electron's energy, and you find that they have more energy as you make the light bluer. Furthermore, if you make the light redder and redder, they come out with less and less energy until you reach a certain color, as you move toward the red, for which no electrons are ejected at all. So if I change the light color to a color that's red enough—no matter how intense the light—no electrons come out whatsoever. Completely at odds with the classical prediction that the color shouldn't matter. Because all the color is determining is the oscillating frequency of the electromagnetic wave. It has nothing to do with the energy.

So we have two real contradictions here with classical physics. Classical physics says it should take a long time before those electrons come out. And classical physics says the color shouldn't matter.

We do the experiment, and we find the electrons come out immediately. And that color matters a lot. If the color is too red, they don't come out at all. If it goes toward the blue, they come out, and they come out with increasingly higher energy. So we got another big problem here.

Now we've got four problems. We've got the existence of atoms. They shouldn't exist because the electrons should spiral into the atoms, into the nuclei, emitting electromagnetic radiation. We've got the spectra of atoms—the individual, discrete colors of light they emit—no explanation for that. We've got the ultraviolet catastrophe that we should all be glowing with infinite amounts of ultraviolet energy. And now we have the photoelectric effect that says, well, this phenomenon should occur, but it shouldn't occur, according to classical physics, in the way it actually does occur. It should occur slowly and in a color-independent way, when in fact, it occurs instantaneously almost and in a way that depends very significantly on color.

How do we resolve these problems? This is the big crisis for classical physics around the turn of the twentieth century. And I'm going to go through the resolution of this crisis in pretty much historical order. That's not the most logical progression, pedagogically, but I think it's important to do it historically because it establishes when concepts were introduced into the idea.

In a way, the most difficult of these problems to understand conceptually is the problem of the ultraviolet catastrophe. In fact, I gave you no hint about why classical physics predicts that this occurs. And I'm going to give you now just the briefest explanation of why the ultraviolet catastrophe—that sense that a hot glowing object should emit infinite energy as you move toward the ultraviolet end of the spectrum—and the reason is a little bit subtle. But the reason is this: What's causing this energy to be emitted from a hot glowing object? Ultimately, it's vibrations of the individual atoms in it emitting electromagnetic waves. Those atoms can vibrate, according to classical physics, with any vibration frequency they want. And there's a law in classical physics—classical thermodynamics, the study of heat—that says when a system has different ways of having energy, it will share all those ways equally. That's called the equipartition theorem.

And it turns out if you look at the ways an atom can vibrate, there are more and more ways, as you get to higher and higher frequency bands. And so if all those share energy equally, two things should happen. First of all, there should be an infinite amount of energy shared by all these possible ways of vibration and most of that energy should be in the very high-frequency range, which, in those days, was known as ultraviolet. That, in a nutshell, is why classical physics predicts the ultraviolet catastrophe.

Well, Max Planck, in 1900, a German physicist, was able to resolve the ultraviolet catastrophe in a way that to him wasn't particularly satisfying. What he did was look at the equation that predicted the ultraviolet catastrophe and he basically said, "Can I fudge this equation in a way that will prevent the ultraviolet catastrophe from occurring or from being predicted?"

And he found a fudge factor he could put in the equation that would get rid of the ultraviolet catastrophe and would, in fact, make the spectrum of light emitted by a hot glowing object, that continuous range of colors emitted by a dense, hot, glowing object, as opposed to a single atom that emits a discrete spectrum. He could match the spectrum of a hot glowing object, which, as we saw last time, gets brighter and also shifts from the red toward the yellow toward the green, and so on, as the temperature goes up.

He could match that with his fudge factor. And after introducing his fudge factor, he looked at what the fudge factor meant physically. First, he put it in mathematically, and then he looked at what the fudge factor meant physically. And what the fudge factor turned out to mean, physically, was it put a restriction on the amount of energy a vibrating atom could have, basically.

And what it did is this: It said, suppose I have an atom vibrating with a certain frequency, back and forth—100 times a second, a million times a second, a billion. I don't care however many times a second that atom is vibrating back and forth—and, therefore, emitting electromagnetic waves at that frequency.

There is a minimum amount of energy that that atom can have at that frequency, a very minimum amount—it's a tiny amount. But it still can't go any smaller. You can't have half that amount. You can't have three quarters of that amount. You can't have five eighths of that amount. You can't have one-and-a-half times that amount. You could have twice the amount or three times or ten times, or one time. But one time is the minimum amount of energy you could have in that particular vibration frequency.

And what that does for the ultraviolet catastrophe is to take those very high-frequency vibrations, the ones that were contributing the infinite energy in the ultraviolet and give them so much energy, because the energy is proportional to the frequency of vibration. The bigger the frequency of vibration, the bigger that minimum energy. And I'll show you that in an equation in just a moment. The bigger the frequency of vibration, the bigger that minimum energy.

For the very highest frequency vibrations, Planck's equation predicts the energy required to get them vibrating is so big that it's very unlikely that it will happen. So the high-energy vibrations don't occur. There isn't an infinite amount of energy. Most of the energy is not in the ultraviolet but lower down, in the visible light for normal temperatures. And the ultraviolet catastrophe goes away.

The formula that Planck used is actually very simple. And, again, this is a non-mathematical course. But it's a very simple formula. And I want you to just appreciate briefly, what it says. So, here's Planck's formula.

What's rescuing us here from the ultraviolet catastrophe is this idea of energy quantization. The energy is now coming in little discrete bundles—a certain, minimum amount for a given frequency of vibration. Here's Planck's equation: It says $E = hf$. It's that simple. E, energy, is something h, some number h, which I'll talk about more in a minute, times f, frequency. That's all it says. E is hf.

What do these things mean? f is the frequency of these atomic vibrations, or as we'll soon see, the frequency of a light wave, or some other process that has a frequency. Frequency means how many times a second does this occur? How many times a second does the atom vibrate? How many times a second does the electron go around the atom? How many times a second does the electric field of the light wave reverse direction? That's what frequency means.

E is the energy, and E, the number you're going to get by applying this formula, is the fundamental unit, the minimum amount of energy, the little bundle of energy that goes with that particular vibration frequency f, for a given vibration frequency f. You apply this formula to f. You multiply f by this number h and you get the minimum amount of energy you're allowed to have at that vibration frequency.

What's h? h is a new constant of nature and a very important one. h is the constant of nature of the twentieth and twenty-first century. It's the new thing that's been introduced. What is h? It's this proportionality that tells you how small or how big are these little, fundamental units of energy that you can have associated with a given vibration. And the name, *Planck's constant*, obviously honors Planck's discovery of this equation.

So there is a new constant of nature in quantum physics. It's called Planck's constant. It's given the symbol h and it represents the proportionality between the frequency of vibration of some kind of system, like a vibrating atom or a circulating electron, or whatever, and the energy associated with that. Planck's constant, very simple equation; energy is Planck's constant times frequency.

So there's a new link here and the link is between energy and frequency. That wasn't there before. Frequency of vibration before was totally independent of energy. In classical physics, you give me a vibrating atom, or a vibrating mass attached to a spring, or a car with bad shock absorbers that's bouncing up and down and it can have any vibration frequency.

No longer true in quantum physics. It can have vibration frequencies and the minimum energy it can have with that vibration frequency is h times f. It can have $2hf$, $3hf$, $101hf$, $1,000,039hf$. But it can't have $.5hf$, or $1.5hf$, or Πhf, or $2,493.97hf$. It has to have an integer number of that fundamental unit of energy; $E = hf$.

So what has Planck done? He's quantized energy. He said energy comes only in little discrete bundles. How big are the bundles? That depends on the vibration frequency associated with this process. The bundles are of size hf. Great. We've got quantization. So, that rescued us from the ultraviolet catastrophe. That's in 1900.

The next rescue came in 1905, that famous year when Einstein published a number of papers, three of which were seminal. Remember, one of them convinced people atoms existed. The other one was special relativity that we spent quite a few lectures on.

The third seminal paper was Einstein's explanation of the photoelectric effect. And Einstein proposed the following. He said, "I think that the energy in light waves is quantized; comes in little bundles. It isn't distributed evenly all over the place, like a wave. It comes in little bundles. It's like it was a particle."

And that explains the photoelectric effect, because in the photoelectric effect we turn on the light and to our surprise, electrons are ejected almost immediately. Why does that happen?

"Well," says Einstein, "it happens because, although in classical physics it would take a long time to extract this spread-out energy from the wave, in fact, the energy isn't spread out in the wave. It's distributed in little discrete bundles here and there throughout the wave. And there's a good chance that as soon as we turn on the light, one of those little bundles will hit an electron and eject it."

So that solves the first part of the photoelectric effect crisis. It's tells us why the electrons are ejected immediately. If the energy is quantized in these little bundles instead of being spread out all over the light wave, we're okay because we can eject electrons immediately, as soon as one of these little bundles hits them. Those little bundles are now called *photons*—little particle-like things of light, little bundles of light energy. (Photon: That's the word. It's a word you hear a lot these days because we have a lot of technological devices using photons.)

"Now," says Einstein, "how big are these photons? What's the energy of the photons?" Well, remember what a light wave is? It's a vibrating system of electric and magnetic fields. It's vibrating with some frequency. For an A.M. radio wave, that's about a million times a second; for an F.M. radio wave, it's 100 million times a second. In your microwave oven, it's 2.4 billion times a second. For visible light, it's about a 1 with 14 or 15 zeros after it times a second. But it's a vibration at a certain frequency.

"What's the energy associated with the photons, with the little bundles of that vibration?" says Einstein. The answer: Let's go back to Planck's equation. It's exactly the same. It's $E = hf$. That is the minimum amount of energy you can have in light of frequency f, says Einstein; $E = hf$.

And that explains the second part of the photoelectric effect, because it takes a certain minimum amount of energy to get an electron ejected from that metal, to tear it free from the electrical forces that are holding it in the metal. And if the light is too red—and red corresponds to lower frequency, blue and ultraviolet, to higher frequencies—if the light is too red, f is too small and hf, the bundle size, the amount of energy, is simply too low to eject the electrons, no matter how bright we make the light. A bundle hits an electron and it just can't knock it out, no matter how bright we make it, no matter how many bundles we have.

So Einstein does the same thing Planck does, but Einstein does it for light waves. He quantizes the energy in a light wave. He said the energy in light comes in little discrete bundles, called photons and they have, lo and behold, an amount of energy, given by $E = hf$. And it explains both aspects of the photoelectric effect. Well, that took care of the photoelectric effect. It took care of the ultraviolet catastrophe. We're still left with the problems of atoms—two problems of atoms. One, they shouldn't exist because the electron should spiral into the nucleus. Two, they give off these discrete spectra, and we have no explanation for that.

In 1913, the Danish physicist Niels Bohr, who was one of the great, towering figures of twentieth century physics, proposed a new model for the atom, a new picture of the atom, an elaboration on Rutherford's picture of the miniature solar system. And you're going to find Bohr's proposal a little bit unsatisfactory. And it is. But it held sway for about 10 years, from 1913 until it was supplanted (10 to 15 years until it was supplanted) by the full theory of quantum physics, which we'll get to in a few lectures.

And Bohr said the following. He said, "I'm going to apply quantization"—this idea that things are quantized—"to the atomic orbits, to the electron orbits, instead of the electron in the atom being allowed to be anywhere, any distance from the nucleus, like a planet in the solar system can be, presumably." "I believe," said Bohr, "there are only certain discrete orbits allowed; only certain distances from the atom are allowed." Now, you might expect me to say Bohr's quantization condition was the energy of those electrons is h times their frequency. But it was a little more complicated than that.

The formula actually involves not the energy directly, but something called the *angular momentum*, which is a measure of sort of rotational motion. And that was what Bohr said was quantized. And he said the angular momentum is simply Planck's constant, divided by something dealing with how fast the electrons are going, times their mass. We don't need to get into the details of the formula. If you want to know it, it's L equals h divided by twice the momentum, mv, also called p—twice the momentum of the particle ($L = h/2p$). But that's not important.

What's important is Bohr quantized the electron orbits. He said only certain electron orbits are allowed, and they are described by a quantization condition that quantizes this property, angular momentum, according to some formula Bohr gave. And no other electron orbits are allowed. And the important point is the quantization condition involves Planck's constant h, once again.

Bohr said something else. He said when an electron is in one of these orbits, it doesn't radiate electromagnetic waves. Why? Sorry. No explanation. This was just a postulate Bohr gave. He said electrons in these allowed orbits do not radiate electromagnetic waves. That solves the problem by fiat of the fact that atoms—classical physics says—atoms shouldn't exist. In Bohr's theory, atoms do exist, because once an electron is in one of these allowed orbits, it doesn't radiate electromagnetic waves.

But atoms can radiate electromagnetic waves. If an electron is in a higher orbit and jumps down to a lower orbit, which it's allowed to do, and in the process, its energy changes by a certain discrete amount, because if the angular momentum and the orbits are quantized, so is the energy, although the quantization formula for the energy is a little more complicated. But the energy is also fixed.

Each orbit has a different energy, and if the electron jumps from a higher orbit to a lower orbit, a certain, precise amount of energy is emitted. Electromagnetic radiation occurs then at a certain precise amount of energy. In fact, electrons do try to fall eventually down to the lowest orbit available to them. But the lowest orbit isn't crashing all the way into the nucleus. There's a minimum orbit they can have—a minimum, quantized amount of this angular momentum and, therefore, of energy.

Now, let's go back to the energy quantization formula for photons, for light, $E = hf$. Okay, what does this say? Well, now we have a certain discrete amount of energy that an atom can emit. As an electron jumps from one level to another, maybe if it jumps from a higher level down, a different amount of energy can be emitted. There are certain discrete amounts of energy that can be emitted by an atom as electrons jump among these orbits.

Well, if we have a certain, discreet E and we have Planck's constant, that means there are only certain, discrete frequencies, or equivalently, colors of light that can be emitted by atoms. And that phenomenon explains the spectra of atoms.

And in fact, Bohr's quantization condition on what orbits were allowed not only explains the existence of atoms and the fact that we should expect discrete spectra, in fact, it predicts spectra that are exactly what we observe when we take light from an atom, run it through a prism to disperse it into its separate colors, and measure carefully what the frequencies or wave lengths of those colors are. Agrees beautifully with Bohr's theory.

So, although Bohr's theory is built on this funny, ad hoc assumption that the energy levels of the atom are quantized and the atoms simply don't radiate electromagnetic waves as long as the electrons stay in their allowed orbits, (and no good physical reason given for that), nevertheless, this ad hoc theory explains a great deal. In particular, why atoms exist, what their spectra are, and in fact, the actual experimentally determined numbers for the wavelengths of the emitted spectra.

What do all the resolutions of all these crises have in common? They have one thing in common. They have Planck's constant h. And what Planck's constant h is doing, in each of these crises—the ultraviolet catastrophe, the photoelectric effect, and the Bohr atom—is causing a quantity that in classical physics was believed to be continuous to be allowed to take on any value.

It's causing some quantity—energy in a vibrating system in the case of the ultraviolet catastrophe, energy in a light wave in the case of the photoelectric effect, angular momentum and, therefore, energy of an electron in orbit. It's quantizing those quantities. It's saying they cannot take on a continuous range of values. They can only take on certain, discrete values, which are predicted by the equation $E = hf$ in the simple case of atomic vibrations and light waves. A little more complicated formula in the case of the atoms, but they all involve Planck's constant. And they all involve the statement that, "Nope, you can't have a continuous range of things. You can only have certain, discrete amounts."

It's again like my egg analogy. You go to the grocery store. And if you didn't know that eggs came in discrete amounts, you could say, "Oh, I'd like a tablespoon of eggs for this recipe and 2.5 teaspoons for this recipe, and 1.4379 cups of egg for this recipe." You can't do it. Eggs don't come that way. They come in discrete amounts. And when you go in the grocery store and see the eggs in cartons, that becomes obvious to you.

The reason quantum physics isn't obvious to us—the reason these discrete energy levels, discrete amounts of energy in the light wave, discrete energy levels of the atom, discrete energy levels of the vibrating atoms in the ultraviolet situation—the reason we don't see that discreteness, the reason we think everything is continuous is because Planck's constant is a tiny, tiny, tiny number.

Here's how tiny it is. Let's take a look. Here's Planck's constant, rescuing us with energy quantization. But how big is Planck's constant? Well, it's tiny. It's 1 over 1 with 33 zeros after it, approximately. Now, I'm doing a very bad thing here, and I've marked my students down for this, because I haven't put any units on that. It doesn't really matter what the units are, this is approximate. But in the kinds of units we use to measure everyday objects, like feet and seconds and things, actually this is its value in the meter, kilogram, seconds, and feet system of units. In fact, it's about 6 times 10^{-34} in that system of units. But we don't need to worry about that. The thing is, it's a very, very tiny number by comparison with anything we do on the regular macroscopic scale of our everyday lives.

What does that mean? It means the gradations in energy levels, particularly in everyday objects. Here's a tennis ball. What if I start vibrating this tennis ball up and down? Well, if we really believe quantum physics, this can have only certain, discrete energies. But vibrating it twice a second, f is 2; h is this tiny, tiny number—2 times that tiny, tiny number is minuscule. The distance between allowed energy levels of this tennis ball is so minuscule, it might as well be perfectly continuous. That's why we don't notice quantization in the world of everyday experience. Even in the world the size of a bacterium or something like that, we don't notice quantization.

But when we get down into the atomic level, where the minimum bundle size of energy becomes comparable to the energies objects have—at electrons going around in atoms, vibrating atoms, the energy it takes to eject an electron from a metal surface in the photoelectric experiment—then we begin, for the first time, to notice energy quantization.

And the key, the essential thing that makes quantum physics different from classical physics is this notion that not only does matter come in discrete little hunks like electrons—we've already found that out—but so does energy, at least in some situations—the situations I've described here. There is a certain minimum amount of energy that you can have for a given frequency. Energy is quantized.

In essence, quantum physics is the theory that says, or the fact that Planck's constant isn't zero. If Planck's constant were zero, the minimum amount of energy we could have would be zero, and we could have any amount, running continuously from zero on upward. There really would be a continuum of energies. Electrons really could be anywhere in the atoms, and the atoms would emit light of all colors. There really would be a continuum of energy spread throughout the electromagnetic wave. And the minimum energy would be so small that it really would be zero. Essentially, that it really would take a long time for the electrons to gain that energy. There really would be ultraviolet frequency vibrations of arbitrarily small size, and there would be a lot of them, and there would be an ultraviolet catastrophe.

But no, it's the non-zeroness of Planck's constant that rescues us from all that. It's the non-zeroness of Planck's constant forcing on us a world in which energy comes in these discrete amounts that makes quantum physics different from classical physics. Quantum physics is the physics of h being not zero. It is, in a way, like relativity. If you think about relativity, our gut sense of the speed of light is it's really so fast, it's essentially infinite. And relativity is the theory that says, "No, it's not quite infinite. It's finite." If you let c go to infinity, relativity would reduce to classical physics, because we'd never be anywhere near the speed of light, if it were infinite. Relativity is really the theory that says the speed of light isn't what you think it is—infinite. It's a little bit less—186,000 miles a second.

Quantum physics goes the other way. Quantum physics is the theory that says, "Nope, Planck's constant, the number that sets the fundamental graininess, discreteness scale of the universe isn't zero, which would give us a continuous availability of energy. It's small, but it's not zero." And that quantization, that fundamental graininess of the universe, is going to be what leads to all the strange results of quantum mechanics that we'll now be exploring.

Wave or Particle?
Lecture 18

Light has, at least, some aspects of particle-ness. It comes in these little, discrete bundles that act pretty much like particles. They hit an electron, and they bounce it out of the metal. So, which is it—wave or particle?

Newton (in the mid-1600s) proposed that light consists of particles. He was able to explain the phenomena of reflection, refraction, and color using his particle model. Christian Huygens (1600) proposed an alternative: that light consists of waves. Thomas Young (1800) provided conclusive evidence that light is a wave. His double-slit experiment showed that light beams interfere, something that is possible only with waves. Maxwell in the 1860s stated that light was a wave. Einstein (in 1905) explained the photoelectric effect by proposing that light behaves as if it were a particle, in that light energy is concentrated in particle-like photons.

The *Compton effect* (1923) showed what happens when light (in this case, x-rays) interacts with electrons. Historically, the Compton effect was for many old-time physicists the final convincing evidence for the reality of quanta. Classical physics predicts that the electron should absorb energy from the light wave, then re-emit at the same frequency. Experiments show that the light scatters off the electron with lower frequency—just as if the light

Thomas Young.

were a beam of particles that interacts with electrons in the same way that two billiard balls collide. An incoming photon bounces off an electron, giving up some of its energy and lowering its frequency (since $E = hf$).

This presents a quantum quandary. If light consists of particles, how can we explain the results of two-slit interference experiments? Try to look at the process in more detail:

- Which slit does a photon go through? Try to find out by covering up one slit—and the interference pattern disappears! How did the photons going through the other slit "know" about the first slit being closed?

- Try putting photon detectors at each slit, to "catch" photons in the act of going through. Again, the interference pattern disappears.

- Dim the light so that only one photon is present at a time. Still, an interference pattern gradually builds up. Somehow, each photon must "know" about both slits.

So is light a wave or a particle? The quantum answer: It's both! If you don't try to detect it, light acts like a wave and exhibits interference effects. The behavior of the waves is governed by Maxwell's equations. A wave is a spread-out thing, and it can "sample" both slits. Thus, one way to answer the question of which slit the photon went through is to say that it went through both. Close one slit, and the wave can't interfere with itself on the other side, so the interference pattern disappears. If you detect light, it behaves as if it consists of particles (e.g., ejection of an electron in a light detector, darkening of a grain on photographic film, and so on).

So is light a wave or a particle? The quantum answer: It's both!

There is a relation between wave and particle, but it is only a statistical one. The *probability* of finding a photon is related to the wave amplitude; the stronger the wave is at some point, the more likely you are to find a photon there. Thus, the wave picture predicts that waves should be strong at certain points on the screen in a two-slit experiment, and quantum physics predicts that's where you are most likely to detect photons.

The *wave/particle duality* at first seems to be a contradiction. How can light be both particle and wave? There's no contradiction, according to Bohr's *Principle of Complementarity*. Rather, wave and particle aspects of light are complementary. Both are needed for a full description of the behavior of light. The two aspects cannot manifest themselves together at the same time. If you do an experiment that involves wave aspects of light (e.g., an interference experiment), you'll find that light acts as if it were a wave. If you do an experiment that involves particle aspects (e.g., a photoelectric experiment), you'll find that light acts as if it were a particle.

Is quantum physics absurd? You may think so, but keep in mind that the theory has been remarkably successful in describing the world at the atomic and subatomic levels. Even some of the pioneers of quantum physics had similar doubts. ∎

Essential Reading

Gribben, *In Search of Schrödinger's Cat*, Chapter 5.

Hey and Walters, *The Quantum Universe*, Chapter 1.

Wolf, *Taking the Quantum Leap*, Chapter 8.

Suggested Reading

Lightman, *Great Ideas in Physics*, Chapter 4, through p. 210.

Lindley, *Where Does the Weirdness Go?*, "Act I," pp. 3–86.

Pagels, *The Cosmic Code*, Part I, Chapter 5.

Questions to Consider

1. You cover first one slit, then the other, in a double-slit apparatus; in each case, you record the pattern that appears on the screen. If you then open both slits, will the resulting pattern be the sum of the patterns you see with only one slit open? Explain.

2. A friend who knows nothing about physics asks you whether light is a wave or a particle. How do you answer?

Wave Interference

A

The two-slit system is an experimiment designed to determine the nature of light. If light were a particle, it would act as it does in *A*. Here, an incoming beam of light particles strikes a barrier with two slits in it. Only those particles that exactly lined up with the slits would pass through the barrier and would then strike the screen at two places directly opposite the slits.

B

If light were a wave, it would act as it does in diagrams *B* and *C*. In *B*, we show only the light passing through the left slit. The light wave hits the barrier and spreads through the left slit in concentric circles.

C

In *C* we see the light waves pass through both slits and interfere with each other. The thick lines mark regions of constructive interference, where wave crests meet crests and troughs meet troughs. Bright bands appear where these lines meet the screen.

Strangely, light has both wave and particle aspects. If we cover up one of the slits in the system, light acts as it does in diagram *A*. The light passes through the slit and hits the area of the screen directly opposite it. But when both slits are open, light acts as it does in diagrams *B* and *C*, forming an interference pattern.

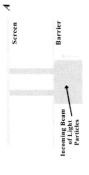

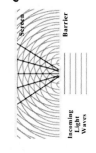

Wave or Particle?
Lecture 18—Transcript

Lecture Eighteen, whose title is, once again, a question—"Wave or Particle?" Here we focus on what, in quantum mechanics, is very much a schizophrenic nature for light. Why? Well, think back to the last lecture, when we saw how Einstein resolved the photoelectric effect problem. And the photoelectric effect, you'll recall—light hitting a metal surface ejects electrons—classical physics says that should happen, but it should happen slowly and in a way that's independent of color. Actually it happens immediately, in a way that depends strongly on color.

And Einstein resolved that by saying the energy in light is contained in little bundles called photons. There's a minimum size for a photon, for a given frequency of light; it's given by Planck's formula, $E = hf$. We're not going to worry about that so much today. We're going to worry about the fact that there is a minimum photon size, a little bundle that light energy comes in, these discrete little bundles. It's almost as if light were a particle.

And that's the reason for my title—"Wave or Particle?" Is light a wave? Well, we've been assuming that all through out this course. But now, along comes Einstein, of all people, on whom relativity theory is built on the idea of light waves after all. And Einstein himself is telling us, no. Light has, at least, some aspects of particle-ness. It comes in these little, discrete bundles that act pretty much like particles. They hit an electron, and they bounce it out of the metal. So, which is it—wave or particle?

We want to explore that today. And to begin exploring it, we want to look briefly at a history of physicists' understanding of light through the centuries. If you're going to explain why, you've got to explain a number of its phenomena. In particular, you've got to explain the fact that it reflects—that is, it bounces off shiny surfaces like metals and mirrors. You've got to explain the phenomenon of refraction, whereby light bends when it goes from one transparent material, like air to glass, or the cornea of your eye or whatever. That's what makes optical instruments possible; it's what lets our eyes focus on things; it's what lets eyeglasses correct your vision, and so on. We've got to explain those phenomena.

We've also got to explain phenomena like color. Isaac Newton, in the late 1600s, laid out a theory of optics in which he assumed that light consisted of particles. And by assuming that different colors corresponded to different kinds of particles and that reflection was obvious—it was like a ball bouncing off a wall—that was easy to explain. Refraction was fairly easy to explain. He explained that by a change in the speed of the particle as it went into a different medium, (although he got it backwards from the way it really is), but he posed a very successful theory of light.

Christiaan Huygens, who was approximately a contemporary of Newton's, had a different theory of light. He said, "I think light is a wave." And for roughly 100 years or so, there was debate about whether light was a wave or a particle. And then, along came a series of experiments that were done in the very early 1800s by Young. And Young did a very simple experiment, which proved conclusively that light must be a wave. So, from about 1801, 1802 on, there was pretty conclusive proof that light behaves like a wave.

Let me show you Young's experiment, because it's a very important one. And it's one that's going to concern us again and again, as we think about quantum physics. Young did a very simple thing. He took some kind of barrier and he made two small holes in it and he made those holes very close together. They don't have to be microscopically close, by the way. You can actually do it, this experiment, yourself, by taking a microscope slide, or some other piece of glass, paint it black, take two razor blades, and hold them right next to each other so the blade ends are very close. Scratch two slits in the black. And you can actually do this experiment, especially if you have access to a laser, which makes it easier.

Young didn't have lasers, but he was able to do this with sunlight actually coming into his laboratory. So, here we have a barrier of some sort, with two slits in it. That's the crucial thing I'm going to talk again and again about. A two-slit system. It's simply a system, with two little pinholes, or two little long slits, and it's going to let light through, and it's otherwise opaque to light some ways away from that barrier. We have a screen, and what we're going to do is shine light on the barrier, in this case, from below. And we're going to ask: What do we see on the screen now, if light consisted of particles?

So here comes a beam of light particles. What would happen? What would happen is all the particles would be blocked, except where they hit the barrier and then the light would go through, basically in straight beams. And we would see on the screen two bright spots opposite the two slits. That would be good evidence that light consists of a beam of particles. Very clear. Very unambiguous. The particles are stopped, except at the two slits. They go straight through at the two slits, and they hit the screen directly opposite the slits. And we would see two bright spots on the screen. And that would be what would happen, if light consisted of a beam of particles.

What, on the other hand, happens if light consists of waves? Coming in, here come the incoming waves. I've represented with these lines the crests of the waves, say places where the waves are highest. In between, they're lowest. And they undulate like that. So I've just drawn you the wave crests, coming in here. What happens? Well, I'm not going to describe, in detail, why this happens. But what does happen when light goes through a very small barrier like this? And you may have seen this, if you've seen ocean waves going through, say, a gap in a breakwater or something like that. You'll see them spreading out, basically, in semicircular wave fronts after the barrier.

So what happens is light goes through these slits. The light waves and each slit, in a sense, acts like a source of circular waves that then spread outward from the slit. So here, from the left hand slit, are the waves that emerge. They form a sequence of semicircular wave fronts that spread out until they hit the screen. And if that were all, we would see the screen illuminated, essentially, uniformly by that light. But, we've got another slit. We've got the right hand slit. And it also acts like a source of semicircular waves. And so now, we have these two patterns of semicircular waves.

By the way, we could do the same thing with water waves, or sound waves, or other waves, and similar phenomena occur. In fact, we could do it if we had two, instead of slits, two sources, that were oscillating right in unison. But making it with two slits makes sure that the light waves are, in fact, oscillating together.

So basically the slit system has simply provided us with two sources of waves doing the same thing at the same time, but they're separated in space.

And you can already see in this picture there's an interesting kind of pattern here. There are places, for example, where wave crests meet. There is such a place. And in this dark spot in between there's a place where two wave troughs meet. Here's another place where wave crests meet; another place where wave troughs meet.

On the other hand, somewhere about here is a place where a wave crest and a wave trough meet. Now, if you think back to our understanding of the Michelson-Morley experiment, that experiment worked by using a phenomenon called wave interference. The fact is, two waves can be at the same place, at the same time, (unlike two particles), and when they are, they interfere. They may interfere constructively, if two crests coincide. They may interfere destructively, if a crest and a trough coincide.

So at the point I've got marked right now, where a crest and a trough coincide, there is destructive interference. And there is essentially no wave there, at this point and in fact, at any point along here. This line, something like this. Wave crests are meeting wave crests; wave troughs are meeting wave troughs; and we have much bigger waves. The waves are enhanced by constructive interference.

So what's this look like? Well, along all those lines (and some more you could draw further out), wave crests are meeting and wave troughs are meeting, and we have constructive interference. In between, we have destructive interference. At the midpoints in between, the interference ranges somewhere between constructive and destructive, changing gradually.

So what are we going to see on the screen? In this case, in this particular case, we're going to see on the screen alternating light, dark, light, dark, light, dark bands, even though there are only two slits. The waves come through those slits. They form these semicircular wave fronts beyond the slits and we see this interference pattern, consisting of light, dark, light, dark, light, dark bands. It's a lot like the bands we saw in the Michelson-Morley experiment, which were caused by a similar but not identical kind of phenomenon. There we set two light beams in different paths and we combined them. Here we have two light beams forming these two sets of interfering, concentric wave fronts in that region between the barrier and the screen. And what we see on the screen is an interference pattern, consisting of light, dark, light, dark, light, dark.

Now, I've got an actual photograph to show you of what such an interference pattern looks like. This was taken by putting a camera film in place of the screen in a two-slit interference pattern. And what you see here are, in fact, two patterns, one in red and one in green. And if you think about that picture I showed you, of how closely those light, dark, light, dark bands are, it's going to depend on things like how far apart the slits were and what the wavelength of the light is in relation to that separation of the slit.

Here I shined a red laser and a green laser through the same double-slit system. And you if you look carefully, you'll see that the spacing of the interference fringes, the bright versus dark areas, is different for the red and green light. And that's because of the different wavelengths of red and green light. Red light has a longer wavelength, a lower frequency. Green light has a shorter wavelength, a higher frequency. So, there's an actual set of interference patterns, taken by putting a camera film in place of the screen of a double-slit experiment and shining both red and green laser lights, simultaneously, through that system. So, we have a very clear experiment that shows that light behaves like a wave. That experiment was first done—it was difficult to do in the 1800s—it was done by Young, and he concluded that light must be a wave.

And then we had Maxwell coming along in the 1860s and showing what kind of wave it is. Light is, as we now know, an electromagnetic wave consisting of electric fields and magnetic fields, continually regenerating each other. Because a changing electric field makes a magnetic field, and a changing magnetic field makes an electric field, and so on. So we're very comfortable with the idea that light is a wave. And that idea figured importantly in Einstein's development of special relativity in 1905.

But now, in the next step in the evolution of our understanding of light, we have Einstein again, in the same year, in 1905, saying, "Wait a minute. Light is a particle." Or at least, it comes in these little, discrete bundles called photons. The wave energy is not spread out. It's bundled in these little bundles that are like photons. They're called photons and are acting like little particles like little billiard balls that can hit an electron and eject it from the metal in a photoelectric effect experiment.

So it's Einstein who tells us light seems to be a particle. And now in 1905, light has acquired this schizophrenic personality. Which is it? Is it a wave or is it a particle? In the photoelectric effect, that explanation requires a particle-like nature for light. In this double-slit experiment, interference is something that happens to waves. It doesn't happen to particles. Two particles can't be in the same place at the same time. You'd get a very different pattern on that screen, if light were a beam of particles. You'd get the interference picture. So, which is it?

I have to describe one other experiment, which is done a little bit later, in the twentieth century. In 1923, there was an experiment done by Compton, and it convinced any skeptics who didn't believe in Einstein's photons. And the experiment consists of taking a basically free electron that isn't tied very strongly to anything—it's actually tied loosely to an atom—and hitting it with a very high-energy photon, an x-ray photon. Remember, we have a whole range of electromagnetic waves, from low frequency radio waves, which, now that we've quantized things, have very low-energy photons—because energy is Planck's constant times frequency—up through infrared, and visible light, and ultraviolet, and then x-rays. Beyond those are gamma rays.

So we're going to hit an electron with a high-energy photon, an x-ray photon, and we're going to see what happens. Or, when x-ray light, with x-rays, classical physics says the following thing ought to happen. Here's my electron. Along comes an electromagnetic wave. This is the same thing that happened classically, in the photoelectric effect experiment. Along comes the electromagnetic wave. The electromagnetic wave sets the electron into vibration. The electron, in that process, is taking a little bit of energy away from the electromagnetic wave. So the wave goes on weaker. Now the electron itself is an oscillating, vibrating, accelerating, electric charge. And we know from Maxwell's electromagnetic theory that it will, itself, be a source of electromagnetic waves. So what the electron will do is radiate out electromagnetic waves in different directions. That describes the process of so-called scattering of an electromagnetic wave by an electron. The electron absorbs the electromagnetic wave and vibrates at the same frequency as the incoming wave, because that's what's making it go up and down. And it radiates back electromagnetic waves of the same frequency—that is the same color as the waves coming in.

That's the crucial point. The classical explanation for this scattering of an electromagnetic wave, or scattering of light off electrons is that the outgoing light gets scattered in different directions. (By the way, a process similar to this, although not off individual electrons, is why the sky is blue. Blue light gets scattered.) That scattering ought to take place at the same frequency. That's the classical prediction.

What does quantum mechanics say? Quantum mechanics says something like this: I'm going to do the demonstration by taking now the light represented, not as a wave, but as a particle. So here's a photon of light, and here's the electron. And we're going to stage a collision between these two. In comes the photon, and it's going to hit the electron. And what happened? Play it again. Rewind if you need to watch that collision again.

What happened? The electron was sitting there. At rest. The photon came in. It hit the electron. The electron went off in one direction and the photon bounced back. That's like the light being scattered. But notice what happened. The photon gave up some of its energy to the electron. And it was quite clear in that collision. And you could see that. That the photon came off moving slower than it was coming in. It had lost energy.

Now here my analogy is a little bit imperfect, because photons can't slow down because they're light, and they go at c. What my stainless steel ball bearing, representing a photon did was to lose energy. It transferred energy to the electron. If it were a real photon, a particle of light that moves at speed c, it can't slow down. But what it does do is lose energy.

How does a photon lose energy? A photon's energy is energy equals Planck's constant times frequency. So the way it can lose energy is by slowing down in frequency, lowering its frequency. Lengthening its wavelength. It goes to lower wavelengths. When you do the experiment, bombarding free electrons basically with x-rays and you measure the scattered white, the scattered x-rays coming off, you find they have lower frequency. And the frequency is lower exactly as predicted.

By doing the simple physics of a simple billiard ball-like collision between two billiard ball-like particles, you have to do the collision using relativistic mechanics, not Newtonian mechanics. But if you do that, and it's a thing a sophomore physics student can do, you will find that the photons, that is, the x-rays, come off at a longer wavelength, a lower frequency. It's called the *Compton shift*. And that is a well-verified thing that happens. And it shows very clearly that the interaction of those electromagnetic waves, "the x-rays," with electrons is particle-like. So we have very clear evidence for particle-like behavior of light—in that case, of x-rays.

So now, we have a real quantum quandary, if you will. Light behaves like a particle and light behaves like a wave. And which is it? And I'm sure you'd love me to give you an answer. It's really waves. It's really particles. But, I can't. The quantum mechanical answer is it's both. It's got to be both. And this is an answer that's forced on us by that fundamental solution to the quantum problems at the turn of the twentieth century, namely, that energy is ultimately quantized in these little bundles. In the case of light, that means light energy is distributed in little bundles, not spread continuously. And those bundles act, basically, like particles. And the photoelectric effect, to some extent, and the Compton effect that I just showed you, to a major extent, convinced you that light has a particle aspect.

On the other hand, we did an interference experiment and we saw that light is a wave. So which is it? Is light a wave or is it a particle? That's the title of my lecture.

Well, let's try to find out. Let's look at our double-slit system again. So here it is. We're looking at the two-slit system again. And we know what happens. If waves come through that barrier, through the two slits, we get an interference pattern on the screen that consists of light and dark and light and dark bands. But now, we're in this quandary. We don't know whether light is a wave or a particle. Certainly, this experiment suggests it's a wave. So, where's the particle-ness of it? Which slit did each photon go through? How did the photons know to go through those slits and then bend themselves in such ways that they would end up at the right places on the screen?

If light's a particle, we ought to be able to give a particle like explanation for this phenomenon. Maybe. Let's try. Well, one thing we could think of doing is blocking one of those slits. Let's block a slit and see if half the photons go through the other slit. What ought to happen? Well, you could speculate on things that might happen. Perhaps we'd see the same interference pattern, but perhaps it would just be weaker, because now only half the photons are involved.

So let's do that experiment. Let's block one slit. And so we'll be looking at only the photons that go through the other slit. So let's see what happens. We're going to block the right hand slit. There we go. And what happens? We see a single spot on the screen. Right opposite the other open slit. We've destroyed the interference pattern completely. What we've tried to do here is catch the photons in the act of behaving like particles rather than waves. And what have they done for us? They've obliged us by behaving like particles. And their wave behavior is gone. No interference pattern left. We tried to catch the photons going through a slit—and no wave behavior. We've now made them behave like particles. We're trying to catch them in the act. We're trying to say, "Okay, which are you, light? Are you a wave or are you a particle?" And we tried to see the particle-like parts of the light sneaking through these slits. And we can't do it. We get particle-like behavior instead of wave behavior. We've destroyed the interference pattern that we were trying to understand.

Now, you could imagine more subtle experiments you might try to do to achieve the same effect. You might put little detectors that sat there at each slit looking at the photons and seeing what they did. Well, the problem with looking at a photon is you can't look at a photon. What you can do is absorb the photon because it's light. But if you absorb the photon—that's what we did by our blockage of one of the slits—then the photon isn't there anymore and it can't go on and make the interference pattern.

If you tried to put anything that will detect the photons going by, if that thing is actually capable of detecting the photons, then it's going to disturb them in some ways—either change their paths or maybe take them out of existence all together. And, lo and behold, you will not get the interference pattern. If you put photon detectors at those two slits and they actually detect the photons going through, then you will find two spots on the screen, just as if light was a particle. You can't do it. You can't catch it being a wave and a particle simultaneously. If you try to catch the particles, they will oblige and be particles, but there will be no interference pattern. Very strange behavior.

So now you try one other thing. You say, "Okay, these photons are doing some weird thing. I mean, one photon, comes through this slit. Another one comes through the other slit. And maybe they somehow interfere to make this pattern on the screen."

So let's turn the intensity of the light down so low, make it so dim that there's only one photon at a time in this whole apparatus. In fact, and this has been done, you could turn it down so low that sometime, once a day, a photon leaves the light source and gets to the screen. And then on the average, it's another day before another photon, making that trip in a tiny fraction of a second, does it. So there's only one photon there at a time. So the photons aren't somehow interfering one photon with another. And then, you must say, "Well, clearly that one photon that's in there, it either goes through the left slit or it goes through the right slit. There's no nonsense about interfering with lots of other photons, because there aren't other photons."

And what happens if you do that? You build up an interference pattern that looks exactly like the one we saw before—the wave interference pattern with bright, dark, bright, dark, bright, dark bands. How did you do it? How did each individual photon, being the only thing in the apparatus in 24 hours— the rest of the time there's nothing going on—and "bing" a photon comes through. How did it get to the screen? Did it go through the left slit or the right slit?

The answer has to be it went through both slits. In some peculiar way, light is both a wave and a particle. Let me show you what would happen, statistically, if you did that experiment very slowly. What I'm going to show you now is a simulation of the build up of that interference pattern, by hits of individual photons on the screen. It's going to look like a random process at first. We're going to show one photon at a time for quite a while. And finally, in the end, we're going to let a few barrages of photons hit the screen. And you'll see, at first, what looks like a random process. But out of it will emerge, eventually, this interference pattern. Somehow, the individual photons know to land on the screen in a way that eventually builds that interference pattern.

Here we go, building up gradually. It looks kind of random. But now, you begin to see a kind of grouping there. They're grouping in certain spots. Those are the places where the bright spots in the interference pattern should be. Somehow, the photons, even though it looks random, are being guided to those spots. And here we are, as more and more photons come through, we're building up and building up something that begins to look substantially like that interference pattern. Here they go—big barrages at the end. And there we have the interference pattern. Alternating bands of light, dark, light, dark, light, and dark. We built up that pattern with photons that went through the apparatus one at a time. There was only one photon in the apparatus at any given time.

What is all this telling us? It's telling us that light is both a wave and a particle and there's no way around it. Light has a dual nature. It's both. If you don't try to detect light, if you don't try to catch the individual photons, what you will always find happening is that light, when left to its own devices, acts like a wave. If you try to do something to detect the individual photons that make up that light, like build a photoelectric effect experiment or put a camera film there (and the way camera film works is individual photons ultimately hit chemicals in the film and cause chemical reactions that make the film brighten or darken at that point) then a particle-like interaction will take place.

But if you leave the light alone and don't try to detect it, you'll see that it behaves like a wave. If you try to do a detection, it ultimately requires interaction with an individual photon—a Compton experiment, where a photon hits an electron; a photoelectric effect experiment, where a photon hits an electron and ejects it from the metal; a photography experiment; a modern charged-coupled device, like is in your video cameras or in an astronomical telescope camera these days, in which a photon hits a semiconductor device and knocks electrons into higher energy levels and creates an electric current. All those things are photon-like interactions.

If you tried to detect the light, if you tried to say, "Oh, I know that light consists of particles. I'm going to find one," you will, but if you leave it alone and don't try to find one, it acts like a wave and it exhibits wave properties. Like interference, it does both things. If you want to insist that there are photons in the wave when you're not looking at it, you can say that. But then you have to say that those photons went through both slits. Somehow, each photon, even when we did the experiment with only one photon in the apparatus each day, that photon somehow knew about the presence of both slits. That's a rather unproductive way of thinking.

I find it much more useful, to say, "Oh, in that case, the light is manifesting itself as a wave." I understand how a wave can sample both slits. A wave is a big spread-out thing. It goes through and it goes through both slits, and that's fine. I understand that. I can't understand that with individual photons. If I try to detect the individual photons, going through the slits to see which slit they went through, then they oblige and act like photons and the wave interference pattern disappears. And I just see a spot on the screen opposite the slit that I didn't try to block, for example.

Very strange schizophrenic behavior for light. It's a wave and a particle both. Is there any relationship between the wave and the particle aspects of light or are these completely disjointed? They're not disjointed. There's a definite, statistical relationship. And this is going to govern a lot of what we say about the quantum mechanics, not only of light, but also of particles.

The relationship is this: In places where the wave is strong, you are likely to find a photon, if you try to detect one. In places where the wave is weak, low amplitude, not much waviness, you're not likely to find a photon. In the case where the wave amplitude is zero, at the darkest spots in the interference pattern, you won't find a photon. There's zero probability. Where the wave amplitude is big, where the waves have undergone constructive interference, there's a good likelihood of finding a photon.

But the association is a peculiar one for a precise, rigorous, science like physics. The association is statistical. It's not that if the wave amplitude is five you will find five photons. If the wave amplitude is big, you are more likely to find a photon. If the wave amplitude is low, you are less likely to find a photon. It's the probability of detecting a photon that depends on the amplitude of the wave.

So what goes on in our two-slit interference pattern? Well, we aren't trying to detect the photons until they hit the screen. Let's not try to catch them on the way. So the light behaves like a wave. It goes through the two slits. It does what waves do. It creates an interference pattern on the other side. There are places in that interference pattern where the waves interfere constructively. They result in the bright spots on the screen. There are places where they interfere destructively. They result in the dark spots on the screen.

If you then go and try to detect the individual photons by putting a photographic plate at the screen, by putting a photoelectric effect experiment at each point on the screen, by putting a charged-coupled device—the modern light detector that we use in video cameras and things—on the screen, then you will detect the photons. And you will detect them in proportion to the amplitude of the wave. So where the wave amplitude is big, where the interference was constructive, you'll detect lots of photons. Where the interference was destructive, where the wave amplitude is low, you will not detect very many photons.

But it's purely statistical. There is no way of saying, "Yes. There will definitely be a photon here." You cannot think of the photon as a little billiard ball that follows a precisely known trajectory through that double slit. If you want to explain the double-slit experiment, you have to say the light is a wave going through that system, but it also has photons in it. I can't talk about the photons as they go through the apparatus. If I try to talk about them and make some sense of it by measuring what they're doing, I destroy the wave aspects of the light. So I don't talk about the photons while they're going through.

I get this wave pattern. I say, "Oh, I hear the waves are strong. I'm likely to find photons." I go to detect them and I find them. But I can't say I am going to detect an individual photon at this point. And at this time, I can't say that with any kind of precision. I can only state, statistically where the wave is big, there's likely to be photons.

So quantum mechanics has given us a strange kind of light. It's neither a wave nor particle. It's in some sense both. And the connection between the two is fuzzy, probabilistic, statistical—a very strange thing for what's supposed to be the most precise, most rooted, most fundamental of the sciences. So there isn't really a contradiction between the wave and particle aspects of light. They're not contradictory. They are, in fact, as Niels Bohr said, complementary aspects of the same reality. They're not contradictory. They're complementary.

And Bohr elucidated a principle called the *principle of complementarity*, which gets us out of any contradictions. We said light is a wave or particle. It's both. And, you say, "Oh, it can't be. That's a contradiction." Nope. Bohr's complementarity principle says it's not a contradiction. Both of those are needed for a full description of the behavior of light. You can't explain the photoelectric effect experiment and the two-slit interference experiment without invoking both aspects of light.

What does the complementarity principle say? Well, Bohr's complementarity principle says, basically, this: It says, wave and particle aspects of light are not contradictory but complementary. You've got to have them both to explain fully the properties of light. Light acts as a wave in some situations, like a particle in other situations. But you'll never catch it acting like both at the same time. You'll never catch light in a contradiction. Sometimes it acts like a wave. Sometimes it acts like a particle.

What determines which it acts like? What determines which it acts like is what you choose to do to it. If you choose to do an experiment in which light is being asked to act like a particle, it will oblige and it will act like a particle. What's an example of such an experiment? A photoelectric effect experiment, where you want light to hit individual electrons. A Compton effect experiment, a two-slit experiment, in which you covered one of the slits—those are experiments in which you're suggesting light ought to behave like a wave. And it obliges you and does so. You do another experiment in which you expect and set up for particle, for wavelength. Did I say wave? Particle, in which you set up and expect wave behavior. Then the light will act like a wave. Example: the two-slit interference pattern experiment, with both slits wide open.

So what determines what light acts like? You do. By setting up the experiment. If you do an experiment looking for waves, like an interference experiment, you get them. If you do an experiment looking for particles, like a photoelectric effect experiment, you see the particle aspect of light. Does this seem absurd to you? If it does, you're in good company.

The physicists who, in the early part of the twentieth century worried about this were really, really baffled. And I want to end this lecture with a quote from Werner Heisenberg, who was one of the real founders of quantum physics and whose uncertainty principle will be the subject of the next lecture. Here's what Heisenberg says. They're in a meeting, in a gathering of physicists, and he's been talking with Bohr. And he says, "I remember discussions with Bohr which went through many hours, until very late at night, and ended almost in despair. And when at the end of the discussion, I went for a walk alone in the neighboring park, I repeated to myself, again and again, the question: Can nature possibly be as absurd as it seems to us in these atomic experiments?"

Quantum Uncertainty—Farewell to Determinism
Lecture 19

Well, it's farewell to determinism and farewell to the clockwork universe once we have Heisenberg's uncertainty principle in place because quantum mechanics is going to eliminate forever the possibility that we live in a completely deterministic universe in which everything is predictable. Why is that? The fundamental reason is the reason of quantization.

Quantization means that we cannot observe the universe without affecting it. This, in turn, limits our ability to make measurements with arbitrary precision. Thus, we must say farewell to the "clockwork universe" of Lecture 3. The least obtrusive way to observe something is to see it—that is, to bounce light off it. First, we will consider how to prepare the light.

Recall that the probability of finding a photon is proportional to the intensity of the associated light wave at that point. If we want to know with precision where a photon is likely to be, then we need a *wave packet*, with the "wiggles" of the wave confined to a small region. We can do this by producing, for example, a very short pulse of laser light. But note that making a localized wave such as this requires a short wavelength and, correspondingly, a high frequency.

Heisenberg's quantum microscope "thought experiment" explores an attempt to measure simultaneously the position and velocity of an electron with high precision, by bouncing light (i.e., minimum one photon) off the electron. To get accurate position information, we need a localized photon. There's a problem, though: The localized photon has high frequency and, therefore, high energy (recall the quantization condition $E = hf$). As it bounces off the electron, the photon transfers a lot of energy to the electron, altering its velocity substantially. The observation destroys some of the information—the velocity—that we sought to measure.

Note the crucial role of quantization here: The requirement for a *minimum* amount of light energy—one photon's worth—causes the problem. We can't observe a system without interacting with it, and when energy is quantized, that means disturbing the system. Surely there's a way out of this problem: We can make the photon energy lower, thus reducing the disturbance. But lower photon energy means lower frequency (again, $E = hf$), longer wavelength—and a less localized photon.

Now our measurement of the electron's position is less precise. The quantum microscope thought experiment reveals a tradeoff between our ability to measure a particle's position and its velocity simultaneously. If you make the velocity measurement more precise, you lose information about position and vice versa.

The *Heisenberg uncertainty principle* is the formal statement of this tradeoff. The uncertainty principle states that it is impossible to measure simultaneously and with arbitrarily high precision both a particle's position and its velocity (actually its momentum, the product of mass and velocity). Quantitatively, the uncertainty principle says that the product of a particle's mass, the uncertainty in its position, and the uncertainty in its velocity cannot be less than Planck's constant h: $m \, \Delta x \, \Delta v > h$. Because h is so small, the uncertainty principle has a negligible effect on measurements of normal-sized objects, such as planets, baseballs, and even bacteria. At the atomic scale, however, where particle masses are tiny, the uncertainty principle severely limits our simultaneous knowledge of particles' positions and velocities.

The uncertainty principle states that it is impossible to measure simultaneously and with arbitrarily high precision both a particle's position and its velocity (actually its momentum, the product of mass and velocity).

What does it mean? Let's consider the philosophical interpretation and implication. Most physicists subscribe to the *Copenhagen interpretation* of quantum physics. This view grows out of logical positivism, with its claim that it makes no sense to talk about what cannot be measured. In the Copenhagen interpretation, not only can one never measure the velocity and position of a particle simultaneously, but it also makes no sense to say that the particle *has* a velocity and a position.

Under the Copenhagen interpretation, such particles as electrons and protons simply can't be thought of as miniature bowling balls, whizzing around in precise orbits. Rather, they're fuzzy, statistical things describing paths that are only vaguely determined. Because precise velocity and position are required to use Newton's laws to predict future motion, the uncertainty principle and the Copenhagen interpretation abolish the strict determinism of the Newtonian "clockwork universe."

Not all physicists accept the Copenhagen interpretation. For all his life, Einstein was among its staunchest critics. His famous remark, loosely paraphrased "God does not play dice with the universe," expresses his rejection of quantum indeterminism. (Einstein's actual words are "But that He [God] would choose to play dice with the world…is something that I cannot believe for a single moment.") Today, a small group of physicists is pursuing alternatives to the Copenhagen interpretation. Among these are *hidden variable theories* that posit an underlying deterministic reality hidden from our measurement by the uncertainty principle. However, recent experiments, to be described in Lecture 21, place severe constraints on such theories. ∎

Essential Reading

Hey and Walters, *The Quantum Universe*, Chapter 2.

Wolf, *Taking the Quantum Leap*, Chapter 7.

Suggested Reading

Lindley, *Where Does the Weirdness Go?*, "Intermission," pp. 87–121.

Pagels, *The Cosmic Code*, Chapters 9–10.

1. Why can't we get around the uncertainty principle by observing the electron first with a high-energy, localized photon to get its position, then with a low-energy, spread-out photon to get its speed?

2. The statistical nature of quantum physics is often cited to explain the possibility of our having free will and has also been used by some in attempts to explain consciousness. What bearing do you think quantum physics has on free will and consciousness?

Wave Packets

Broad

- Photon not localized
- Long wavelength
- Low frequency

Narrow

- Photon well localized
- Short wavelength
- High frequency

Heisenberg's "Quantum Microscope"

Heisenberg devised this thought experiment to show how interactions between an observer and the system under observation result in unavoidable and unpredictable disturbances in the system—a phenomenon that underlies Heisenberg's uncertainty principle. The experiment attempts to measure the position and velocity of an electron by shining light on it and detecting the scattered light. But light has both a wave and a particle nature, and to know precisely where the light is we need a "wave packet" of short-wavelength light (A, above). Using such a packet, we can determine the electron's position to high precision. But a short wavelength corresponds to a high frequency and, by the quantization equation $E = h f$, to a high energy for the light photons. High-energy photons scattering off the electron impart momentum to it in an unpredictable way, thus disturbing its velocity. So this experiment measures the electron's position precisely, but provides little information about its velocity.

To get around the problem of high-energy photons disturbing the electron's velocity, we try instead to use low-energy photons. But these correspond to low frequency and thus long wavelength—and long wavelength "wave packets" aren't precisely localized. Thus the experiment measures the electron's velocity with precision, but its position remains uncertain. It's impossible to measure simultaneously both position and velocity with arbitrary precision. The standard interpretation of quantum physics goes further to say that it is meaningless to talk about a subatomic particle's having simultaneously both a well-defined velocity and a well-defined position.

Quantum Uncertainty—Farewell to Determinism
Lecture 19—Transcript

At the end of the last lecture, we left Heisenberg himself baffled about the peculiar nature of the world at the quantum level. What did Heisenberg do about his bafflement? Well, ultimately he went on to build a principle that embodies the fundamental bafflement of quantum mechanics, and I want to explore that principle today. It's the principle of uncertainty, Heisenberg's famous *uncertainty principle*.

If you think back, way back—you probably watched the early lectures, a long time ago—think back to Lecture Three. Its title was "The Clockwork Universe." Well, it's farewell to determinism and farewell to the clockwork universe once we have Heisenberg's uncertainty principle in place because quantum mechanics is going to eliminate forever the possibility that we live in a completely deterministic universe in which everything is predictable. Why is that? The fundamental reason is the reason of quantization. We solved the problems that classical physics had with the subatomic world by pointing out that things had to be quantized: atomic vibrations, the energy in light waves, the orbits of electrons in atoms. These things are quantized. They come in certain discreet units.

How does that force on us an end to classical determinism? Well, the fundamental idea is this: If there's a minimum size to the interaction we can have—if there's a minimum amount of energy—then if we want to understand the world, we have to interact with it, and the impact of that interaction on the world cannot be made arbitrarily small. Why? Because there's a limit to the size, the bundles of energy, we use in interacting with the world. If there were no such limits, we as observers of the world, we as physicists, or we as people doing experiments, we as people wanting to understand how the world works, we could do experiments. We could probe the world. We could ask questions of it without having to disturb it in any way at all.

But nature doesn't work that way. Nature gives us a certain minimum size to the bundles of energy we can use when we try to interrogate the world and, therefore, we, in interrogating the world, inexorably change it. We become a factor in the very experiment we're trying to perform. No longer are we some objective, disconnected observer observing things. We become part of what was observed. That's fundamentally, philosophically, deeply what quantization does to our ability to interact with the world and to understand it. We become part of the world we're trying to interact with. We can't observe the universe without affecting it.

Well, think a little bit about experiments we might do to observe the world to try to understand it. How do you observe things? Well, the least obtrusive way probably is simply to watch them, to look at them. What does it mean to look at something? Well, it means to detect light that has shined on that thing and come back to your eyes or your measuring instruments or whatever and allowed you to tell what's going on. So let's think about a very simple experiment we might want to do to observe something with light. Before we do that, we have to talk a little bit about what it means to observe something with light.

Here's what we want to do. Here's our outline. We'd like to take some object—and we'll take a really simple object like an electron, for example—we want to see what it's doing. We want to know how it's moving, that is, how fast it's moving and in what direction, and we want to know where it is. Ultimately, why do we want to know those things? Because, according to classical physics, anyway, if we know its position and its velocity and we know the laws of motion, which we do, then we can predict what it will do in the future. It's knowing the position and velocity of a particle that allows us to predict its future behavior and gave rise to this idea of a clockwork universe that's completely determined if, in principle, the positions and velocities of all the particles that make it up are known.

So let's focus on something very simple, a single electron, and let's ask, how can we go about measuring its position and velocity? And I just said the simplest way to do that is to try to see it, to shine light on it and see where it is. So that's what we want to do. Before we do that, let me make a point, though, about the light we might choose to shine on it because light, after all, has this particulate nature to it. It consists of waves, yes, but it consists of photons. If we want to know where this electron is, we're going to have to shine on it photons that themselves are going to act pretty particle-like. How do we do that?

Well, remember about the statistical relationship between waves and particles. The relationship is where the light wave is strong, where it has a high amplitude, we're likely to find the photon. Where it has a low amplitude, we're not likely to find the photon. If we want to make a localized photon, if we want to know where the photon is so it acts particle-like—as in our Compton experiment, where we shot an x-ray photon, and it acted like a billiard ball—if we want to localize the photon, we've got to make a wave situation in which the wave has a strong amplitude in a relatively localized region and otherwise doesn't have much amplitude. If we have a wave that's spread out over all space, that isn't going to help us much because we don't know where the photon is likely to be. It would be just as equally likely to be anywhere.

Well, to give you an example that is far from that extreme, I have here a picture of two possible *wave packets*. What's a wave packet? It's a wave that's been put together in such a way that instead of being a continuous undulation that goes on forever, it's negligible in most places and then has a strong oscillation, a strong undulation, in just a limited region of space. How you put together a wave packet is a little bit interesting. Mathematically, to make a wave packet, you actually have to combine a bunch of waves of different frequencies. A wave packet doesn't have a single frequency or a wavelength well-defined. It's made up of a bunch of different frequencies, and therein lies part of our problem here.

The wave packets I'm showing you consist of two different types. There's a broad wave packet. It's spread out over space, as you can see. This is the upper one that's red. It's spread out over space. It's broad. The photon is somewhat localized but not very localized. If I were to try to detect a photon in this wave—the photon, remember, is likely to be found where the wave is strongest—but here the wave is pretty strong over a fairly broad region, and, therefore, the photon is likely to be found somewhere in a fairly broad region. It's not very well localized. The wave packet has essentially long wavelength. Although, as I've just told you, it is made up of a lot of different waves. But this broad wave packet—they can be fairly long wavelengths, and it can be fairly well defined, and correspondingly, it's a fairly low frequency. When you have a wave, a high frequency corresponds to a short wavelength and a low frequency corresponds to a long wavelength. So in this particular case, I have a broad wave packet.

If I try to find photons, where they are is not terribly well localized because there's a broad region where the wave packet has high amplitude, where the wave is big. It's got long wavelength and, therefore, essentially low frequency. On the other hand, if I really want to make a wave where I can say with pretty good assurance that I'm going to find a photon here, what I can do is make the lower, blue wave packet. That one is much more localized. It's very narrow. There's only a limited region of space where the wave has a high amplitude. So the photon, if I find one, is going to be with very high precision within that limited region. To do that, I had to use waves of relatively short wavelength and, therefore, high frequency. Again, that's a slight approximation to what's really going on.

To make these wave packets, I have to build up a lot of different frequencies. But to make the narrower one, it turns out the high frequencies are more prominent, the short wavelengths. So, in a crude approximation, the narrow one has a well-localized photon. If I detect the photon, it has a short wavelength, but it has a high frequency. So those are two different wave packets I might try to use to shine on my electron to see what it's doing.

The experiment I'm going to describe is actually a thought experiment. Remember we talked earlier about thought experiments. They're experiments that, in principle, one could do but in practice they would probably be difficult, at least at the time the thought experimenter thought them up. They help you to illustrate an important idea if you think about what would happen in this thought experiment. So this is Heisenberg's famous thought experiment, sometimes called the quantum microscope experiment, and it consists of shining light on an electron in an attempt to determine the position and velocity of the electron simultaneously so you can predict what the electron is then going to do.

So, how do we do the experiment? Well, here's a very simple picture of it. I'm going to have a light source. I'm going to imagine shining a single photon out of that light source. I'm going to represent that photon by one of these wave packets I've been talking about. I'm going to shine the light, the photon, on the electron. The photon is going to bounce off the electron and go into a detector, and I'm going to use that to tell me where the electron is and how it's moving. Now, in practice, to do this I'd really have to shine a beam of light and extract the information, basically statistically, from how all the photons come off the electron, and probably I'd do it with a bunch of electrons and so on. But, in principle, we can imagine the minimum interaction would be to look at the electron with just a single photon. So, that's what our impractical thought experiment consists of. We're going to shine light on the electron.

Now here I show us doing that. Okay, the electron is moving along. I've indicated the electron's position by where I've put it in the picture. I've indicated the electron's velocity by an arrow. In this case, the electron is moving to the right and slightly upward on the screen. This is before the photon comes in and attempts to measure it. Now, what happens? The photon I've chosen in this particular case is the localized photon. I need that because I want to know where the electron is. So I'm going to localize the photon because, if I don't know where the photon is, I can't tell where the electron was that it bounced off of. So I'm going to localize the photon by generating this wave packet.

How would I do that in practice? Well, I might take a laser beam and I might make a very, very short pulse of light—very, very short in time—a short, short pulse of light, and that would be a very compact wave packet where I knew essentially where the photon was. So I know essentially where the photon is, represented by this wave packet. Statistically, I'm very likely to find the photon somewhere in this little region. With this packet of waves, there's going to be a photon somewhere in it—there's some statistical chance of finding it—and the chances are it's very close to that localized region. That wave packet is traveling along at speed c, as electromagnetic waves do, and there's a photon somewhere there if I tried to detect it. That photon is going to interact with the electron.

So what happens? Well, the photon interacts with the electron. It bounces off, and I get it in my detector. Now, actually, I'd have detectors all around so I could be sure I knew where it came off, at what angle, and so on, but in my simplified experiment it goes right into my detector. Great. Because I had a localized photon—I knew where it was, I knew that wave packet was very narrow—I've got a very precise measurement of the electron's position. So I've measured very precisely where that electron is. Unfortunately, there's a problem. And the problem is that localized wave packet. That's the one that had a narrow distribution in space. It corresponded to short wavelengths or high frequencies.

High frequencies—remember the solution of all the quantum problems or all the problems classical physics had at the end of the nineteenth century. The solution was quantize—quantize energy. In particular, as Einstein said, quantize the energy in light. It comes in little bundles. How big are those bundles? E, the energy of one of those bundles, is Planck's constant h, times the frequency. Well, here we have a high-frequency photon. We needed the high frequency in order to localize it so we could measure precisely where the electron was. The problem is, high frequency, through quantization—and this is where quantization comes in—means high energy. E equals hf. So this photon, necessarily, has high energy. If you want to work with light at that frequency, if you want to have a localized wave packet, then there is a minimum amount of energy you can be sending out there to interact with the electron, and that minimum amount of energy is relatively big because E is hf, and f is big. It's a high-frequency situation. E is hf. E is, therefore, relatively big.

So this photon had high energy. It hit the electron. Remember what happened when I did the demonstration on this table? When the photon hit the electron, the photon recoiled in some direction, and the electron took off in some direction. Whatever the electron was doing, however, it was moving before the photon hit it. That high-energy photon disturbed its motion in a significant way, which I can't exactly predict because, even though the photon is fairly well localized, I don't know exactly where it is in relation to that electron. So I don't know exactly how it hit it. I've represented that on this picture by showing you a bunch of possible arrows that represent the possible states of motion of the electron after it's been hit by the photon and a big fat question mark that tells you we don't know the electron's velocity after this experiment has been done. Why not? Because it's been hit by a high-energy photon that has disturbed its velocity.

That's the problem. Quantization tells us there's a minimum size to this photon in terms of energy. I can't interact with the electron with any less energy than that. If I do interact with the electron with that relatively large amount of energy, I disturb its motion. I change its velocity. I've measured its position to high precision, but I know nothing about what it was doing before I hit it. So in measuring its position to high precision, I've destroyed any information about how it was moving, about its velocity. That's not good. I set out to try to measure the electron's position and its velocity to high precision because that's what I needed to predict its future behavior. I ended up measuring its position to high precision, but I destroyed any information I had about its velocity. Its velocity is now not what it was before, so even if I could measure its velocity precisely, at this point it's not what it was before because the electron has been disturbed. Why? By the fundamental bundle of energy, the minimum size I could have for energy of a photon of that frequency, that wavelength, that compactness, that localized-ness.

Well, surely there's a way out of this problem. Surely the way out of it is to use a lower energy photon. Remember, quantization doesn't say there is an absolute minimum energy any photon can have. It just says that if I have light of a certain frequency—the waves are oscillating at a certain rate—then, for that frequency, there's a minimum energy hf. But I can make the energy lower by making the light redder, making the frequency lower, or making the wavelength longer. If I do that, I'll have a lower energy photon. Its interaction with the electron will be less violent. It will disturb the electron's velocity less. In principle, I could take the frequency of the photon, and, therefore, its energy, down as close to zero as I want. I can't do that for a fixed frequency, but I can do it by taking the photon frequency down, and, therefore, I can minimize the disturbance of the electron's velocity.

So, I can measure the velocity. Great. Let's do that. So here's the microscope experiment again. Now I've put in not my localized high-energy, short-wavelength, high-frequency photon but my wave packet that is spread out. It's broad. It has long wavelength. It has low frequency. Therefore, it has low energy. E equals hf. Therefore, it's not going to disturb the electron very much. And, again, in principle, I can make it as low frequency as I want and make that disturbance as small as I want, and, therefore, when the photon bounces off the electron, I can get a precise measurement of its velocity. I'm confident that's still the velocity because the photon carried very little energy, and, therefore, it disturbed the electron as little as I want it to. Great. I've measured the velocity of the electron.

But there's a problem. Because I used a low-energy, low-frequency, long wavelength and, therefore, broad and spread out wave packet, I don't know where the photon was. The photon was not localized in this case, and, sure, it bounces back into my detector, but I don't know exactly where it was, so I don't know where the electron was. I've measured the electron's velocity accurately, but I can't tell you anything about its position or very much about its position, and I've represented that in this picture by a big fat question mark and a lot of possible positions for the electron after the interaction. It's got the same velocity it did before the interaction this time, because I've minimized the disturbance by lowering the energy of the photon. But lowering the energy of the photon necessarily spread it out in space, and I couldn't localize it, and, therefore, I don't know where the electron is.

Now, what's the problem? Well, we've got a kind of a trade-off. I can measure the position accurately with a high-energy photon, but then I disturb the velocity. I can measure the velocity accurately with a high-energy photon, but then I don't know anything about the position. Or I can do something in between; I can kind of compromise. I can take a medium-energy photon that will disturb the velocity of the electron somewhat but not too much. I won't know exactly where it is, but I'll know somewhat where it is. There's a trade-off here. If you try to measure the position accurately, you destroy information about the velocity. If you try to measure the velocity accurately, you destroy information about the position.

There's a compromise place in between. You can measure the velocity with some reasonable degree of precision and the position with some reasonable degree of precision, but they'll both be off a bit. You can't, at least with an experiment like this, measure simultaneously both the position and velocity of a particle. Why? Ultimately, again, it's because of that point I made at the beginning. To measure something about the world is to interact with it, and quantization sets a fundamental, minimum size to that interaction for a given frequency of light that you want to use. We've seen how that results in a disturbance of the electron, and if we try to get around that by lowering that frequency to lower the minimum energy of the interaction, then we lose information about position. So we can't simultaneously measure the velocity and position of an electron or, in principle, any other particle.

That's basically what the Heisenberg uncertainty principle says. The Heisenberg principle is basically a formal statement of that quandary, that trade-off, that you can't measure simultaneously the position and velocity of a particle with arbitrary precision. You can't measure simultaneously the position and velocity. Actually, technically, Heisenberg's uncertainty principle says not velocity but momentum, and momentum is the product of the particle's mass with its velocity. You can't measure them both with arbitrarily high precision. You can measure one with perfect precision and know nothing about the other. You can measure the other with perfect precision and know nothing about the first. Or you can measure them both with some degree of precision, but neither will be perfect. Strictly, mathematically, Heisenberg's uncertainty principle is stated by saying that the product of the uncertainty, how uncertain you are about the position with how uncertain you are about the momentum—again, the mass times the velocity—that product can't be zero.

In classical physics it could be zero, and we could measure both with arbitrary precision. In quantum physics, it can't be zero. In fact, it has to be bigger than a number, which is, approximately, guess what? Planck's constant h, that fundamental number that sets the fundamental graininess, the fundamental quantization of the universe. Remember, I said quantum mechanics is really the physics about the fact that Planck's constant isn't zero. If Planck's constant were zero, that product of uncertainties could be zero, and that means I could measure them both with arbitrary precision. Both uncertainties could be zero. If the uncertainty is zero, it means I know the quantity precisely. If Planck's constant were zero, the product could be zero and both uncertainties could be zero, and I could measure with arbitrary precision.

But Planck's constant is not zero. It sets that fundamental graininess. And, in the quantum microscope thought experiment, it's that graininess, or the imprecision associated with the wave particle duality that led to my inability to measure both the velocity and the position simultaneously. Mathematically, then, Heisenberg's uncertainty principle reads something like this (h: $m \, \Delta x \, \Delta v > h$): The uncertainty in momentum, which is the mass times the uncertainty and velocity, times the uncertainty and position, is greater than h. In this expression, m is the mass. Delta x (Δx) is the uncertainty in position. That Greek letter Delta—capital Delta, a triangle-like symbol—to a physicist or a mathematician means a change in or a range in some number, a variation. So Delta x means a range in x. The position x is the variable that stands for position; v stands for velocity.

So this says, take the mass of the particle and multiply it by the uncertainty in its position—how unsure you are. Do I know it to within plus or minus 2 meters? Do I know it to within plus or minus 10 light years? Do I know the electron's division within plus or minus a nanometer (a billionth of a meter)? What's the uncertainty? Take that number. Multiply it by the uncertainty in velocity. Is the velocity 25 meters a second plus or minus 1 millimeter a second, or is it half the speed of light plus or minus a third of the speed of light? What is the velocity and how uncertain is it? The important number is how uncertain it is plus or minus how many meters a second, miles an hour, or whatever. That's the Delta v (Δv). Multiply them together. Those numbers cannot come out smaller than approximately h. Whether it's exactly h or h divided by 2Π or different things depends exactly on the details of your wave packet, but it's approximately a number on the order of h, and h is a very small number. That means the product of the uncertainties can be very small.

Also, look at the mass in there. I could divide both sides of that equation by the mass, and I'd have Delta x Delta v greater than h over the mass. That means for massive objects like a tennis ball or me or a car, the product of the uncertainty in position and uncertain velocity is miniscule. That's why the uncertainty principle doesn't affect macroscopic-sized objects like tennis balls, or people, or cars, or the Earth. Do I know what this tennis ball is doing? To an enormously high degree of precision, I can measure that. I can measure its position very accurately and its velocity very accurately because it has a pretty big mass, and it's the mass times the uncertainty in position times the uncertainty of velocity that has to be bigger than that tiny, tiny little number h.

So the uncertainty in position times the uncertainty in velocity, which has to be bigger than that tiny number h, divided by the relatively big mass of this tennis ball—that's a pretty tiny number. So I don't have to be very uncertain about either. For all practical purposes, I can measure the position and velocity of macroscopic objects to a tremendously good degree of accuracy, and I can predict their subsequent motion very, very precisely. In principle, though, I can't predict it with infinitely good precision. But in the case of objects like basketballs and tennis balls and stars and spacecraft and cars and airplanes, the uncertainty principle has a negligible effect on our ability to predict what's going to happen.

On the other hand, for things like electrons that are not very massive at all, again divide both sides of that equation by m. Then h, which is a small number, gets divided by another small number, and that makes things a bit bigger. For things the size of electrons, for things on the atomic scale, the product of the uncertainty in position with the uncertainty in momentum becomes significant. That's why it's only at the atomic scale that we begin to notice quantization. That's why quantum physics wasn't obvious. If we lived in a world where h was not 1 over 1 with 33 zeros after it, but in a meter, kilogram, second, or pounds and inches and seconds system of units it had value around 1, then all this would be obvious to us.

When you turn on a light bulb, you'd be bing, bing, bing, bing, bing, hit by photons, because they'd be much bigger. Quantization would be obvious to us, and we would live intuitively with the uncertainty principle. We wouldn't really be able to play basketball very well because we wouldn't even know where the basketball was, and could it go through the hoop, and maybe it went through the hoop, and maybe it didn't go through the hoop, and maybe it interfered with itself. If all those things would happen at the macroscopic world and quantum mechanics would be obvious, it would be a very strange world to live in. It's the smallness of Planck's constant, again, that makes quantum effects important almost always only at the microscopic, subatomic kind of level. Again, it's because h is so small, so the product of these uncertainties is small.

Well, that's the physics of Heisenberg's uncertainty principle, but what does it mean? Well, there's where things get interesting. According to most physicists, but not all, the philosophical interpretation of the uncertainty principle says this: It follows the school of philosophy called logical positivism, and one of the tenets of logical positivism is that, if you can't measure something, it makes no sense to talk about its existence. I could say, there's a big, purple hippopotamus standing on the stage with me, but it's invisible, and you can't weigh it, and it doesn't reflect electromagnetic waves, and it doesn't emit heat, and all these other things, and anything you try to do to detect it, I'm going to tell you, "No, you can't do that." Well, that's nonsense. Maybe it's there. Maybe it's not. But there's no way we can ever verify that, so it makes no sense to talk about it. That's what logical positivism says.

Well, along comes Heisenberg, saying, "Look, I can't measure the velocity and position precisely." Logical positivism and the so-called Copenhagen interpretation of quantum mechanics, which flows from it, which most physicists subscribe to, says, "Well, if you can't measure both the velocity and position simultaneously, it makes no sense to say the electron or any other subatomic particle, or actually the Earth or a bowling ball, for that matter, has a precisely defined velocity and position at the same instance. It makes no sense to talk about those. If you can't measure them, and the uncertainty principle says you never can, then it makes no sense to say it has them." That's the Copenhagen interpretation.

Most physicists believe that or else they simply do quantum mechanics and they try not to think about the philosophy of it. Because, believe it or not, quantum mechanics takes wonderfully accurate predictions about some kinds of things and allows us to build computers and semiconductor devices and lasers and all kinds of things that would be impossible without it. It works, so a lot of physicists sort of forget about the philosophical interpretation. But it's there, and, when pressed, most physicists would say, "Yep, I agree with the Niels Bohr and the Copenhagen interpretation that says a particle simply doesn't have a velocity and position." That's a meaningless statement to say an electron has a velocity and a position but I can't measure them. Because I can't measure them, it's like the purple hippopotamus. It doesn't exist. That's the Copenhagen interpretation.

Remember, I told you our language would fail us when we came to quantum mechanics. Here's how it's failing us. We want to think of an electron as a little billiard ball-like thing. I grab this steel ball. I say, here's my electron. It's got a position and a velocity. It doesn't, according to quantum mechanics. It's not like a little billiard ball. It's a kind of fuzzy, statistical thing. According to the Copenhagen interpretation, this picture of the atom is way wrong. The atom can't consist of electrons in precise orbits the way the Earth is in a precise orbit around the sun because, at that scale, Planck's constant h, comes into play and is big enough, tiny as it is, that it says, "No, you cannot describe an atom by saying that electron is right there at this instant moving that way."

The electron is a fuzzy, smeared-out, statistical kind of thing. You can say, "Oh well, it's got a velocity in this range and a position somewhere in this range," and about the most you can say about the position is, it's somewhere within about a tenth of a nanometer. That's about the size of an atom—a tenth of a billionth of a meter. You can say, "Well, I'm pretty certain it's in that region. Maybe it's a little further out. Maybe it's way far out, but there's not much probability of that." You can't talk about electrons and protons or atoms or anything else as having precise velocity and position. If they don't have precise velocity and position, then the whole clockwork universe idea dies because the clockwork universe was based on the idea that every particle had a precise velocity and position and, even though we might not be able to know it, it was sufficient to know that it had them because then the laws of physics would take over, and they would determine the future behavior of that particle and all the other particles in the universe, and everything would be completely determined.

But once you let go of the notion that particles have unique velocities and positions, well-defined at all times, then that idea is gone. So, farewell to determinism. Farewell to the clockwork universe. Now, does everybody believe this? No. The most famous physicist who refused to believe this was Einstein. Einstein went to his grave very, very troubled with quantum mechanics, and he had a series of wonderfully famous debates with Bohr. Einstein's famous statement about quantum mechanics is summed up by saying, "God doesn't play dice." He just could not believe that the universe was fundamentally probabilistic and statistical. He didn't quite say, "God does not play dice." He said, instead, "It is hard to sneak a look at God's cards, but that he would play dice with the world is something I cannot believe for a single moment." That's Einstein on quantum physics. It looks like Einstein was wrong.

There have been other schools of physicists who tried to pursue this idea, who tried to say, "Look, we can't measure these things simultaneously, but really they still exist," and they searched for so-called hidden variables operating at a lower level that we could never detect but that would determine exactly what was happening. Those theories, although there are still physicists and philosophers who are working on them, were dealt a very serious blow with a series of experiments, done in the 1980s particularly, and we'll get to those in Lecture Twenty-One. For now, it really looks like quantum mechanics is giving us a very strange and new philosophical picture of a world that, at the fundamental level, is fuzzy and probabilistic and imprecise in a very, very fundamental way.

Particle or Wave?
Lecture 20

"Particle or Wave?" Did I say that right? Didn't I have a title just like that before? Well, if you think about Lecture 18, the title was "Wave or Particle?" This one's very, very similar, for reasons that will soon become obvious.

L ecture 18 showed that light has a dual nature; it is both wave and particle. In 1923, the French prince Louis de Broglie (pronounced "de Broy") put forth, in his doctoral dissertation, a remarkable idea: If light exhibits both wave and particle behavior, why not matter as well? Then, there should be "matter waves" associated with material particles.

De Broglie proposed that the wavelength of a matter particle's associated wave depends on the particle's mass and velocity: *wavelength* $= h/mv$. Because h is tiny, so is the matter wavelength, especially for normal-sized objects such as planets, people, baseballs, and bacteria, whose mass m is also substantial. With the wavelength much less than the size of the object or the systems with which it interacts, we don't notice the wave aspect

Because wavelength also depends on velocity, it can become significant, even in macroscopic systems, when particle velocities become very small— something that happens only at temperatures close to absolute zero.

of ordinary matter. For subatomic particles, though, the mass m is small and, therefore, a particle's wavelength can be comparable to the size of the systems with which it interacts. In particular, the wavelengths of atomic electrons are comparable to the sizes of atoms. Because wavelength also depends on velocity, it can become significant, even in macroscopic systems, when particle velocities become very small—something that happens only at temperatures close to absolute zero.

De Broglie's matter-wave hypothesis has been verified in experiments involving electron beams similar to the double-slit experiment of Lecture 18. In practice, the closely spaced atoms of a crystal serve as the slit system, and the electrons exhibit the same interference effects as the photons in an optical double slit. Which slit did the electrons go through? The answer is the same as in Lecture 18. When we're not trying to detect it, the electron acts as a wave and "samples" both slits.

We have seen how quantum physics describes the behavior of light. Light waves (electromagnetic waves) are governed by Maxwell's equations of electromagnetism. In classical physics, the solutions to those equations describe the electromagnetic waves, and that's the end of it. In quantum physics, the solutions to Maxwell's equations describe waves that give the *probability* of detecting photons. The link between the wave equation and the particles is a statistical one.

The quantum description of matter is similar. For each particle, there is a *wave equation*, which in nonrelativistic quantum physics is the *Schrödinger equation* first proposed by Erwin Schrödinger in 1926. The solutions to this equation don't directly describe actual events. Instead, they give the probability that the associated particle will be found at a given place and time.

Matter waves explain Bohr's atomic theory. The allowed electron orbits are those in which a *standing wave* can fit—just as the notes played by a violin string are those that can fit on the string. Because wavelength is related to frequency and, thus, to energy ($E = hf$), the atomic energy levels are quantized. So are energy levels in any confined system.

Waves are continuous, spread-out entities—and here "waves" includes the solutions to the Schrödinger equation that describe the behavior of matter particles. This leads to unusual new phenomena, such as quantum tunneling, wherein a particle "tunnels" through a barrier from which classical physics says it doesn't have enough energy to escape. Does it really occur? Yes, radioactive particles appear outside atomic nuclei from which classical physics says they do not have the energy to escape. In another example, hydrogen nuclei in the Sun's core overcome the "barrier" of electrical repulsion when classical physics says they can't. They fuse to make helium—and in the process, release the energy that keeps us alive! Increasingly, microelectronic devices make use of quantum tunneling, including devices that may be at the heart of computers thousands of times faster than those we have now.

A long, continuous wave has a well-defined wavelength, hence velocity by de Broglie's relation *wavelength = h/mv*. But the associated particle can be found anywhere, so the position is completely uncertain. A short, localized wave is built up of lots of different wavelengths. You know where the associated particle is, but its wavelength and velocity are very uncertain. The uncertainty principle and wave-particle duality are inherent features of nature. There is no way around them.

Nevertheless, quantum physics does make some exact predictions. For example, it predicts precisely the energy levels of atomic electrons. The atomic spectra predicted from these energy levels agree precisely with experimental measurements. What quantum physics can't predict is the precise path and behavior of individual particles. ∎

Essential Reading

Hey and Walters, *The Quantum Universe*, Chapter 3.

Wolf, *Taking the Quantum Leap*, Chapter 5.

Suggested Reading

Lightman, *Great Ideas in Physics*, Chapter 4, from p. 210.

Questions to Consider

1. Whenever a particle is held in a confined space, as between rigid walls or by the electric force in an atom, its energy levels are quantized. Use de Broglie's matter-wave hypothesis to explain how this energy quantization arises.

2. Planck's constant can be expressed as $h = 10^{-33}$ in the standard meter-kilogram-second system of units. The mass of an electron is about 10^{-30} kilograms, and an atom is about 10^{-10} m in diameter. Use de Broglie's formula *wavelength* $= h/mv$ to estimate the wavelength of an electron moving at 10^6 meters per second and compare with the size of an atom. Repeat for a baseball (mass 0.1 kilograms) moving at 10 meters per second and compare with the size of the baseball. Your answers will show why the wave nature of matter is crucial at the atomic scale but completely irrelevant on the scale of everyday objects.

Standing Waves

On Strings...

This diagram shows a string fastened at both ends to rigid supports; an example would be a violin or piano string. The string can be set into vibration, but because the ends are fixed the only vibrations it can sustain are those wave patterns have fixed points the same distance apart as the ends of the string. At the left is one such vibration; for this case two full wavelengths fit between the ends. At right is a vibration that is not allowed, because at the right-hand support the wave is not at a fixed point. In a stringed musical instrument, the allowed vibration frequencies determine the notes that can be played.

In Atomic Orbits...

A similar effect occurs in atomic electron orbits, where the wave function for the electron must fit around the orbit. An allowed orbit is shown at left; a disallowed orbit on the right. The allowed orbits correspond to the discrete energy levels of atomic electrons.

Quantum Tunneling

This diagram illustrates quantum tunneling, a remarkable phenomenon that becomes evident on subatomic scales. The sequence above shows a particle trapped between two barriers. Classical physics asserts that the particle can move back and forth and might be found anywhere between the barriers, but that it will never be found beyond the barriers.

In quantum physics, however, a particle is not localized precisely. Instead, the probability of finding the particle at a certain place depends on the size (amplitude) of the wave function at that point. The mathematical description of the wave function requires it to have nonzero amplitude outside the barriers. As suggested by the amplitude of the wave function shown here, the particle is most likely to be found between the barriers (left frame). But there's a small probability that it will be found outside (right frame)—having "tunneled" through the barrier. Such quantum tunneling plays an important role in many phenomena, including electronic devices and the nuclear processes that keep the Sun shining.

A Quantum Corral

Courtesy of
IBM Corporation, Research Division
Almaden Research Center

391

Matter Waves and Uncertainty

Continuous Wave:
Wavelength, hence velocity, precisely known.
Position completely uncertain.

Wave Packet:
Position well known.
Wavelength, hence velocity, uncertain.

Particle or Wave?
Lecture 20—Transcript

Lecture Twenty: "Particle or Wave?" Did I say that right? Didn't I have a title just like that before? Well, if you think about Lecture Eighteen, the title was "Wave or Particle?" This one's very, very similar, for reasons that will soon become obvious. Where are we at this point? Well, we've studied the wave-particle duality for light. We asked the questions: Is light a wave? Is light a particle? And we ended up with this somewhat unsatisfying answer: It's both. But we recognized, through Bohr's principle of complementarity, that that was not a contradiction, that we need both aspects of light to describe fully its behavior. Well, that's true for light.

What about matter? In 1923, a French prince—his name was Louis de Broglie—finishing his Ph.D. thesis in physics, had a brilliant insight. He said if light, which we thought was a wave, exhibits particle-like aspects, why shouldn't matter, which we think is composed of particles, exhibit wavelike aspects? Why shouldn't there be a wave-particle duality for matter as well as for light? In his Ph.D. thesis in 1923, de Broglie proposed a wave-particle duality for matter. He said, "I think with each matter particle there is associated a wave of a certain wavelength." "How do you calculate that wavelength?" asked de Broglie. Well, it depends on the particle's mass, one of its fundamental properties. It also depends on how fast it's moving. And it depends on that fundamental number of nature that tells us about the quantum physics, tells us about quantization, tells us about that fundamental graininess of the universe at the subatomic scale.

I'm going to show you the equation that de Broglie wrote to talk about the wavelength of matter waves. Again, we aren't getting very mathematical here, but I want to give you a sense of what that equation looks like so you can understand how it works. Let's take a look at the equation. Here it is: *wavelength = h/mv.* It says wavelength, that is, the length of a matter wave associated with a given particle, is equal to a quotient, Planck's constant *h*, the same Planck's constant we've seen before, divided by the particle's mass times its velocity—same as in all of quantum physics. If *h* were zero, the wavelengths of all particles would be zero and they wouldn't have any wave aspect. But quantum physics is about the fact that the fundamental graininess, the fundamental divisions, of nature are not zero, that there is a fundamental limit to how much you can subdivide things, and it's set by the size of that constant h.

So here we have wavelength equals h over mv. The wavelength of a matter wave is Planck's constant divided by the mass times the velocity of the particle we're talking about. I've drawn here a little picture of a wave to remind you what the wavelength means. The wavelength is the distance between two wave crests. It's the distance along the wave over which the wave repeats itself once. So that's what I mean when I say the wavelength of a matter wave.

Now, what does it mean to talk about matter waves? Do I have a matter wave associated with me? Does a car running down the road have a matter wave associated with it? Does the moon in its orbit have a matter wave associated with it? Well, in principle, yes, but in practice we don't notice the wave properties of things like people or baseballs or the moon or other large, macroscopic objects. We wouldn't even notice it for something like a bacterium. It takes very small objects to notice the wave properties. Why is that?

Well, let me give you an example. If you walk out of the room in which you're now watching or listening to this course—and don't do it if it's your car—but otherwise, walk out of the room a minute, and you'll notice something. You can probably still hear me talking, but you probably, unless you're right in the line of sight to your TV screen, can't see me talking about this anymore. Why not? Well, light seems to travel in straight lines, and light does not go around corners, and light goes out of the door of the room you've walked out of, but it doesn't bend around the corner. Sound does. And yet light and sound are both waves. Why the difference?

Well, it turns out that the wave properties of a wave are most noticeable when it interacts with systems—other things in its environment—that are themselves roughly comparable to the size of a wavelength. If those systems are much, much bigger than a wavelength, you don't really notice the wave aspect of the phenomenon you're talking about. That's one of the reasons why it wasn't obvious from the outset that light is a wave—why Newton and Huygens could disagree about whether light was a wave or a particle long before quantum physics came along—because most of the objects with which light interacts in our everyday experience are much bigger than the wavelength of light.

The wavelength of light is not infinitesimally tiny. It's about a thousandth of a millimeter. A thousandth of a millimeter—it's a pretty small amount, but, in the scheme of sizes physics measures, it's still not that small. It's actually a rather big distance now in the scale of, say, integrated circuits we use in computer chips, for example. So the wavelength of light is not infinitesimally tiny, but it's pretty small. That's why you couldn't just take a couple of holes in a wall and do a double-slit interference pattern, although you could do it, as I described, by taking two razor blades and putting them very close together. If you do that, you get a spacing that is many times the wavelength of light still, but at least it's close enough that you can begin to see the interference.

Here's the point. Wave interference occurs, and other wavelike phenomena occur, when the wave interacts with a system that is roughly the size of a wavelength. So you get a good double-slit interference pattern if the slits are close together or approximately the size of a wavelength or a few wavelengths. If they're further apart, you still get an interference pattern, but the light, dark, light, dark, light, dark come so often they smear together and you don't notice it. Similarly, if a light wave goes through a small hole comparable to the size of a wavelength, then we get that spreading out effect that we saw when we had our light in the interference experiments going through a slit. On the other hand, if light goes through a very large hole, like the door to your room, then it basically goes through in straight lines as if it were a particle.

That's why, when you walk out of the room, you don't see me talking on your video screen anymore unless you happen to be looking in a direct line of sight to it. Why do you hear me talking? Because the wavelength of sound waves is comparable to the size of the door of the room. So the sound waves exhibit wave behavior quite dramatically in that situation, and part of that behavior is bending, analogously to the way the light waves in our double-slit experiment spread in circular paths after they went through the slits. So the important point there is a wave is noticeable, or the wave behavior of some wavelike phenomenon is noticeable, when it interacts with a system that's roughly a wavelength in size or maybe a little bit bigger. But if it interacts with a system that's much, much bigger than a wavelength, then you don't notice the wave aspect of it.

Well, let's go back to de Broglie's equation. Wavelength is Planck's constant, which is a tiny little number, divided by the mass times the velocity of the wave. Mass of a big object like me or the moon or the Earth or this tennis ball is a pretty big number compared to the mass of something like an atom or an electron. So the wavelength of this tennis ball, as it's moving along like this, is Planck's constant (that tiny, tiny number, 1 over 1 with 33 zeros after it) divided by the mass (which might be a good fraction of a kilogram) divided by its velocity, which might be a meter a second or something like that—and that comes out to a tiny, tiny, tiny, tiny, tiny number, like a 1 over 1 with 33 or 34 or 30-something zeros after it, of a meter. That's infinitesimally tiny. We'll never, ever notice the wave properties of this tennis ball even though, according to quantum physics, they exist.

So we don't notice the effect of de Broglie's hypothesis on ordinary matter, the kind of matter we deal with, for two reasons. One is that Planck's constant is so small, and, second, ordinary matter that we deal with is sufficiently massive that we don't notice the wave aspects because the wavelength is too small to be of any comparable size to the object itself or anything with which the object might interact, like in the case of the tennis ball, a tennis racquet, for example. On the other hand, if we look at subatomic particles, back to de Broglie's equation again, m is very small, and the wavelength can become large, and, when the wavelength becomes comparable to the size of the systems that are important for those objects—those microscopic objects, those subatomic particles—then the wave aspects will become quite dramatic.

For example, for a typical electron in a typical atom, if you take Planck's constant divided by the mass of the electron times the typical velocity an electron might have in an atom, you'll find the wavelength associated with that electron is roughly comparable to the size of the atom itself. Well, there's a situation in which the electron is interacting with a system, the atom of which it's part, in a situation where the size of the system is comparable to the wavelength. Then the wave properties ought to be noticeable. So, if de Broglie's hypothesis is correct, we ought to notice wave aspects at the subatomic world, the wave aspects of matter, but we won't notice them in the macroscopic world.

There is one interesting exception to that last point, and I'm not going to dwell on it in this course because it's kind of an aside, but it's worth mentioning. Look once again at de Broglie's hypothesis. Wavelength is Planck's constant divided by the mass times the velocity. If I could make the velocity of a particle very small, even if it had a substantial mass, I could make the wavelength big. If I could bring this tennis ball truly to rest, I could make its wavelength very big. I can't bring it to rest because it's composed of zillions and zillions of atoms and things bouncing around, and the tennis ball is always really jostling around. But if I could bring it to rest, I could see quantum effects even in macroscopic-sized systems, that is, systems that are the size of ordinary, everyday objects that we deal with.

There is one case where I can actually do that. If I cool matter to temperatures very, very close to absolute zero, then that frenzied vibration that basically is heat comes as close as it can to a stop. Quantum mechanics doesn't let it stop altogether, but it comes as close as possible to a stop. Then even macroscopic-sized systems will exhibit very bizarre behavior that is directly attributable to the fact that matter has a quantum wave aspect to it. That's one reason low-temperature physics is such a fascinating field. It's used to confirm quantum mechanics, sometimes with systems that are not tiny little microscopic systems.

So, that's one way we can look at de Broglie's matter wave hypothesis. It's not the most obvious way. The most obvious way, for example, is to say, "Okay, de Broglie, you claim that particles, like electrons, have a wavelength, have a wave associated with them." Well, one way to prove that would be to do a double-slit experiment with electrons. Send a beam of electrons like you generate in the back of your TV set through a double-slit apparatus and see if you get an interference pattern. Those experiments have actually been done. They were done as early as the 1920s. De Broglie's hypothesis, remember, is 1923. By the end of the 1920s, those experiments had been done. They aren't done quite like the double-slit experiment I described. They're physically different but conceptually similar.

What one typically does is send a beam of electrons at a crystal of some metal or some other substance. A crystal is, simply, a regularly ordered structure with atoms at particular locations, and the spaces between the atoms act like the slits in a double-slit interference experiment. The electrons go through, and you get them onto a photographic plate or a screen or something, and you look at what happens, and lo and behold, you see patterns that are identical to the same patterns you would get if you sent light or x-rays or something through that same structure. You get interference patterns with electrons. If you say to yourself, "Okay, which slit . . .," let's pretend it's just a double-slit experiment instead of one of these crystals; it's conceptually the same thing, you say, "Which slit did the electron go through?" The electron is a matter particle.

You run into exactly the same situation you did with light. It went through both slits. It went through neither slit. It somehow sampled both slits. It wasn't an electron while it was going through the slits; it was a wave. Exactly the same questions we asked about light—is it a wave or a particle?—can be asked about electrons or other subatomic particles, and the answer comes out the same. There is a wave-particle duality for matter as well as for light. Is matter solid particulate stuff, and light is ethereal waves? No. We already know light isn't just ethereal waves. It's got waves, but it's also got particle aspect. Now we're finding out that matter, which we thought of as solid little things like little, tiny billiard balls—huh uh—it's got a wave aspect as well.

Both matter and light have wave and particle aspects to them, and there's no getting around that. Don't try to ask which slit did the electron go through any more than you can ask which slit did the photon go through. Somehow the electron is acting as a wave when it's going through that crystal or through that double-slit apparatus, and it is sampling all the slits or both slits or whatever there are. And only when you try to detect the electron, only when you say, "I want to find an electron," a little, hard, sphere-like ball, then you'll find a little, point-like object, and you'll say, "Oh, the electron is here." But if you leave the electron alone and let it go on its own, its wave nature determines what it does. Very strange.

Well, this is, in one way, baffling. In another way, it's a beautiful kind of symmetry. What's true for light, that it's both wave and particle, is now true for matter. It's both wave and particle. It kind of parallels Einstein's matter and energy being in some sense equivalent. It's not the same thing, but it's got an esthetic symmetry to it. Both matter, which we thought of as hard, solid particulate, and light, which we thought of as ethereal and wavy, both of them have aspects of what we thought the other was. Both have a wave-particle duality.

How are we going to develop a physics that can describe the behavior of a universe in which both matter and light are both wave and particle? I want to spend a few minutes showing you how that, in fact, was done. We already know the answer for light, and, after I review the answer for light, then the answer for matter will become more obvious. In the case of light, we have equations, Maxwell's equations of electromagnetism, that we showed, not mathematically but conceptually, how they led to the prediction that there should be light waves—because these equations lead to electric fields changing, which make magnetic fields changing, which make electric fields changing, and that whole structure propagated through space, and we found it did so at the speed we called c. Out of that we developed the idea of special relativity.

The conceptual framework for doing that was we had some equations. In fact, Maxwell's equations reduce to what's called a wave equation, an equation that describes the waves. We had this wave equation, this equation that describes how waves behave, and we solved that wave equation, and it tells us how electromagnetic waves behave. It tells us what they look like. It tells us how fast they go. It tells us everything we'd want to know about them. So the conceptual framework, if you will, of classical physics explaining wave behavior is to say, "Okay, we have a wave equation ultimately coming from the theory of electromagnetism, and it predicts the behavior of the waves and all their properties." That's it.

In our wave-particle duality quantum description of light, we modified that, but we didn't just throw out the wave equation. We didn't throw out Maxwell's equations all together. What we simply said was this: We have this theory that leads to a wave equation, Maxwell's theory of electromagnetism.

It leads to a wave equation. The wave equation predicts the behavior of waves. In classical physics, we said, those waves are the reality. In quantum physics, we make that one step removed. Okay, we say, the wave equation predicts what the waves will do. Then there is a statistical link between the wave equation and the possibility of finding the associated particle. Where the wave equation predicts a strong wave, we predict that we're likely to find photons. Where the wave equation predicts a weak wave, we predict that we are unlikely to find photons.

We do the double-slit experiment. We theoretically think about the double-slit experiment. We say, "Oh yes, I see what happens." We predict that there should be a strong wave at this point on the screen, a weak wave here, a strong wave here. That doesn't tell us exactly where we'll find photons, but it tells us we are more likely to find photons at the place where the wave is strong. There's that statistical link. So we have equation predicts wave behavior, but wave behavior does not determine exactly what happens. Instead, it gives us a statistical determination of what is most likely to happen.

Well, the situation for matter turns out to be exactly the same. For every particle there's a wave equation that describes the behavior of its associated matter wave, the wave that de Broglie talked about, the wave that de Broglie hypothesized. There's a wave equation. For electrons and other simple particles in the approximation where they're moving at much less than the speed of light, so sort of a Newtonian situation, a nonrelativistic situation— we do have relativistic equations, but I'm not going to talk about those yet— in that situation, in 1926, Erwin Schrödinger came up with a wave equation that describes the behavior of the "matter waves" associated with simple particles like electrons. The equation is a little bit more complicated than the wave equation that follows from Maxwell's equations, but it's pretty simple, and people in first- or second-year physics in college regularly solve this equation and understand how it predicts the behavior of matter in quantum mechanical-sized systems with considerable accuracy.

So, Schrödinger came up with this Schrödinger equation. What does the Schrödinger equation do? Well, it does what Maxwell's equations did. It gives us the behavior of a wave. What wave? In this case, the matter wave that de Broglie associates with the particle. Then how do we determine what the particle itself actually does? In the same way. Once we know the solution to the equation that describes the matter wave, we know how the matter wave behaves. But we don't say the matter wave is reality. Instead, we say the matter wave gives us the probability that we'll find the particle. So this sounds a little involved. If you want to know how an electron behaves in quantum mechanics, you do the same thing you would if you want to know how a photon behaves. You solve the associated wave equation that describes the waves, and then you ask yourself, "Okay, where am I going to find this electron? What's it going to be doing?" It's going to be found where the wave amplitude is big, and it's unlikely to be found where the wave amplitude is small.

Let me just summarize that with a little conceptual diagram here that gives you a sense of how this all works first in classical physics and then in quantum physics. In classical physics, we could predict the behavior of a matter particle very simply. How did we do that? If we had a classical particle of matter, we had some equations. And, again, I'm not getting mathematical here. I'm just telling you what a physicist would use. We had equations. In that case, they were Newton's laws of motion that told how particles move in response to forces. If we solve those equations of motion, we immediately get from that the behavior of the particle. So the solution to the equation of motion leads directly to the behavior of the particle.

In classical physics, if we wanted to know how a wave behaved—a typical example of a wave is light—we would solve a wave equation. In the case of classical light waves, the wave equation ultimately comes from Maxwell's equations of electromagnetics, magnetism, and the solution to that wave equation gives us the behavior of the wave. So in classical physics, whether we're dealing with matter or waves, we solve one equation. In the case of matter, it gives us the behavior of the particles. In the case of waves, it gives us the behavior of the waves. That's reality and we're done with it.

What happens in quantum physics? In quantum physics, we have this three-step process I've been describing, and the quantum concept here holds for both matter and light because now, with de Broglie, we have this complete symmetry between the two. We solve a wave equation, the Schrödinger equation, if it's a matter particle; the wave equation coming from Maxwell's equations if it's light, if it's a photon we're talking about. We get a solution for wave behavior, but that isn't reality itself. That's only this kind of probabilistic guide to reality, and that tells us the probability of finding the particle, whether it's a photon or an electron. So there's a conceptual look at how this all works out.

What do we gain with de Broglie's hypothesis? What we gain immediately are some explanations for some aspects of quantum mechanics that before we couldn't explain, and I'm going to give you two examples of that. We gain, first of all, an understanding of Bohr's atomic theory. Remember, Bohr said the electrons were stuck in these orbits and they couldn't jump between them and so on, but he gave no explanation of why. Now we have a very good explanation for these allowed orbits in the electron, and here's why. I'm going to do this with an analogy with, say, a violin string. Suppose I have a violin string and I want to make a noise, set it vibrating with the bow of the violin. It can make certain notes, but it can't make every note.

Why not? Here's why not. Take a look. These are called standing waves, the kind of waves that oscillate up and down on things like violin strings or piano strings. On the left, you see a wave. The ends of the string are held fixed. So a wave that can vibrate on that string has to be a wave where it fits, actually, a certain number of wavelengths or half wavelengths, an integer number of them, between the two ends. That's what the wave on the left is doing. Look at the wave on the right. It doesn't work on a string of this length because the right end of it isn't pinned at rest like it's supposed to be in the violin. So you can't make the note corresponding to the right-hand wave on this string. That's not a possible allowed vibration on this violin string.

Well, the explanation of the allowed electron orbits, using de Broglie's matter wave hypothesis, is exactly analogous except these orbits are circular orbits around the nucleus and so we're talking about how many waves can fit in this circular kind of orbit. In the atomic orbits on the left, we have a situation where an integer number of these half wavelengths, of these de Broglie matter waves, fit around that particular orbit. That's an allowed orbit. But on the right we have a situation where the number of wiggles is a little bit more, and the wave doesn't come back on itself. That's not a possible situation.

And, in fact, if you work out the details of this, you find that if you apply de Broglie's hypothesis, you indeed get the allowed orbits in the Bohr theory of the hydrogen of the atom. You get the orbits that are allowed because they're the only orbits on which these waves can close back on themselves, in which these standing waves can exist. So matter waves explain Bohr's atomic theory. The one on the right is ruled out. The one on the left is allowed. And since the energy is related to the frequency of vibration and so on, the energy levels in the atom are indeed quantized, and we have an explanation for all that.

The second phenomena can be understood if you think about what a wave is. A wave is kind of a continuous, spread out thing. If you have a wave going along, it can't just stop abruptly. It has to kind of spread continuously, and the mathematics that comes out of the wave equations describes that. Let me give you an example where that's important. In the very first lecture, I talked about time travel, tunneling, tennis, and, tea. Let's go to that second word, tunneling. Remember, I had you in a concrete prison cage and you were pacing back and forth, and then all of a sudden you found yourself on the outside. Here's why that's possible. It's a phenomenon of quantum tunneling. Here I've got a barrier, and I've got a ball that's moving back and forth between the walls of that barrier. It can't get out. So there it is. A little while later, I might find it moving that way. A little while later, I might find it moving that way.

If this were a classical particle and obeyed the laws of classical physics, I'd never find it outside of there. But this is a quantum mechanical particle in a quantum mechanical box, maybe some silicone structure we've fabricated in making a microelectronic circuit. Maybe it's just an electron in an atom trapped there by the electric forces or something. But it's got some kind of associated matter wave. The way we described its particle, its behavior, is to solve the matter wave equation. Look at the matter wave. Where the matter wave is big, the particle is likely to be found. Where it's small, it's not likely to be found. I'm not going to go through the solution procedure, but there is, in fact, what the solution, what the matter wave looks like for that particular case. It's very big and of basically constant amplitude inside that barrier. That says we're equally likely to find the particle anywhere inside that barrier.

Again, it's the overall height of the matter wave, not whether we're locally at a high point or a low point that's important. It's the overall amplitude. So anywhere between those pieces of barrier, the matter wave is equally likely to be found. The mathematics of the Schrödinger equation, the equation that describes these matter waves, requires that the wave be continuous, so it can't suddenly end abruptly at that barrier. Instead, what has to happen is the wave amplitude tapers off as we go through the barrier, as you can see it doing. Then, when we get outside, there's still a little bit of waviness left, and the waviness outside the barrier has a much lower amplitude than it does inside the barrier. That means we're very unlikely to find the particle out there, but the probability is not zero.

So it may happen that sometime we suddenly find the particle outside the barrier. Don't ask how it got there. Don't try to follow it through the barrier because, if you try to follow it through the barrier, you'll be doing just what we tried to do when we tried to follow the photon through the slit in the double-slit apparatus—you'll destroy the wave phenomenon you're looking at. So you can't ask that question. But it is possible that the wave will be found outside the barrier. Does that really occur? It sure does. Those radioactive alpha particles that Rutherford used to find the atomic nucleus, in fact, escaped from the nucleus of uranium atoms by quantum tunneling through an electrical barrier that classical physics says they don't have enough energy to get through.

As I mentioned the very first day—I think the very first lecture—quantum tunneling also occurs as two hydrogen nuclei come together in the sun. There's an electrical repulsion because they're both positive, and, classically, they typically, at even the 20-million-degree temperature at the center of the sun, don't have enough energy to overcome that. Yet they quantum tunnel through that barrier, come together, fuse, and release energy that makes the sunlight that keeps us alive. Finally, we are increasingly using microelectronic devices that use the principle of quantum tunneling. In fact, we may be building computers in a few years that are thousands of times faster than today's computers because they rely on a very cold—near absolute zero—quantum tunneling device. That one happens to be called the Josephson junction, and people are working on computers that may use these devices.

Here's a picture, in fact, from our newfound ability to manipulate matter. This is a picture taken from a scanning tunneling microscope, which is itself a device that uses quantum tunneling to image systems as small as individual atoms. What you see here is a so-called corral of 48 iron atoms on a copper surface, and inside there is an electron. What you're seeing are ripples that represent the wave function of the electron, the wave that describes the electron, and the electron is sort of trapped. You can just see this beautifully in this picture, trapped inside that circular barrier. But there is a chance it will be found outside, and you can see that in the slight hint that those rippley waves continue outside the barrier.

So this is a so-called quantum corral. It was made at IBM. They manipulate individual atoms into that place and put 48 iron atoms down on that copper substrate with an electron trapped inside, but it's not completely trapped because there's a finite chance that it will tunnel out. Wow. The wave-particle duality for matter is giving us a richness that we never imagined possible.

Finally, now that we understand that matter also has a wave and particle aspect just like light, let's take another look at the uncertainty principle because it's now entirely possible to understand the uncertainty principle just from the wave-particle duality. As long as you acknowledge that matter, like light, has both a wave and a particle aspect, you have to see immediately that it makes no sense to talk about the position and velocity of a particle at the same instant. Why is that? Let's just take a look. Suppose I have a wave that extends forever in both directions, and I've potentially made this wave go right off the edges of your screen because I want you to get the sense that this wave extends forever in both directions. Now, that wave is continuous. Its wavelength is absolutely precisely known.

Remember, in the last lecture we talked about wave packets and how you made a little localized wave packet by combining waves of different wavelengths, and so the packet was slightly undefined in terms of its wavelength and frequency. There was some ambiguity. Well, with this continuous wave, there's none of that ambiguity. The wavelength is absolutely known because the wavelength is absolutely known and because de Broglie says wavelength is Planck's constant over mass times velocity. That means the velocity is absolutely known. So this wave represents a particle for which we know exactly what its velocity is. There is no uncertainty in its velocity.

On the other hand, remember how to interpret the matter wave. The matter wave gives you the probability of finding the particle. This wave extends forever with equal amplitude everywhere. Where's the particle? It could be anywhere. We know nothing about this particle's position. The uncertainty in its position is infinite. The uncertainty in its velocity is zero. This is the extreme case where we know nothing about position and everything about velocity. So this is a situation where the position is completely uncertain.

On the other hand, we could make a wave packet. To make a wave packet, we combine waves of different wavelengths. So in this packet the position is pretty well-known. In this particular case, it isn't perfectly well-known, but it's quite well-known. It's localized. Certainly the particle is not going to be found out here, where the amplitude is zero. You might find it here, but it's unlikely. You're almost certain to find it in this region. The position is well-known, but, because we had to combine a lot of different wavelengths to make that wave, the wavelength is uncertain, and, because the wavelength is Planck's constant over mass times velocity, the wavelength, and hence the velocity, are also uncertain.

So quantum physics tells us that the uncertainty principle follows directly from the wave-particle duality. If you admit that waves and particles are both aspects of matter, then you have to acknowledge that you can't measure the velocity and position of a particle at the same instant because you invariably mix the wave and particle aspects. There's no way around that. Now, I don't want to leave you with the sense that quantum mechanics is all fuzzy and uncertain. There are some things it predicts exactly. It predicts the energy levels of atomic electrons. It predicts the spectra precisely, and the things we measure are exactly in accord with those predictions. What quantum mechanics can't do, though, is give us a billiard ball-like, clockwork universe prediction for the behavior of particles.

Quantum Weirdness and Schrödinger's Cat
Lecture 21

This lecture, which is new to this version of the course, is a digression into a couple of topics that are of considerable interest in the popular study of quantum physics. They are also matters of debate among philosophers and physicists who are interested in the interpretation of quantum physics.

We've seen that electrons or photons in a double-slit experiment act as waves. In a sense, an electron can be said to pass through both slits. More precisely, until we detect the electron, quantum physics says it's in a *superposition state*, neither here (slit 1, say) nor there (slit 2) but with a probability of being found here and a probability of being found there. The existence of superposition states is an inherent result of the wave-particle duality, because it is the wave behavior of an unobserved particle that makes the ambiguity of superposition possible.

Quantum superposition is a real phenomenon, and recent experiments have succeeded in creating an atom that is in two places at once—i.e., in a superposition of "here" and "there."

Measurement of position, velocity, or another property of a particle in a superposition state always gives a definite value. At the instant of measurement, the superposition ceases to exist and the particle is instead in a definite state. This is called the *collapse of the wave function* and represents an inherent involvement of the measuring apparatus and/or observer in the system under observation. In quantum physics, unlike classical physics, observer and observed are inextricably intertwined.

Quantum superposition is a real phenomenon, and recent experiments have succeeded in creating an atom that is in two places at once—i.e., in a superposition of "here" and "there."

Can quantum superposition affect the everyday world? To show how it might, Schrödinger devised his famous cat example. Place a cat in a closed box that also contains a quantum system, in this case a radioactive atom that has a 50% chance of decaying each hour. The decay of the atom is a random event; quantum physics can predict only the probability of decay. Also in the box is a geiger counter that senses the decay of the radioactive atom. The geiger counter is connected to a diabolical apparatus that disperses poison into the cage when the counter detects the decay.

After the cat has been in the box for an hour, is it dead or alive? According to the Copenhagen interpretation, a quantum system is in a superposition state until a measurement forces the collapse of the wave function. Until we look in the box, the cat is in a superposition of dead and alive! Doesn't the geiger counter constitute a measuring system, collapsing the wave function at the instant it detects a decay? The answer is yes only in classical physics, where the measuring system can be considered distinct from what it is measuring. In quantum physics, we must consider the quantum state of the measuring apparatus as well. According to the Copenhagen interpretation, the box and its entire contents are in a superposition until we look in the box.

The many-worlds interpretation says that the universe splits. In one universe, the cat is alive; in the other, it's dead. The quantum state of a macroscopic object such as a cat or a geiger counter is more complicated than that of a single atom. The behavior of its individual atoms is not coherent or coordinated, and the entire cat remains in a superposition state for only an infinitesimal time. After that, it's really alive or dead, and we just don't know it. There is still much to learn about the boundary between quantum and classical systems.

In 1935, Einstein, Podolsky, and Rosen proposed a "thought experiment" that they felt revealed an underlying, objective reality independent of measurement. This is termed the EPR experiment from the initials of their last names. A simple example of an EPR experiment involves a particle that decays into a pair of electrons (not EPR's original example, but easier to grasp). The electrons fly apart in opposite directions. Electrons have *spin*, a microscopic, quantum version of *the angular momentum* of a spinning top or wheel or planet. Spin has two possible directions: "up" and "down" or "left" and "right" or, generally, opposite directions along any line you care to test it on. Angular momentum is conserved, and the original particle has none. So the two electrons have opposite spins. Therefore, if you measure the spin of one electron, you immediately know the spin of the other—even though it's far away and you haven't interacted with it!

EPR claimed that the fact that the state of the second electron could be determined without interacting with it meant that its spin had an objective reality regardless of measurement. Spins of the two electrons are determined together, at the moment they are created. Otherwise, the result of the measurement on the first electron would have to be communicated instantaneously to the second—something Einstein called "spooky action at a distance" and that appears to violate special relativity. Bohr said no. The uncertainty principle still holds for the second electron as well, so EPR does not prove that quantities exist independent of measurement. Copenhagen rules quantum physics!

John Bell (1964) considered an experiment in which the electrons are tested again after a first spin measurement. He showed that the statistical distribution of the results would be different depending on whether the states of the electrons were really determined when they were created or were truly indeterminate until measured. EPR experiments done in 1982 by Alain Aspect of the University of Paris yielded statistics confirming quantum indeterminacy. Since then, ever more sophisticated experiments—including some in which the measuring apparatus is switched randomly after the particles are created—have confirmed that quantum indeterminacy is an inherent feature of EPR-type phenomena. By the late 1990s, experiments in Switzerland had confirmed the EPR effect in particles separated by several miles.

What does it mean? There is a strange "quantum connectedness" between the two particles in an EPR experiment. Somehow each "knows" what's happening with the other. Quantum physics is an inherently *nonlocal* theory; particles in different places can be "entangled" in a way that precludes a strictly local description of each. However, this "spooky action at a distance" (in Einstein's memorable phrase) does not violate relativity because it is impossible to use the effect to send information. ■

Gribbin, *In Search of Schrödinger's Cat*, Chapters 8–11.

Lindley, *Where Does the Weirdness Go?*, pp. 88–132.

Davies and Brown, *The Ghost in the Atom*, especially Chapter 1.

Lindley, *Where Does the Weirdness Go?*, pp. 133–148.

1. Does the "uncertainty" of quantum physics, with its seeming measurer-measured interconnectedness, have any parallels in other areas of modern thought? (This is an epistemological question—what can we know and how can we know it? Is anything certain and on what basis can we assert it is?)

2. Review newspapers and news magazines (not specialized scientific publications) for articles on quantum physics, say for one month. How many articles did you find? How well did they explain things (based on what you know from this course)? Did they offer any "practical" application of the latest findings? Do you think that quantum physics is important in everyday life?

Einstein-Podolsky-Rosen Experiment

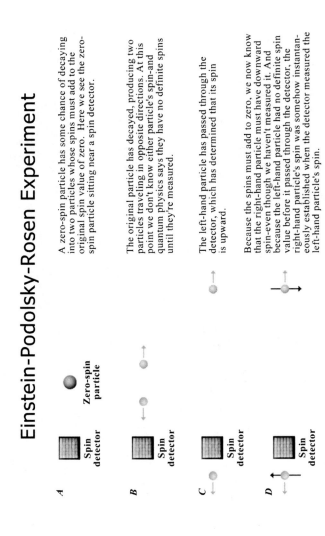

A

Zero-spin particle

Spin detector

A zero-spin particle has some chance of decaying into two particles whose spins must add to the original spin value of zero. Here we see the zero-spin particle sitting near a spin detector.

B

Spin detector

The original particle has decayed, producing two particles traveling in opposite directions. At this point we don't know either particle's spin-and quantum physics says they have no definite spins until they're measured.

C

Spin detector

The left-hand particle has passed through the detector, which has determined that its spin is upward.

D

Spin detector

Because the spins must add to zero, we now know that the right-hand particle must have downward spin-even though we haven't measured it. And because the left-hand particle had no definite spin value before it passed through the detector, the right-hand particle's spin was somehow instantaneously established when the detector measured the left-hand particle's spin.

411

Schrödinger's Cat

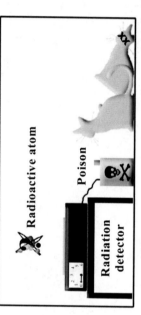

Schrödinger's famous thought experiment shows how quantum ideas applied at the macroscopic level can lead to disturbing conclusions. The experiment consists of a radioactive atom with a 50 percent chance of decaying in one hour. A radiation detector signals when the decay occurs, and is connected so as to open the lid on a bottle of poison. This diabolical apparatus is sealed up in a box that also contains a live cat. Thus the fate of the cat is determined by a random, subatomic event.

If we seal up the box and wait an hour, what state will the cat be in? Classical physics says it will be either dead or alive, even if we haven't opened the box. But in the standard interpretation of quantum physics it makes no sense to talk about quantities that can't be measured. As long as the box is closed, we can't "measure" whether the cat is dead or alive. Therefore, according to the standard interpretation, the cat is in a "superposition" of dead and alive. After one hour there's a 50-50 chance that the atom will have decayed, so the cat is in an equal mix of dead or alive. Opening the box constitutes "measuring" the cat's state and thus leads to a definitive answer to the question "dead or alive?"

Quantum Weirdness and Schrödinger's Cat
Lecture 21—Transcript

Welcome to Lecture Twenty-One: "Quantum Weirdness and Schrödinger's Cat." This lecture, which is new to this version of the course, is a digression into a couple of topics that are of considerable interest in the popular study of quantum physics. They are also matters of debate among philosophers and physicists who are interested in the interpretation of quantum physics. One of them goes back to the early years of the twentieth century. Another is still a matter of very much contemporary research. So we're taking again a digression into a couple of interesting questions that arise in the study of quantum physics. Before we do, let me discuss a particular concept called a *superposition state.*

In quantum physics, because systems are described ultimately by waves—remember, it's ultimately a wave equation you solve; you get a wave solution and that wave solution then determines, but only statistically or probabilistically, what particles are doing—the wave equation and the wave nature of a quantum system are important. And waves have the property that one can form a superposition of more than one wave. Two waves can exist at the same place at the same time. That's a superposition. That's what leads to interference, if you have constructive interference or destructive interference. Because of the possibility of superposition, you can have quantum systems whose state is somewhat ambiguous. For example, when you ask, "Does the electron go through slit one or slit two?" or "Does the photon go through slit one or slit two?" One way of posing that is to say, when I answer to you, "Well, you can't answer that question. We don't know. It goes through both."

One way is to say, "Well, the electron is in kind of a superposition of two states. One state is the state of going through slit one, and another state is the state of going through slit two." With a pure matter particle, that's absurd—nonsense—but when we talk about waves, that's entirely possible. The wave nature of matter and of light makes possible these superpositions. To give you another example, I could have an atom in which an electron is definitely in one of the allowed orbits. I could have an atom in which it's definitely in another of the allowed orbits. I can't have an atom where it's in between one or the other, but I can have an atom in which an electron is in the superposition. It's got a 25 percent chance of being in an upper orbit and a 75 percent chance of being in a lower orbit. Or maybe it's 50/50. Or whatever. That's a superposition state.

When I go to measure, when I go to detect or determine what's going on, what I've actually got, I'll always find the electron in either the lower orbit or the higher orbit. If the wave equation is telling me I have a 75 percent mix of the lower orbit wave and a 25 percent mix of the upper orbit wave, then, if I do the experiment 100 times, I'll find roughly—not exactly, because this is all statistical—roughly 75 percent of the time I'll find the atom in the lower orbit state. Twenty-five percent of the time I'll find the atom in the upper orbit state. That's what a superposition state is. It's a mix of two possible wave solutions, and it's entirely possible because of the wave nature of matter and of light in quantum physics.

So we can have these superposition states. The problem is, once we go to measure the system, if we ask, which slit did the electron go through, we can ask that question, and we can ask, did it go through the left-hand slit? We can determine the answer to that question. At that point, the electron ceases to be in the superposition state, and it becomes in the definite state associated with going through the left-hand slit. If I look at this atom that's in this mixed, superposition state and I say, "Okay, is this particular atom in the lower level or the upper level?" I do a measurement. Even though it's originally in this superposition, I'm going to find either it's in the lower level or it's in the upper level. I'm not going to find both. I'm not going to find somewhere in between. In other words, what I'm going to do by measuring the system is force it into one of the two possible states. The superposition collapses. The uncertainty associated with knowing which state the system was in goes away. The wave behavior also goes away, and I find the system in one definite state.

That phenomenon of making the measurement, of taking a quantum system that may be in an ambiguous or superposition state and doing a measurement on it, is the process of measurement, and measurement itself is a matter of philosophical debate in quantum physics. Exactly what does measurement mean? And, again, the reason that's a matter of debate and interpretation goes back to this original situation in quantum mechanics that, because of the quantization, because there's a minimum interaction we can have when we interrogate nature, when we do experiments, the act of measurement becomes an act that involves, disturbs, or interacts with the system in question in a way that is not reducible to zero interaction. Consequently, the theory of measurement—what it means to measure—becomes a major part of quantum physics.

One thing it means to measure, though, is to force a system into a definite state. I've got a wave going along, and there's a photon associated with that wave, say it's a light wave, and we say, "Well, the wave is spread out all over space, so we know the photon could be anywhere." The act of measuring the photon, the act of catching the photon, the act of detecting the photon with a photoelectric experiment or with a Compton-type experiment or with some other experiment like a photographic film or an electronic detector, the act of localizing the photon and saying, "Ah ha, here it is." That act collapses that uncertain wave function into one where we know exactly where the photon is. It's called the *collapse of the wave function*.

So we have these superposition states. We have quantum measurements that then collapse the wave function and result in a definitive value for whatever quantity we're trying to measure. In the process, they may destroy any evidence about some other quantity we might have wanted to measure. For example, in Heisenberg's quantum microscope, we had an electron. We didn't know its momentum; its velocity state, that is, what velocity it had; what direction it was moving; and how fast. We didn't know exactly where it was. We set out to measure them. If we did an experiment that collapsed the wave function to one of definite position, we lost information about velocity. If we did an experiment that collapsed the wave function to one that had information about a definite velocity, we lost information about position. If we collapse the wave function, we find the quantity we're trying to measure—quantum physics gives us a definitive value—but, in the process, we lose the properties that the system had before.

Now, we've known about this quantum superposition idea for a long time. The idea, for example, of an atom that isn't necessarily in one definitive state but has a chance of being found in this state or that state—that's been with us for a long time. There are more bizarre instances of quantum superposition that have only recently been realized experimentally. For example, experimenters have recently created a single atom in a superposition state that involved being in two quite different places, separated by many times the size of an atom, at the same time. What does that mean? Well, it meant when you weren't looking at the atom—remember, when you're not looking at a quantum system, it's the wave behavior that governs how it evolves—when you're not looking at it, there's a probability that the atom will be here and another probability that it will be here, and they created the wave function in such a way that there's not much probability in between.

But which of those two places is it? You can't say. You go to do a measurement, and you look here, and you find the atom here. The wave function collapses, and there's no longer any chance you'll find it here. You've now put it in a state where it's definitively here. You measure over here and see if it's here and, if you find it, there it is, and you've collapsed the wave function to a state where the atom is definitively here and not here. So, by doing the measurement, you influence the system, you take away the ambiguity, you force the quantity you're looking for—position of atom, in this case—to have a definitive value. But, before you did the measurement, you had created a state in which the atom, in that case, was equally likely to be found in either place. You do 100 measurements and for roughly 50 of them—not exactly, but roughly 50 of them—it comes up here, and for roughly 50 of them, it comes up here.

That experiment has recently been done, and a single atom was created literally in two places at once. So it was in two places at once, but you never caught it in two places at once. You knew you created a wave function that had that property. When you go to measure it, you find it in one place or the other with a certain probability in either place. Remarkable. Can this have any bearing on the macroscopic world?

Well, Schrödinger himself posed a very, very famous example to get us to think about this question of whether quantum superposition could exist in the macroscopic world, the world of everyday-sized objects. The famous example is called Schrödinger's cat. There's a book, by the way, in my bibliography for this course called *The Search for Schrödinger's Cat*, which is a good general introduction to quantum physics and especially some of these quandaries it presents. Here's how Schrödinger's cat experiment works. Schrödinger's cat is, again, a thought experiment. I hope nobody has carried out this experiment because it's a little bit gruesome.

So, let's take a look at what the experiment looks like. Here's the experiment. We take a nice, friendly, live cat, and we put it in a closed, sealed box. In this box, we put some kind of quantum mechanical system. Now, remember, the main aspect of quantum mechanics is that the behavior of a system is governed by its wave equation, but what actually happens ultimately is only statistically linked to what the wave solution says. So we're going to put in this box a radioactive atom, and I'm going to choose a radioactive atom that has a half-life of one hour. What that means is, if I take 100 of these radioactive atoms, in an hour, 50 of them, roughly—again, it's statistical—it may be 49 or 47 or, a rare chance, it might be 62 or something, but, on the average, it's roughly 50, will have decayed away, spewed out a radioactive particle, and decayed into something else. The other 50 won't have. That's what it means to have a half-life of an hour.

Another way to put it is, if I have just a single atom there and I wait an hour, there's a 50/50 chance that I'll find the atom there or not. Now, you can debate what statistics means when I have only one thing. Ultimately, I think what it means is, if you repeated the experiment 100 times, 50 percent of the time it would do this and 50 percent it would do that. That's my interpretation of it. But, anyway, we've set up this system. We've got a radioactive atom. It's got a half-life of one hour, meaning it has a 50/50 chance of having decayed sometime in the next hour after we start the experiment. We have a radiation detector, a Geiger counter or some other radiation detector, looking at this radioactive atom. If the radioactive atom decays, the Geiger counter detects that high-energy particle that spewed out in the radioactive decay. The Geiger counter makes an electrical pulse.

If you've ever heard a Geiger counter, it goes, "click." It makes a little, loud pulse in a loudspeaker every time a high-energy particle comes through its detector system, and that's entirely doable. We've known how to do that for a century. So the radiation detector works. It detects the radiation, and it's hooked up, not to a loudspeaker that goes, "click," but, rather, to a flask of poison. And, if the radioactive atom decays, a mechanism connected to the Geiger counter opens the flask of poison, and the cat dies. That's the Schrödinger's cat experiment. It's cruel and gruesome, and that's why we don't want to do the experiment. But it's an interesting thought experiment to think about whether the probabilistic nature of the quantum world can affect the macroscopic world. And, in this case, it clearly can.

The first thing we can say is, because the decay of the radioactive atom is truly a random event—remember, it's got a 50/50 chance of decaying—if you will, its wave function after an hour has evolved to a superposition that says we're 50 percent likely to find the atom intact and 50 percent likely to find it in the form of a high-energy particle that spewed out and a remnant that's something else other than the original atom. That's what that means. If we calculated the quantum mechanical wave function for this system, it would evolve from a system that started out as pure this atom into a system that was a superposition of this atom and a decayed state. And, if we did a measurement, if we looked at the atom, we would find either one or the other. We would collapse the wave function, and it either would have decayed or it wouldn't have decayed.

So this is a random event. This random event, microscopic atom-scale event, quantum mechanical random event, is linked through the Geiger counter—which might just make a benign click, but, in this case, it's going to open a vile of poison—to a macroscopic system. So whether the cat dies or not is truly random—truly random. It's not deterministic. There is no clockwork universe at work here. We set up this experiment, and we cannot predict what's going to happen. There's a 50/50 chance the cat is going to die. You can do 1,000 of these experiments, and in roughly 500 of them the cat will die, and in roughly 500 it won't. But in any single one, there is nothing—nothing—that gives you the ability to predict that. That statistical link from wave equation to wave behavior is definitive, but, from wave behavior to what we actually find in the particle material world, that's statistical, and there is no going deeper than that, according to the standard interpretation of quantum mechanics.

Okay, so we set up the apparatus. We put the radioactive atom there. We turn on the Geiger counter. We set the mechanism for the flask. We drop the cat in. We close the box and we wait an hour. The fascinating question is not so much whether probability rules the life or death of the cat. We already know that's a probabilistic event. The fascinating question, according to Schrödinger, is, what state is the cat in an hour later, before we open the box? Now, surely your answer is going to be, sometime during that hour, the atom may have decayed. If it's decayed, the radioactive particle has gone into the Geiger counter, the Geiger counter has sensed it, and the Geiger counter has operated the mechanism that opens the vile of poison, and the cat is dead. Or maybe the atom didn't decay, and the cat is just fine. But surely it's either alive or dead.

That interpretation turns out to be inconsistent with at least the strictest interpretation of quantum mechanics, according to the Copenhagen school. The Copenhagen school would say, until you do the measurement, until you collapse the wave function, until you do something that forces that superposition state into one of its possible concrete outcomes, the system is in a superposition state. It's like that atom I just described that is in two places at once. It literally is. It's got a wave function that says it's likely to have a 50 percent chance of being here. Until you ask where it is and actually do the experiment to measure where it is, it is in both places at once or neither place. If you ask the question, "Which place is it?" I guess I'd answer, "It's both." If you measure it, there's a 50 percent chance you'll find it here and 50 percent chance you'll find it here.

The strict Copenhagen interpretation of quantum physics says, in the case of Schrödinger's cat, you wait an hour and the cat is in a superposition state of half dead and half alive. You wait half an hour and it's in a super state, a superposition state, of more alive than dead, but there's still a chance either way. That is the interpretation Copenhagen's school imposes on Schrödinger's cat experiment. This macroscopic-sized object, this cat, is, according to people who go by the strict interpretation of quantum mechanics, in a state that is neither dead nor alive. It's a 50 percent superposition of dead and alive. You then do a measurement. You open the box, and, in the process, you collapse the wave function to one of those two possible states, and you find a cat that's dead or alive. But the strict interpretation of quantum physics says the cat is neither dead nor alive until the moment you open the box, and your act of doing that is what determines what happens.

Wow. You might say, wait a minute. Didn't the Geiger counter measure whether the atom decayed, and, therefore, wasn't that the thing that collapsed the wave function so the cat really is dead or alive? Well, the strict interpretation of quantum physics says, no, the Geiger counter and the cat and the flask, they're all part of the sealed system. They're all one system subject to laws of quantum physics. It's the wave function of that system which has evolved itself to the superposition state, and it's really you opening the box that collapses the wave function and creates which state it is, whether the cat is dead or alive.

The obvious retort to that is, what on Earth about the cat? What does the cat feel all this while? I don't want to get into that. That raises the whole question of consciousness, which is a fascinating question that some people are trying to link with quantum physics, I don't know yet how successfully. I don't know what consciousness is. It's a fascinating question. I'm not even going to think about that right now, but I'll let you speculate about what the cat feels, especially if it's in a superposition of dead or alive or worse. Put yourself in there instead of the cat.

Are there ways out of the strict Copenhagen interpretation? Yes, there are. Some of them are rather bizarre. There's a logically consistent but extremely unusual interpretation of quantum physics. It's called the Many Worlds Theory, and it says every time a situation occurs where a wave function collapses—for example, I have the two atoms here—every time I do a measurement to see if the atom is here, 50 percent of the chance I find it is and 50 percent of the chance I find it isn't. The Many Worlds Theory says, every time I do a measurement like that, both possibilities are realized, but the universe splits—bifurcates—and suddenly where there was one universe there are two universes, in one of which the atom is here and in one of which it's there.

Or, in Schrödinger's cat, when I open the box, the universe bifurcates in two, and there's a universe in which the cat is dead and a universe in which the cat is alive. At every place an interaction is done that forces a superposition to collapse—a wave function to collapse—to one definitive state, the universe splits into multiple universes, depending on how many outcomes there are, and each outcome is realized in a different universe. Wow. Talk about richness of the universe. That is a logically consistent interpretation of quantum mechanics, bizarre as it sounds.

I think a more reasonable way of looking at the Schrödinger cat situation is one that has been discussed in the context of really how much can we apply quantum physics to macroscopic systems like the cat. Well, in principle, we can apply quantum physics to all the individual atoms and molecules, and as long as they retain a kind of quantum coherence—their waves all sort of know about each other and are in step with each other—then we can talk about quantum mechanics applied to the whole macroscopic object like the cat. But if you take that interpretation, you can show that, in a very, very, very short time—a miniscule, almost instantaneous, time for something as complex as a cat—the individual wave functions of the individual parts will become incoherent, and the system will no longer be a quantum system. It will be a classical macroscopic system to which our everyday common sense applies, and the cat will, after a very short period of time, in fact, be either alive or dead and not in some silly superposition.

Well, I don't know the answer that makes the most sense to me. But what it points to is that one thing we don't understand about quantum physics is where the boundary is between quantum physics. We know that things the size of atoms, even some of our microscopic electronic circuits, are governed almost completely by quantum physics. Things like baseballs and our cells and even bacteria are governed by classical physics. Where's the boundary in between?

Bohr has another principle, the Correspondence Principle, that says quantum physics has to go over smoothly to classical physics in the limit as the size of the quantum system gets bigger and certain numbers in quantum physics get bigger. So it's a smooth transition, but where does it occur, and what's going on in this mid-scale region? That's the region where some people are looking, for instance, for explanations of consciousness. There's a lot to learn still about that boundary between classical and quantum physics.

Well, let's move on from Schrödinger's cat and look at one other example of quantum weirdness, in this case something called the EPR experiments, about which a lot has been in the philosophical debate lately. This goes back to 1935, when Einstein, Podolsky, and Rosen came up with a "thought experiment" that they claimed showed that the Copenhagen interpretation of quantum physics couldn't be right. I'm going to give a description of the EPR experiment that is rather different than what they proposed, but it's rather easier to understand conceptually.

Before I can give that, I've got to give you a little exercise in particle physics. It turns out that many of the elementary particles like electrons can be thought of very crudely as being in a state of spinning, like a little ball spinning around an axis. Now, this is a very bad way to think about it, but it's the only way we can really picture. I've added a little pencil to this ball, representing an electron, to show you the direction it's spinning in. In fact, what I'm going to say is, if it's spinning this way, counterclockwise, if I look down from above, then I'm going to have the pencil pointing up. If it's spinning this way, clockwise as viewed from above, I'm going to point the arrow down. So I'm going to use the direction of the arrow to represent which way it's spinning. If I curl my right hand around the pencil, let my right thumb point toward this point, that's the way the thing is spinning.

So that's what the pencil represents, the direction of spin. I don't care what these particles are. I'll just call them particles. But I'm going to imagine I have another particle that has the property that its spin is zero. Spin, by the way, is closely related to angular momentum, rotationalness. This one has no spin and it's radioactive. Now, spin, or angular momentum, is conserved. It means it can't change. This one has no spin but it's radioactive, and it's going to decay into a couple of these. If it decays into a couple of these and it had no spin, what's got to be true? The pair of particles it decayed into has to have no spin either. But this one's got a spin, and this one's got a spin. So, if this one's spin is this way, this one's spin has got to be down to cancel it.

That's the basic idea we're going to be using in the EPR experiment. We're going to start with a particle that has no spin, and we're going to end up with two particles that have spins in opposite direction because they have to conserve that quantity. Now, there's another thing you need to know about spin. It's a funny quantum concept. Its direction is not uniquely determined until you measure it, just like any quantum quantity, as we've now found, isn't really determined until you try to measure its value, and then you collapse the wave function. So here's a particle with spin, but I don't know which way this arrow is pointing. I send it through a detector, and the detector can detect either spin up or spin down but nothing in between. I send the particles through the detector, and I will always find that it either has spin up or spin down, and I'll find, unless some weird situation is giving it a symmetry, that it has a 50/50 chance of being up or down. But before it went into that detector, I can't say it has any particular direction at all. It could be anywhere. And if I made a detector aimed at detecting spins that were either this way or this way, and I sent the same particle through that, I would find its spin was either this way or this way.

That's the way quantum physics works. I asked a quantum question, and I'm going to get an answer that's consistent with the question I asked. Is the spin up or down? Yep, it's either up or down. Is the spin sideways or backwards? Yep, it's either sideways or backwards. You'll find one or the other. That spin property is what I'm going to use to probe this unusual paradox that EPR talked about.

So here's the EPR paradox situation. What we're going to do is start with a spin detector, and for our spin detector we are going to use some kind of apparatus that can tell us whether the spin of a particle is up or down. So there's my spin detector. I'm going to put somewhere to the right of the spin detector that zero spin particle, the one that has no spin and that is radioactive and that is going to decay. What is going to happen? The particle is going to decay. It's going to send out two particles going in opposite directions. They're going to have spin, but the spins are undetermined until I make the measurement. After the left-hand particle has passed through the spin detector, I can look at the spin detector, and it will tell me the direction of its spin. Perhaps it measured a spin up. Fifty percent of the time it will measure up. Fifty percent of the time it will measure down.

If I determine the spin this way, of the left-hand particle, and I'm in the situation shown in the picture right now, if I determine the spin of the left-hand particle is up, what do I know about the right hand particle? Well, because the original particle had zero spin, and spin is conserved, if the left-hand particle has spin up, I know immediately the right-hand particle had spin down. Well, so what, you say. That's no big deal. It is a big deal. The reason it's a big deal is because, before the left-hand particle went through the spin detector, its spin was undetermined. That doesn't mean nobody knew what it was or I hadn't measured it yet. It means it really was indeterminate. It could take on any value.

When it went through that detector, it was going to take on a positive value 50 percent of the time and a downward value 50 percent of the time, and you can't know ahead of time which. It really was indeterminate, and the value of the spin of the right-hand particle was also, at that instant, indeterminate. So when I pass the left-hand particle through the spin detector, that act determined, immediately and instantaneously, what the spin of the right-hand particle was even though it was way far away. Einstein, Podolsky, and Rosen argued from that that this is absurd. They said, "No, the spins had to have been established beforehand, and the interpretation of quantum mechanism, that these quantities are undetermined until you measure them, can't possibly be right." That's what they argued.

Bohr said, "Nope, the Copenhagen interpretation of quantum mechanics is what really reigns, and it really is true that the act of measuring the spin of the left-hand particle fixes the spin of the right-hand particle." How are we going to resolve this? Well, in 1964, John Bell developed a famous theorem that looked at the statistics of doing this measurement. In fact, he particularly said, "What happens if I put a second spin detector after the first one?" Bell was able to show, using some mathematics I'm not going to go through, that, if you take that spin that's now up and run it through a spin detector that looks for spins—horizontal or vertical, or back and forth, rather than up and down—you'll get some statistical chance of finding it one way versus the other way. Bell showed that the statistics of the outcome of that experiment depends on whether or not the spins were established at the instant the two particles were created or whether they were really established by the measurement at the spin detector. Bell's theorem—*Bell's inequalities*, they're called—gives a very definite prediction that those statistical outcomes should be different, depending on which of those two cases is correct.

Remember, Einstein, Podolsky, and Rosen said, "Well, this proves that, really, the spins must have been created up and down right when the particles were created." Bohr says, "No." Bell says, "I know how to tell the difference—you do this experiment." Now, it's a very difficult experiment to do. It was a thought experiment for Einstein, Podolsky, and Rosen, but it was, in fact, done in the 1980s by Alain Aspect at the University of Paris first and then repeated by many others, and the first experiments that Aspect did, indeed, confirmed the statistics that suggest quantum indeterminacy reigns here. The statistics were consistent with the Copenhagen interpretation that the spins didn't have actual values until the left-hand spin was measured, and then the right-hand spin instantaneously acquired its value.

Those experiments have been done with increasing sophistication since then. There have even been experiments set up in which the spin detector essentially didn't exist or didn't have an orientation at the moment the particles were created, and, in the short time it took the particles to fly to it, the electronics was done fast enough to switch that spin detector in different orientations. Nevertheless, the experiment comes out the same. Recently, the experiments have been done with such sophistication that the particles were piped by fiber optics—they were photons, in that case—to different suburbs on different sides of Zurich, Switzerland, and the measurement was made in one suburb of the one particle, and that immediately determined the state of the other particle. Weird.

Einstein was really bothered by this. He called it "spooky action at a distance," and you can imagine he didn't like it, because what we have happening is here a measurement over here of the one particle determines instantaneously—instantaneously—the state of the other particle. How does that jive with relativity? It doesn't sound too good. Actually, it isn't a problem for special relativity, and the reason is there is no way we could send information using that mechanism. There's no way because we don't know what the spin is going to be. We can't say, one if by land and two if by sea like Paul Revere did. We can't use that to signal. We can't say, if you measure a downward spin of the second particle, that means the British are coming. We can't do that.

Why? Because we don't know what the spin of the second particle is going to be, so we can't send information. Nevertheless, somehow the first particle has—has it sent information? Well, I don't know. But somehow the first and second particles are connected in a kind of strange, quantum way, and it appears that that connectedness, that so-called *nonlocality*, is a part of the quantum world. You can imagine that philosophers, maybe even theologians, all kinds of people, are having a heyday with that. There does seem to be this strange quantum connectedness. There is some sort of connection between something happening over here and something happening over here. Now, it doesn't mean we're all connected to some kind of giant, holistic web necessarily, but it does mean there are strange situations, and the EPR experiment brings them out, in which quantum weirdness is so weird that what happens over here instantly affects what happens over there. In some way, those two particles were intimately connected from their creation, and they remain connected even as they spread to significantly large distances throughout the universe. That's quantum weirdness. Weird.

The Particle Zoo
Lecture 22

[W]e're going to resume our descent into the heart of matter. We're going to resume a search for the ultimate constituents of everyday matter—bizarre, exotic matter—all the matter that makes up the universe.

By the early 20th century, physicists were aware of several atomic particles and, over the next decades, using "atom smashers," they learned of many more. This created a complexity that by the end of the century scientists had largely "simplified."

We move from macroscopic object to molecule (e.g., water) to nucleus and electrons of the atom (think of Rutherford's model). Electrons appear to be truly elementary, indivisible constituents of matter. The atomic nucleus is made up of *protons* and *neutrons*, collectively called *nucleons.*

Nucleons are made of still smaller *quarks*, believed to be truly elementary or fundamental. Quarks carry fractional electric charges ($\pm 2/3$ or $\pm 1/3$) of the electron charge. They are described as "up" and "down" quarks, depending on the charge. Quarks combine in threes to make protons, neutrons, and a host of other particles, collectively called *hadrons* (heavy particles), that were once thought to be elementary. Hadrons other than protons and neutrons are unstable, eventually decaying into other particles. Quarks also combine in twos to make another class of particles called *mesons.*

Today, physicists recognize just three fundamental forces that are responsible for all interactions in nature. Most believe

Gravity is the weakest of the forces. It is at once the most obvious to us and, in many ways, the least understood.

that the three will someday be understood as aspects of a single, underlying force (more on this in Lecture 24). *Gravity* is the weakest of the forces. It is at once the most obvious to us and, in many ways, the least understood.

Gravity is universal; every bit of matter gives rise to gravity and every bit of matter responds to gravity. The general theory of relativity is our description of gravity; i.e., gravity is the geometrical structure of spacetime.

The *electroweak* force comprises the electromagnetic force (as described by Maxwell's equation) and the so-called *weak nuclear force*. The electromagnetic force is responsible for the structure of everyday matter from the scale of atoms on up. The weak nuclear force mediates certain nuclear processes, including nuclear reactions that make the Sun shine. The *color force*, also called the *strong force*, acts between quarks, binding them together to make hadrons and mesons. Unlike gravity and the electroweak force, the color force does not decrease in strength with increasing distance between the particles. For that reason, it appears impossible to separate quarks, and isolated quarks have never been observed. The residual color force between quarks in different nucleons provides the *nuclear force* that binds atomic nuclei together.

All matter is composed of two basic types of particles, which interact via the fundamental forces. One of these basic types is the quark. The other is the lepton, or "light" particle. Leptons include electrons and related particles and the elusive neutrinos.

There are three families of particles, each including two quarks, an electron-like particle, and a neutrino. Everyday matter is composed of particles from the first family. These include the up quark, the down quark, the electron, and the electron neutrino. The second family includes the charmed quark, the strange quark, the muon (as in the time-dilation experiments from Lecture 9), and the muon neutrino. The third family includes the top quark, the bottom quark, an electron-like particle termed the tau particle, and the tau neutrino. Discovery of the top quark in the mid-1990s completed verification of the three-family structure of the standard model.

Particles in the second and third families are more massive and, therefore, require more energy to create. All are unstable, which is why they aren't normally found. They can be created in particle accelerators and in high-energy astrophysical processes and were important in the early history of the universe. Experiments show that there probably cannot be additional families of matter.

In addition to the matter particles are particles called *force carriers*. In the quantum description, a force between two particles involves an exchange of a third particle—the force carrier. The force carrier for the electromagnetic force is the familiar photon; for the weak force, the carrier is the W and Z *bosons*; for the color force, it's the *gluon*. For gravity, the force carrier would be the *graviton*—although this awaits a successful quantum theory of gravity (more on this in Lecture 24).

Finally, a massive particle called the *Higgs boson* should exist and, if it does, is responsible for the other particles' masses. Particle accelerators becoming operational around 2005 may be able to produce the Higgs particle. ∎

Essential Reading

Kaku, *Hyperspace*, Chapter 5.

Kane, *The Particle Garden*.

Spielberg and Anderson, *Seven Ideas that Shook the Universe*, Chapter 8.

Suggested Reading

Lederman, *The God Particle*.

Pagels, *The Cosmic Code*, Chapters 3–11.

Riordan, *The Hunting of the Quark*.

Questions to Consider

1. If we can never isolate an individual quark, how can we possibly say that these particles exist? For that matter, we can't see electrons, protons, and neutrons either, and we have only the most rudimentary images of atoms. How do we know these particles exist?

2. How can atomic nuclei stick together? They contain only positively charged protons and neutral neutrons, so the electrical repulsion of the protons should tear them apart.

The Structure of Matter

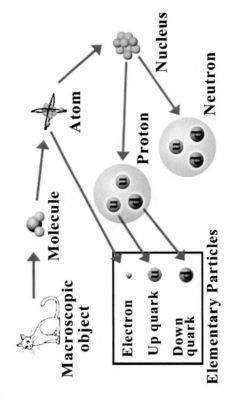

Force in Quantum Physics
(exchange of a "virtual particle")

Example: the electromagnetic force:

Electron   Virtual photon   Proton

Force carrying particles:

Electromagnetic force	photon
Weak force	W , Z boson
Color force	gluon
(gravity)	(graviton)

The Particle Zoo
Lecture 22—Transcript

Welcome to Lecture Twenty-Two: "The Particle Zoo." We got ourselves into quantum physics by looking at experiments that were done largely around the turn of the twentieth century to try to determine the ultimate structure of matter. We started with Democritus in 400 B.C., who argued there were indivisible atoms, but by the time we reached the year about 1900, we recognized that there were atoms, but they weren't indivisible. And, by the time we had Rutherford's solar system model, we at least knew of two kinds of subatomic entities, electrons, which had been discovered by Thompson in the late 1800s and comprised the outermost part of the atom; the electrons being in orbit then around the very massive, positively charged nucleus. So we had nuclei and we had electrons. We also had the subatomic particles, alpha particles, particles called beta particles, which turned out mostly to be electrons, and other particles that were emitted by radioactive atoms.

So there already, at the turn of the twentieth century, were a host of subatomic particles that physicists were discovering. At that point in this course, we stopped and we looked at the evolution of the theory, quantum physics, that describes the behavior of matter at the subatomic scales, and we found some very strange and bizarre happenings. But we're done with that for now, and we're going to resume our descent into the heart of matter. We're going to resume a search for the ultimate constituents of everyday matter—bizarre, exotic matter—all the matter that makes up the universe.

Throughout the twentieth century, particle physicists did increasingly sophisticated and increasingly powerful experiments that were essentially like Rutherford's experiment that discovered the atomic nucleus. Namely, they bombarded matter with high-energy, subatomic particles. Again, as I said before, the drive for ever higher energies to probe ever deeper into the heart of matter is what gave rise to the building of ever larger particle accelerators, to the point where today we have accelerators that are many miles long or miles in diameter and cost many billions of dollars and don't get built all that often for that reason. But we need to go to higher and higher energies to probe ever deeper into the heart of matter.

433

Well, by the middle of the twentieth century, it was evident that there were hundreds and hundreds and hundreds of different kinds of supposedly elementary particles that had been created in these particle accelerator experiments. You smash one high-energy particle into matter, and all kinds of other particles come out, and you can characterize them by things like their mass and their charge and their spin, like we talked about for electrons and a few other fundamental properties that matter seems to have at the subatomic scale. But why should there be hundreds and hundreds of particles? Why should nature be so complex? I thought physics was about nature being simple.

So physicists, in particularly the second half of the twentieth century, tried to piece together some kind of overriding structure that would explain the plethora of particles that were found, that hopefully would explain them in terms of a few smaller, more fundamental entities. That result has been largely successful. By the 1990s, a complete picture was in place—it's called the *Standard Model of Particles and Forces*—that seems to explain all of the particles we've observed and is basically a complete theory, although it, too, requires some explanations that aren't contained within that theory. Rather than give you the history and all of the hundreds and hundreds of particles in my talk here on "The Particle Zoo," I'm simply going to tell you what we now know in the twenty-first century about particle physics and about what the essential particles are that make up matter.

So let's take a look at the structure of matter. Let's probe into ordinary matter with a microscope that can go ever more deeply into the heart of matter. Let's see what we find. We'll start with a macroscopic object, something the size of a person or a tennis ball or a typical-sized object we deal with. In this case, it's a cat. It looks a lot like Schrödinger's cat, doesn't it? So we're going to take this cat, and we're going to look inside it at ever and ever smaller scales and see what we find. Now, we're not biologists here, so we'll skip the fact that we'll come to organs and then cells and then organelles within the cells and so on. We'll get right to the chemistry of this cat, and the first thing we see at the level of chemistry are molecules.

Molecules are the basic, essential particles making up the chemical substances. Molecules can be simple, as the molecule I've shown here, which is a water molecule, (H²O, two hydrogen atoms and one oxygen atom), or they can be complex like the DNA molecules that carry the heredity of this cat. So we've looked with our microscope and come across a water molecule. That's the fundamental piece you can have of water. Anything smaller than that is not water anymore. Water is a chemical substance. It's not a fundamental element like hydrogen or oxygen. It's a chemical compound made up of those things. So here's a molecule of water. What happens if we look with our powerful microscope at the molecule of water? What do we see?

Well, we see that it's made of atoms, and I show here a typical atom. It happens to be a helium atom because it's got two electrons. It's not one of the atoms of water, but it's an atom like the oxygen atom that makes up water and the hydrogen atoms that make up water. In fact, here's a picture of an oxygen atom. So here we have an atom, and this is the point we were at with Rutherford's solar system model of the atom in the early 1900s. This atom consists of a nucleus, as far as we know, or as far as Rutherford knew, a tiny, dense object carrying virtually all the mass of the atom but occupying only about, in diameter, 1/10,000 to 1/100,000 of the overall diameter of the atom. Again, these atom pictures are very much symbolic. They're very much not to scale.

So here's our atom. Now, what's the atom made of? If we apply our microscope to the different parts of the atom, what do we see? Well, one thing we see are the electrons. The electrons comprise the outermost part of the atom. They're in orbit around the nucleus in the Rutherford picture. In the Bohr picture, we know things get a little more complicated, and, because of quantum mechanics, we know we really can't think of them as little balls whizzing around. They're kind of fuzzy, statistical objects. Nevertheless, we're going to picture them in this crude way as little, tiny, point-like particles. There's the electron. What if we probe into the electron?

Well, all attempts to probe into electrons have never found any structure inside the electron. The electron appears to be a truly structureless, point-like entity when we try to probe it. When we ask questions of the electron—what kind of particle are you?—trying to make it act like a particle, it does not have any structure. It's a structureless point particle. We believe the electron is truly one of the elementary entities that make up the universe. There's no way to take an electron apart. There's no way to find anything inside an electron. An electron appears to be truly elementary.

So we've now analyzed the outer part of the atom, which consists of electrons. We find these truly elementary-looking, point-like particles that have a few simple properties. They have a certain amount of mass. It's about 2/1,000 the mass of the other particles that make up the atom. It's rather small. Electrons comprise only a small fraction of the mass of an atom. They carry a negative electric charge of one unit of charge, and they have a few other properties, spin and so on. We understand electrons very well. They are also stable particles. If we have an electron, it seems to last forever. It never disappears. It has nothing else it can decay into. Well, that's the electrons in the atom.

What if we now probe at that nucleus, which to Rutherford was simply a small, massive object carrying positive charge? Well, again, these probings are done with a sequence of experiments in which you bombard the nucleus with ever more high-energy particles, and you see whether the nucleus flies apart, you see what particles come out, and so on, and you begin to understand the composition of the nucleus. What happens if we look in the nucleus?

Well, here's the nucleus enlarged in detail, and we find the nucleus has a lumpy structure. It has a real size. Nuclei are approximately, but not exactly, spherical. They have a measurable size. They're not point-like particles like the electrons. They're definitely not elementary. They seem to be composed of other particles. The particles that compose the nucleus are called, collectively, *nucleons*, and they come in two kinds. There are *protons*, which are positively charged particles, and *neutrons*, which are negatively charged particles. When protons and neutrons were discovered, they were assumed to be probably fundamental particles like the electron, elementary particles with no further structure.

So the nucleus is comprised of protons and neutrons. The protons carry plus one electric charge. There are as many of them as there are electrons in the atom. The atom is, therefore, electrically neutral. The neutrons are neutral.

They carry no electric charge. They are very similar to protons otherwise; they have the same mass or almost the same mass, in particular. What if we try to look in more detail at these two constituents of the nucleus, the protons and the neutrons? Well, if we probe protons and neutrons by bombarding them with high-energy particles, we find, in Rutherford-type experiments, that occasionally we get particles scattering off what looks like a tiny, point-like, massive object inside the proton and the neutron.

So protons and neutrons themselves appear not to be elementary. They appear to be made up of more fundamental entities. We can actually measure a size for the proton and a size for the neutron. It's about 10^{-15} meters, if you want to know what it is numerically. It's a tiny, tiny size, but it's a measurable size, and these things appear to have internal constituents. What are the internal constituents? Well, early in the second half of the twentieth century, as physicists were trying to make sense of these hundreds and hundreds of particles, several physicists came up independently with the notion that maybe we could describe a lot of these particles if we postulated the existence of some fundamental entities called *quarks*, which were even smaller and which would combine in threes to make protons and neutrons and a lot of the other particles in twos or threes.

Well, for a while that was just kind of a bookkeeping device for keeping track of all the different kinds of particles, but when experiments probing the structure of protons got to high enough energy, we found, indeed, there seemed to be particles within the proton, and those particles are the quarks, and those particles are in the neutrons. So the proton and the neutron are themselves made of smaller particles called quarks. The quarks are kind of unusual. They carry fractional electric charges. When Millikan did his famous experiment in the early 1900s and found that electric charge was quantized in units of the electron charge, he was slightly wrong. Electric charge is, in fact, quantized in units of a third of an electron charge, and some of these quarks carry a third of an electron charge and some carry two-thirds of an electron charge.

In particular, the proton is composed of *up quarks* and *down quarks*, two up quarks and one down quark. The quarks have these very whimsical names, as you'll see. I'm going to introduce you soon to more quarks, and they all have slightly whimsical names that really have no bearing to their meanings in ordinary reality. There's nothing up about an up quark. There's nothing down about a down quark. Those are just names. The proton is composed of two up quarks. Each one of those has two-thirds of an electron or proton's worth of electric charge, positive two-thirds. That adds up to positive four-thirds, and the down quark has negative one-third, so that subtracts to make the proton's charge of plus one. The neutron is made up, instead, of two down quarks—minus one-third charge, minus one-third charge, that's minus two-thirds—and plus two-thirds in one up quark, and that makes the neutron neutral.

So the proton and the neutron are very similar. They're each made of three quarks: the proton, two up quarks and a down quark, with a net positive electric charge of plus one unit; and the neutron, two down quarks and one up quark, with a net electric charge of zero, making the neutron neutral. So, again, if we probe deep inside one of these nucleons, a proton or a neutron, we find up quarks and down quarks. We believe that the quarks are structureless, although some people have proposed theories where maybe they're not. But it looks right now like the quarks are truly structureless point particles like the electrons, like the quarks are fundamental constituents of matter.

So, in this picture, anyway, of the ordinary matter that the cat is made up of, we've gotten down to what we believe at this point is the fundamental level. We believe that three kinds of particles in two broad classes—quarks and electrons—comprise all ordinary matter and make up all ordinary matter. It just takes three kinds of particles, to do it—electrons, up quarks, and down quarks—and those are the truly elementary particles of nature. If you'd taken physics a few decades ago, we probably would have been saying, "Well, the elementary particles are electrons, protons, and neutrons." Nope. The elementary particles are electrons and the two kinds of quarks that appear in protons and neutrons.

But things are going to get more complicated than that. Before they do, though, let me mention that the combination of up-up-down or up-down-down aren't the only combinations you could put together of quarks. In fact, you could put together many combinations of three different quarks, and all those combinations make up possible *elementary particles*, particles that were once thought to be elementary but are now known to be composed of quarks and aren't really elementary. All the others, except the proton and neutron, are highly radioactive and unstable and decay in a relatively short time, so we don't see them around. The neutron itself is unstable if you take it out of the atom. A neutron decays—half of it will be gone—in an average of about 20 minutes. The neutron is an unstable particle also. The protons appear to be stable over very long terms, although there are some theories that predict they ought to decay in times on the order of a 1 with 33 zeros after it numbers of years.

Experiments looking for proton decays in large numbers of protons—huge tanks of water, tens of thousands of gallons—have not yet discovered that protons decay. So protons seem to be stable particles. Neutrons are not, and all the other particles that you can make from three quarks are not stable. Those other particles, including the neutron and proton as well, are called *hadrons*. It means "heavy particles" because they are made up of quarks and they are pretty heavy. The quarks are massive, and they combine to make fairly massive particles. The proton and neutron are each about 2,000 times as massive as the electron. Quarks can also combine in twos, a quark and its anti-quark. Remember, the electron had an anti-particle, the positron. Every quark has an anti-quark, and, if quarks combine in pairs, a quark and an anti-quark, they make a whole other class of particles called *mesons*. So there's quite a bit of possibility here with these quarks.

So, what is it that holds all these things together? What sticks the quarks together to make protons and neutrons? What sticks the protons and neutrons together to make nuclei? What holds the nuclei and electrons together to make atoms? What holds the atoms together in molecules? And why doesn't the cat simply fall apart into a blob of molecules and ultimately quarks and electrons? Well, the answer is the same in all those cases. The answer is some kind of force interacting between the particles to hold them together. We had a lot to say about force earlier in this course. For example, there was the force of the push as I shove on this table. There was the force of gravity, pulling on a ball and dropping it to the floor, or pulling on the moon to keep it in its orbit.

There was the force, when I stretched that bungee cord, to show you how $E=mc^2$ applied even to that kind of motion, all kinds of forces you can think of. There's the force of friction we worried about in Galileo's early studies of motion. How many kinds of forces are there? Well, actually, physicists today believe there are at most three fundamental forces, and most physicists believe there's probably only one, but we haven't yet discovered how the three are related into one. More on that in Lecture Twenty-Four. So, we believe there are only three fundamental forces, and together they account for all the forces. Friction, everything else that happens, all the forces, all the interactions in the universe, we believe, can be explained in terms of these three fundamental forces.

If I had been giving this lecture series a few decades ago, I would have said four fundamental forces. If I'd been giving it a few centuries ago, I would have listed many more. But we've achieved a lot of unification of previously separate entities, and now we're down to three fundamental interactions. I want to describe them briefly. What are they? What are the fundamental forces? Well, the first is probably the most familiar but probably also the least understood, and it's by far the weakest of the forces. That's the *force of gravity*. What is gravity? Well, we know that, from our study of general relativity, gravity is ultimately the geometrical structure of spacetime, and it has the interesting property that it acts on all matter. No matter is exempt from the gravitational force. Gravity acts between every piece of matter in the universe. Protons act on neutrons. Neutrons act on neutrons. Electrons act on protons. Electrons attract electrons. It's just that they repel much more, with a different force.

Gravity is a force that acts between all particles in the universe. It's a very, very weak force, and so the only time we see it being significant is when there are enormous accumulations of matter. That's why the gravity of the Earth or the sun seems significant to us. Enough matter gets together to warp spacetime enough to make significant changes in the motion of particles. So gravity is a very weak force. It doesn't seem that way to us simply because we live very near a very large accumulation of matter, namely, the Earth, which warps spacetime in its vicinity. So gravity is a weak force that's really, in a way, the least understood of the forces, even though we have the general theory of relativity.

Then there's a force called the *electroweak force*. This used to be two separate forces, but now it's one. What's the electroweak force? Well, it comprises what I've talked about before, the electromagnetic force as described by Maxwell's equations of electromagnetism. The electromagnetic force itself is composed of two forces that were once thought to be separate, but, as we saw in the lecture where we looked at electromagnetism, electricity and magnetism are really aspects of the same thing. So the electric force and the magnetic force, once thought to be distinct, are, in fact, aspects of the electromagnetic force.

Then there's another force, somewhat obscure, called the *weak force*, the *weak nuclear force*, and it's responsible for a certain limited class of nuclear interactions that occur. We see them in nuclear experiments. They occur on the sun and are partly responsible for the generation of solar energy that keeps us alive. It's called the weak nuclear force as opposed to a different nuclear force, the *strong force*. The weak nuclear force and the electromagnetic force are now understood to be aspects of the same underlying force called the electroweak force. That understanding came about, theoretically, in the 1970s and was confirmed experimentally in the 1980s. So that's a fairly recent addition. Before that time I would have told you about the electromagnetic force and separately, independently, the weak force. Now we know they are combined into something called the electroweak force.

Finally, there's a force that is sometimes called the strong force, although a probably better name for it is the *color force*. This is the force that acts between quarks. This is the strongest of the forces we know about, and it's responsible for holding quarks together to make protons and neutrons and the other hadrons and the mesons. The color force has an unusual property. Remember how the gravitational force falls off with distance? That's why we can have a finite escape speed to get objects infinitely far from Earth. The electric force also falls off, as does the magnetic force, with distances. As they move electric charges farther apart, the force gets weaker. The weak force does the same thing. The color force, the force that acts between quarks, remarkably, does not fall off with distance.

If you try to separate quarks, you're pulling always against a constant force, and, therefore, it would take infinite energy to get two quarks separated.

That can't happen. In fact, if you try to separate two quarks, when you get so much energy buildup, you may have enough energy to create a pair of particles by that pair creation process we described and, instead of getting the quarks apart, you simply start creating more particles. You cannot separate quarks because of that feature of the color force, the force that binds quarks together. It doesn't decrease with distance, and, therefore, we've never been able to isolate a quark. Most physicists believe we will never find an isolated quark because to isolate quarks is impossible. Yet, by probing the structure of nucleons, protons and neutrons, we know that quarks exist, and we know that they can combine with each other in different ways. But we can never have a separate, isolated quark. One will never see an isolated quark.

By the way, if you've ever studied nuclear physics at all, you were probably told there's a strong nuclear force that binds neutrons and protons together to make an atomic nucleus. That's true, but the nuclear force isn't a separate force. It's simply a manifestation, a kind of a residual manifestation, of the color force between the quarks that is acting even more tightly to bind the quarks into the nucleons, and that color force is felt a little, small distance beyond the nucleon itself, beyond a proton or neutron, and it's that small, residual effect that is, nevertheless, the very strong nuclear force that binds the protons and neutrons together to make an atomic nucleus and is responsible for nuclear energy conversions, nuclear weapons, nuclear power, and nuclear reactions of all kinds.

If we go up in our hierarchy back to the cat, then the forces become more familiar again. What binds the nucleus to the electrons? Well, that's the familiar electric force. What binds the atoms together to make molecules? Well, that's a residual of the leftover electric force that is felt between neutral atoms because different parts of the atom with different electric charge may be in different places, so there's a residual electrical effect; just like the nuclear force is a residual of that color force. Then the molecules are held together, ultimately, by electrical interactions to make the cat. If they weren't held together, gravity would pull the cat to the floor in just a blob of stuff. So there's how the forces work to hold things together.

We put together these forces with the known particles, and we get what's called the *Standard Model of Particles and Forces*. This is the model that governs our best understanding of the way the universe works to date. We know it's not complete, but it does a pretty good job of explaining all the observed particles and their interactions. The particles in the Standard Model come in two general classes. There are quarks. We've talked about quarks before. I've talked about the up quark and the down quark, but there are actually several other quarks, which I'll introduce in a moment. So, there are quarks. Then there are particles called *leptons*. This means light particles, particles that have very little mass. You've met one lepton already. It's the electron. You've actually met another lepton back in Lecture Nine.

More on that—the leptons come in two varieties. There are particles like the electron, the electron and some close cousins of it, and then there are very illusive, ghostly particles called *neutrinos*, which hardly interact with matter at all. They can barrel right through the Earth with very little likelihood of anything happening to them. They're generated in copious amounts in the sun. They're very hard to detect. They were thought, until very close to the end of the twentieth century, to have zero mass-like photons and, therefore, to travel at the speed of light, but, in experiments that were done at the end of the twentieth century, it was determined that neutrinos almost certainly have a very, very small amount of mass. We're still studying that today. That's an active area of research.

What do these particles look like in the Standard Model? Well, remarkably, they come grouped in families, and here's how it works. The particles in the Standard Model, the first family, consist of the up quark and the down quark. Each family has a pair of quarks. The up quark and the down quark are the quarks in the first family of particles in the Standard Model. Then there's the electron and the electron neutrino, which is this ghostly, illusive particle associated with the electron. That's it. That's the first family. The first family of particles—two quarks, an electron, and an electron neutrino, to a first approximation—are all that's necessary to make all the ordinary, everyday matter of which we're composed.

I say, "to a first approximation" because very recent research is showing that, although a proton is, indeed, composed of two up quarks and a down quark and a neutron of two down quarks and an up quark, there are, in fact, coming into existence and going out of existence, quickly and ephemerally, pairs of quark/anti-quarks of heavier varieties and that those ephemeral, so-called virtual particles that come into existence and disappear play a significant role in establishing properties of the proton and neutron, like their mass and that spin that elementary particles have. So you may hear it said that all ordinary matter is described by this first family of particles. That's approximately, but not entirely, true. It looks like particles from the other families are also playing some role.

What are the other families? Well, the second family includes a quark called the *charm quark* and another quark called the *strange quark*, a particle called the *muon*, which is like a heavy electron and which is unstable. The muon is the particle we studied in Lecture Nine, when we had these muons created in cosmic rays and we saw time dilation occurring with the muons. Then, with the muon, there is a neutrino associated with it. As we go from one family to another, we go to particles that are more and more massive, and, consequently, they require higher energies to create. So it takes higher energies, bigger particle accelerators, to study the more massive particles.

The third family includes the so-called *top quark*, the *bottom quark*, a very heavy electron-like particle called the *tau*, and a particle called the *tau neutrino*. The top quark was only discovered in the mid-1990s, and that's what completed this picture. So we now know that all these particles exist. We've detected them all. By the way, there's a relationship between these families, and this is found out in the experiments that study the masses of neutrinos. It appears that electron neutrinos can change into muon neutrinos, can change into tau neutrinos. Neutrinos can oscillate back and forth from one form to another. Now, you might say, "Okay, this looks good, but I'm sure next year's edition of this course is going to have a fourth family as we discover them." Well, no. There are experiments you can do that tell you what the number of families of particles that can exist to be, and these experiments seem to confirm, very precisely, that it's no more and no less than three. So we believe these are all the families of particles there are. Why? Who knows? We don't know why there should be three families of particles, but that's the way nature appears to have made things.

Again, I want to emphasize that only the first family appears regularly and routinely and continuously in ordinary matter. Particles in the second and third families are all unstable, and they quickly decay to other particles. So protons, neutrons, ordinary matter, are made of electrons and up quarks and down quarks. Although, as I've just described, the other ones do play a modest role in the existence of protons and neutrons and maybe a role that's not so modest. But they aren't there all the time. They're coming into existence and going out of existence. They did play a major role in the early universe when things were so hot there was enough $E=mc^2$ energy around to create even these heavier particles, and they do play a role in some astrophysical processes, and they play a role in our particle accelerators.

Now, there are other particles. We aren't quite done. There are other particles, and the reason there are other particles is because, in quantum physics, the description of a force is not somebody pulling or pushing nor is it necessarily that field picture, although it's a kind of modified field picture. In quantum physics, the description of a force between particles is that particles interact by exchanging yet another particle, and that particle is called a *force carrier*. I'm going to give you an example of how that works for a familiar force, the electromagnetic force. In this case, I'll show you a proton and an electron. They're interacting. It's an attractive force in this case, the electric force that's attracted between those two.

What happens is that one of them emits a photon. That's the force carrier for the electromagnetic force, the familiar photon, the particle that is also the particle in visible light or electromagnetic waves of any kind. What happens is that the electron emits the photon, the proton absorbs it, and, in that process, they've developed that attraction that goes on between them. Then the proton emits a photon. It goes back and forth. Photons go back and forth. This is called a *virtual photon* because it comes into existence at the electron, travels to the proton, goes out of existence at the proton, and we never see it. So it's called virtual. But that process, the exchange of a virtual photon, is a quantum physics description of how forces actually act.

So there are other particles. There's the photon, the particle that carries the force, the electromagnetic force, and there are other force-carrying particles. What are they? Well, for each of the forces, there's an associated force-carrying particle. For the electromagnetic force, that particle is the photon. Electromagnetism is mediated, if you will, by the exchange of photons between charged particles. For the weak force, there are two force-carrying particles called *W* and *Z bosons*, and it was their discovery that confirmed the electroweak unification that made the weak force understandable as being related to the electromagnetic force.

For the color force, the force that binds quarks, there's a particle called the *gluon* because it glues the quarks together, and two quarks interact by exchanging gluons. Gluons are fascinating because they themselves interact, and they can make particles, called glue balls, which are big clumps of gluons. These are, again, areas of very active research. Finally, for gravity, which doesn't fit into the Standard Model, we don't know how to incorporate gravity with these other forces. That's going to be the subject of my last lecture, attempts to do that. If gravity could be described in terms of a quantum theory, and we don't yet know how to do that, then there would be a particle called the *graviton*, which would mediate the gravitational force. So, in addition to the quarks and the leptons, the quarks and the electrons and the neutrinos and the muons and the muon neutrinos and so on, there are these additional force-carrying particles, the photon, the W and Z bosons, the gluon, and maybe the graviton.

Finally, there's one more particle you'll probably hear about. It's a particle called the *Higgs boson*. Nobody's ever seen a Higgs boson because we haven't got particle accelerators that reach the energies we think are going to be needed to create it. The Higgs boson is some sort of a master particle whose existence is believed, in the Standard Model, to be responsible for determining the masses of all the other particles. There's a particle accelerator going on-line in the first decade of the twenty-first century. It's not there yet. It's going to be in Switzerland at CERN, on the French-Swiss border. It's called the Large Hadron Collider, and it probably will reach energies that are capable of producing the Higgs particle.

Again, it takes a lot of energy because $E=mc^2$. If that happens, we will, by that time, have completed that Standard Model of Particles and Forces. What good does this do us? Well, I'm going to show you in the next lecture how our understanding of the particles and forces leads us partway to an understanding of the evolution of the entire universe.

Cosmic Connections
Lecture 23

The bottom line is we know now that we live in a universe that is
expanding and that appears to have had a definite beginning. There
are three main lines of experimental evidence that lead us to that
conclusion, and I want to describe those for you in some detail.

In the 1920s, Edwin Hubble (for whom the Hubble Space Telescope is
named) discovered, using the Doppler shift, that distant galaxies are all
moving away from us with speeds proportional to how far away they
are. Hubble had discovered the *expansion of the universe*. It's important to
keep in mind that this does not imply that we're at the center. Every observer
sees the same thing. Each is like a raisin in a rising loaf of raisin bread;
each raisin sees all others moving away with speeds proportional to their
distance. An obvious implication of cosmic expansion is that the matter in
the universe was once much more densely packed. Extrapolating back in
time from the observed expansion suggests that the universe began in a *Big
Bang* explosion some 15 billion years ago.

In the 1960s, Arno Penzias and Robert W. Wilson discovered the *cosmic
microwave background radiation*. This was radiation left over from the
time the universe first became transparent, about half a million years after
the Big Bang. Cosmic background explorer (COBE) satellite studies of the
microwave background in the 1990s showed it to be remarkably similar in
every direction, but with tiny "ripples" that may be the seeds of the large-
scale structure (galaxies and clusters of galaxies) that we see today.

The standard model of particles and forces allows us to explore conditions
that would have held in the first instants after the Big Bang. The model
predicts particle interactions that would lead to a distribution of matter
similar to what is, in fact, observed. Thus, particle physics helps confirm the
Big Bang concept.

The main theme of cosmic evolution is that, as the universe expands, it cools,
allowing ever more complex structures to form. Here are some highlights of

the story of the universe during its expansion: At about 10 microseconds, nucleons formed from a "quark soup." Within the first three minutes, helium nuclei formed. At about half a million years, electrons surrounded nuclei to make atoms, creating a transparent universe and producing the cosmic microwave background in the process. When the universe had reached 100 million to several billion years of age, galaxies formed. Later came the production of heavier elements, including carbon, oxygen, and others needed for life by nuclear fusion in stars. Supernova explosions spewed heavy elements into space. New stars and planets formed. Eventually life and intelligence evolved.

Change is essential to this story, and general relativity shows that the universe must be expanding or contracting (although the 1998 discovery that the expansion is accelerating rather than slowing muddies the simple picture). There are essentially two possibilities: either the universe will expand forever, or it will eventually contract in a "big crunch"—just as a ball thrown upward may escape Earth forever if its speed exceeds escape speed but will otherwise return.

We're literally "children of the stars"; the elements that make up our bodies were forged in the cores of stars that have long since exploded.

Current theories suggest an *inflationary universe*, in which a period of very rapid expansion very early in the Big Bang (at 10^{-34} seconds!) smoothed out large-scale curvature to produce a universe whose overall geometry is flat—and, thus, barely able to expand forever. This inflationary scenario solves several outstanding problems with the Big Bang theory. However, it requires a higher density of matter than we've yet detected in the visible glow from stars and luminous gas or even in the gravitational influence of *dark matter* in galaxies. A sobering thought is that cosmologists now believe that most of the mass in the universe cannot even be in the form of ordinary matter but must consist of hitherto unknown forms of matter and/or energy. The universe we see and detect may be just a tiny fraction of what's really there.

We're literally "children of the stars"; the elements that make up our bodies were forged in the cores of stars that have long since exploded. Physicist Freeman Dyson imagines that intelligence may persist to the infinite future, even as the universe evolves through an unimaginable richness of new forms and structures. ■

Essential Reading

Kaku, *Hyperspace*, Chapters 13–15.

Lightman, *Ancient Light*.

Weinberg, *The First Three Minutes*.

Suggested Reading

Barrow and Silk, *The Left Hand of Creation*.

Hawkins, *Hunting Down the Universe*.

Mather and Boslough, *The Very First Light*.

Padmanabhan, *After the First Three Minutes*.

Questions to Consider

1. If we see all the distant galaxies receding from us, why can't we conclude that we're at the center of the universe?

2. Why couldn't atoms, or even nuclei, exist at the very earliest instants of the universe?

3. Freeman Dyson's vision of intelligence in a forever-expanding universe imbues intelligent life with a kind of immortality. Discuss this concept and examine how cosmic expansion and Dyson's vision affect your feelings about your own mortality.

For more treatment of the topics covered in this lecture, we recommend The Teaching Company course *Understanding the Universe: An Introduction to Astronomy, 2nd Edition* by Professor Alex Filippenko of the University of California at Berkeley. Professor Filippenko led the team that demonstrated the accelerating expansion of the universe in 1998.

A Cosmic Timeline
(approximate and <u>not</u> to scale)

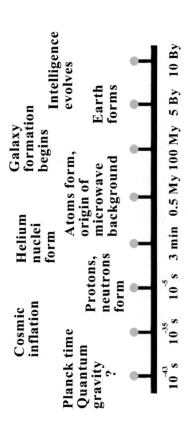

Cosmic Futures

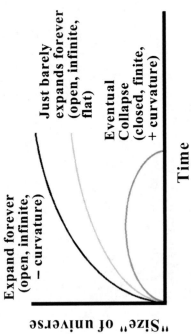

Expand forever
(open, infinite,
– curvature)

Just barely
expands forever
(open, infinite,
flat)

Eventual
Collapse
(closed, finite,
+ curvature)

"Size" of universe

Time

Cosmic Connections
Lecture 23—Transcript

Lecture Twenty-Three: "Cosmic Connections." Since Lecture Sixteen, we've been exploring the universe at the smallest scales until, in the previous lecture, we were down to the scale of quarks and electrons, the really fundamental things that make up the universe. Before that, we dealt with general relativity, the theory that describes, among other things, the large-scale structure of the universe on the largest possible scales of galaxies and the whole universe itself. Now, in our next to the last lecture, we're going to combine the largest and the smallest in a look at what we know about the universe that general relativity and our theories of subatomic particles, the Standard Model of Particles and Forces, can tell us.

The bottom line is we know now that we live in a universe that is expanding and that appears to have had a definite beginning. There are three main lines of experimental evidence that lead us to that conclusion, and I want to describe those for you in some detail. The first goes back to the year 1920, roughly. Before 1920, astronomers, looking through what were then the most powerful telescopes in the world, would see faint, fuzzy patches in the sky, and there was some debate about these patches. Some astronomers believed they were clouds of gas or dust or something within the general system of stars that constitutes our Milky Way. Others, though, took a more visionary approach. They said perhaps these are distant "island universes," whole assemblages of stars, like the Milky Way, maybe containing billions or hundreds of billions of stars that are so vast we can't see the individual stars.

Well, before the 1920s, that was the debate. But, in the 1920s, large, new telescopes became available, and with those it became very clear that these distant, fuzzy, so-called nebulae, at least some of them, were, indeed, island universes, were distant galaxies containing hundreds of billions of stars. The whole human idea of the scale of the universe expanded enormously, unfathomably largely, at that point in human history. Well, during the 1920s, an astronomer named Edward Hubble undertook a series of observations of these distant, now-discovered galaxies, and he found something remarkable. He found all the galaxies, beyond any reasonable distance from our own, seemed to be moving away from us. And, the further away they were, the faster they were moving.

How did Hubble find this out? Well, he needed to know the distances to these galaxies, and he needed to know how fast they were moving. Distance is not terribly hard. If you look at an object, the further away it is, the dimmer it appears. If you make the reasonable assumption that these galaxies are all approximately the same kinds of objects, then, by looking at how bright they seem in your telescope, you can give a rough sense of how far away they are. How do you get a sense of their speed? That actually turns out to be fairly easy. If you stand by the roadside and listen to trucks going by, you hear something like this [humming] as the truck goes by. What's going on there?

As the truck is approaching you, it emits sound waves at successive intervals, but, because the waves are being emitted as the truck is moving toward you, the frequency that you hear for the waves appears to be increased. You hear the waves, sound waves, at a higher pitch. After the truck goes past you— the truck is receding from you—the sound it's emitting is being emitted at successively greater distances, and that results in a lowering of the pitch of the sound waves. [Humming], that's the kind of thing you hear. That's called the Doppler effect. It's well-known, and it happens for all kinds of waves, including light waves, although the description of physically what's going on is slightly different because light waves, unlike sound, have no medium, but they still exhibit the Doppler effect. The speed doesn't change, but the pitch changes, that is, the frequency of the light wave, depending on whether objects are moving toward you or away from you.

Well, in the case of astronomical objects, galaxies, they're all moving away from us. That means all their light is shifted toward the red. Do they look redder? Well, most of them don't. At least in Hubble's day they didn't. But what one can see, very easily, are the spectral lines emitted by atoms in those galaxies. Remember how atoms emit these discreet spectral lines. That was one of the pieces of evidence that led us to quantum physics—very discreet colors, very particular colors. When one looks at the light from distant galaxies, one can analyze these spectral lines, and one finds that, indeed, they are shifted toward the red, and one can use that shift to measure the speed of recession of the galaxies. So Hubble found that all the galaxies were receding from us, and they were doing so with speeds that depend on their distance from us.

Now, I have to give you an immediate caution. It sounds like we've broken the Copernican paradigm. It sounds like we're at the center of this cosmic expansion, going off in all directions. Well, in a sense, we are at the center, but so is everybody else. There's nothing special about the Earth. We've known that throughout this whole course, and Hubble's observation doesn't change that at all. How is that possible? Let me give you two analogies. One analogy would be a loaf of raisin bread. You mix up your dough. You knead it. You knead in the raisins. You've got these raisins all spread throughout this big lump of dough, and, to make life easier, let's imagine the dough is infinite. Well, if you don't like infinite, don't worry about it. The dough is there, it's got raisins in it, and it's expanding.

What happens? Every raisin moves apart from every other raisin. If you were sitting on one of those raisins and you could see through the opaque dough, you'd see all the other raisins moving away from you. The further a raisin is from you, the more dough is expanding in between you and it, and the faster it seems to be going away. Yet, no raisin can claim to be at the center, especially if I make the assumption that this particular weird loaf of raisin bread is infinite in extent. So, we don't have to be at the center.

Let me give you another analogy, which I'm going to show you here on the screen. I have a picture showing a whole lot of little, elliptical-shaped objects spread randomly all over the screen, in random direction. Those are galaxies. I'm going to take this same picture, and I'm going to enlarge it by about 25 percent, and I'm going to put the enlarged picture on the same screen, so we're going to see what the universe looks like at a certain time and what it looks like a bit later when it's expanded by about 25 percent. Here's what happens. Now, I happen to have placed this particular expanded universe so that this galaxy right here is viewing the expansion. This galaxy doesn't see itself going anywhere, so the bigger version is right on top of itself. By the way, the galaxies themselves do not expand. My picture shows that they do, but that isn't what really happens. Simply the distance between them grows.

So this galaxy, which is "at the center," but it isn't really at the center, looks out, and it sees nearby galaxies that have moved slightly from there to there. This one's moved from there to there. If it looks further away, it sees galaxies that have moved a little bit further. This one's moved from there to there. Further out, this one's moved from there to there, and so forth. So this particular galaxy, whose viewpoint we're taking in this picture, sees all the other galaxies moving away from it at speeds that are proportional to how far away they are. Let's look at the perspective, though, from another galaxy—this one. It sees exactly the same thing. It sees all the galaxies expanding, moving away from it, at speeds that are proportional to how far they're away. Here are some close-by ones. They haven't moved much. Here's one that's further away. It's moved from here to here. Here's a rather distant one. It's moved from here to here. No matter which galaxy I look at this picture from, I see the same thing. I see an expanding universe, expanding in all directions.

One more case, now looking from the point of view of this galaxy right down near the bottom of the screen. It, again, sees the same thing. Nearby galaxies have moved slightly. More distant galaxies have moved further. Way distant galaxies have moved quite a bit away. Every single galaxy sees the universe expanding in all directions at speeds proportional to the distance, and not one of them is at the center. Now, there's an obvious implication of this universal expansion. If the galaxies are all rushing apart at speeds proportional to their distances, what happened in the past? Well, the obvious implication is, in the past, they were closer together and closer together and closer together, and, if you measured the present expansion rate and make some simple assumptions, in fact, if you apply general relativity (the best thing to do), you can extrapolate back in time and ask yourself, was there a time when all the galaxies were close together?

The answer is yes, and the answer is that time seems to be somewhere on the order of 12 to 15 billion years ago. That time is called the age of the universe, and the measurement of it comes from looking at the expanding universe and exploring how the galaxies' motion depends on their distance from us. So it appears that the universe began in some kind of cosmic explosion. That explosion is called the Big Bang, and that explosion took place, we believe, something like 12 to 15 billion years ago. Now, Hubble made this discovery in the 1920s. To honor his discovery, of course, the Hubble Space Telescope is named after him. But, through much of the twentieth century, there was debate about whether cosmic expansion really implied a beginning, a Big Bang. There were other competing theories.

For example, there was a theory that the universe has always been expanding and new matter is created to fill the void as it expands, so it really always looks the same and never had a beginning. But most of those theories were put to rest in the 1960s with the discovery by two scientists, Penzias and Wilson at Bell Laboratories, of what's called the cosmic microwave background radiation. These guys were actually doing communications kinds of experiments with a radio telescope. They had a peculiar static in their telescope, and they couldn't get rid of it. They tried and tried and tried. They shoveled some pigeon dung out from pigeons that were nesting in the telescope because that produces some electromagnetic radiation as it decomposes. It didn't help. Nothing helped. They pointed the telescope up. They pointed it down. This mysterious static was coming from everywhere.

Then they learned of some theorists at nearby Princeton University who were arguing that, if you look at a scenario of a universe expanding from a Big Bang explosion, there ought to come a time when a burst of electromagnetic radiation is released, and that radiation ought to travel throughout the expanding universe and manifest itself today as a kind of cosmic background everywhere, most of which is now in the microwave region of the electromagnetic spectrum. Microwaves are very short wavelength radio waves, basically, that are used to cook in a microwave oven. They're used for communications. UHF television signals are often in the microwave region of the spectrum. So Penzias and Wilson and the Princeton theorists got together and realized that Penzias and Wilson had discovered the cosmic microwave background radiation, a fossil remnant from a time about 500,000 years after the Big Bang. I'll show you why that is a little bit later.

So they discovered the cosmic microwave background. It's very difficult to explain with other theories. By the way, if you'd like to see the cosmic microwave background, sometime when you're watching some boring soap opera or something on television, turn instead to a channel where there's no signal in your area and you see just a fuzzy, snowy pattern on the screen. The cosmic microwave background is strong enough that, for a UHF television channel, a few percent of the signal you're getting—that fuzzy, snowy static—is actually cosmic microwave background. You are receiving, in your TV antenna, photons of microwave radiation that have not interacted with anything since a time 500,000 years after the Big Bang when they were created. So, take a look at that. It's a lot more interesting than other things you might find on television.

Today, we look at the cosmic microwave background not with antennas at Bell Laboratories but with a satellite, particularly the cosmic background explorer satellite (COBE), which has discovered that the background radiation is very, very uniform to a high degree of precision, but it's also discovered very, very faint ripples in the microwave background, ripples that suggest the formation of a large-scale structure in the universe. This has solved one of the major outstanding problems, which is, how did we get from a universe that might have been smooth and homogenous at the beginning to a universe which is very clumpy and shows us galaxies and clusters of galaxies and so on? In the 1990s, we discovered the structure, these very faint ripples, in the cosmic microwave background radiation. So the cosmic microwave background provides a second major clinching of the theory that we live in a universe that is (a) expanding and (b) had a definite beginning. It's very difficult to explain the universe and what it's doing now and having a cosmic microwave background radiation without invoking some kind of Big Bang beginning.

The third piece of evidence—and this is where we tie in previous lectures and tie in the very small—comes from advances in particle physics in the last half, particularly the last quarter, of the twentieth century. I outlined for you those advances in the previous lecture, culminating in the Standard Model of Particles and Forces with which we could predict, with reasonable accuracy, the interactions of the fundamental particles. What we can do with that is go back to a time very, very early in the expansion. We can say, what did the universe look like when it was literally a billionth of a second old? We know it was very hot then, and we know the energies involved, so we know the kinds of interactions that occurred among the quarks, and we know what the constituents of the universe looked like then. And, we can say, what if it expanded a little bit more? What, then, would be going on? We can ask ourselves, what would this original, primordial soup of quarks and things evolve into?

When we follow that, using our Standard Model of Particles and Forces, we find that it evolves to a universe whose constituents, mostly hydrogen and a little bit of helium and a little, tiny bit of everything else, are, in fact, what the universe today seems to contain. Remarkable. So the advances in particle physics that I've described in the last lecture, the advances that took place particularly toward the end of the twentieth century, led us to a further confirmation of the Big Bang picture because they allowed us to go back to conditions that prevailed very early in the Big Bang. In fact, one way of describing the energies of our big particle accelerators is by saying how far back in time they can take us. The new particle accelerator I mentioned that's going to come on-line in the first decade of the twenty-first century can take us back to about a trillionth of a second after the Big Bang occurred. Why? Because it can create energies that were typical of the particle energies that existed in the universe at that time.

So we have these three pieces of evidence that give us a picture of a universe that has been expanding from a Big Bang explosion. What exactly does that expansion look like? What's the history of such a universe? Well, I want to give you the main theme in that history, and then I want to describe, in detail, some of the major events in that highlight. The big theme is this, though. The universe starts out incredibly dense, maybe infinitely dense. It expands. As it expands, it does the same thing that happens if you let air out of a tire. Let air out of your car tire. You'll feel it's cooler. As the gas expands, it does work, it loses energy, the vibrations of the constituent particles slow down, and that's cooling.

So, as the universe expands, it cools, and as it cools, the forces that hold matter together in a very, very hot universe are not strong enough—if I have two particles stuck together by one of those forces and a third particle comes along with high energy, it can knock them apart, and it's hard for things like atoms, or, at early enough stages in the universe, even nuclei, and, at even earlier stages, even protons and neutrons, to stay together because they get knocked apart by high-energy particles in the very hot, early universe. But, as the universe cools, things can stick together more and more. So part of the main theme is expansion—cooling—and, therefore, more complex structures can form. That's the main theme.

I now want to look at some of the details of that evolution. I have a cosmic time line. I'm showing you a cosmic time line. It's very approximate. It's not to scale. There's not an even division or even a logarithmic even division, for those of you who are into such things, between these numbers on my bottom scale. It goes from a time of 10^{-43} seconds. That's 1 over 1 with 43 zeros after it. We don't know what happened before that time, and there's a very good reason for that. We do not have a theory that would allow us to understand the behavior of matter at those extreme densities. The reason we don't is because at that point we need to take into account both quantum physics and general relativity because we have extreme spacetime curvature because of the high density of matter and we're dealing with a very tiny region.

So we have quantum physics, and we don't know how to handle that. That's called the *Planck time*, and we don't know what to do before the Planck time. But, beyond that, our Standard Model of Particles and Fields can give us at least some sense of what's going on. So, what happens before Planck time quantum gravity, we don't know. After that, at about 10^{-35}, 10^{-34} seconds (10 to the minus something means a 1 over 1 with that many zeros after it)—these are incredibly small times—we believe the universe went through a period of enormously rapid expansion. I'll have a little more to say about that a little bit later. That happened about 10^{-34} seconds after the Big Bang. Sometime around a 1/1,000,000 to 1/100,000 of a second—1/100,000 is 10^{-5}—for the first time, things got cool enough that, out of what had been a soup of quarks and very elementary particles, for the first time, quarks could stick together without being knocked apart, and they could form protons and neutrons.

So we have protons and neutrons forming in the first 1/1,000,000 or 1/100,000 of a second. Again, these numbers are all approximate. As things cool, in a period from about 3 to 30 minutes—roughly in that interval—for the first time, protons and neutrons could stick together, and they wouldn't be knocked apart because the energy of the particles whizzing around was low enough now, and helium nuclei formed. When we do calculations based on our understandings of nuclear physics, we find that, during that time, given the conditions we think existed, the universe should have evolved from pure protons and electrons—that is, hydrogen—but it wasn't hydrogen atoms yet, it was just protons and electrons. A proton is a hydrogen nucleus. Hydrogen is the simplest atom, one proton—that's it—in its nucleus.

At the end of that roughly 30-minute period, enough protons and neutrons should have come together to make the universe approximately 25 percent helium nuclei and the rest hydrogen nuclei, and after that the building of these nuclei essentially stopped. So, at the end of that time, there should have been about 75 percent hydrogen; 25 percent helium; a very tiny, tiny amount of a few other, very light elements—lithium, beryllium, a heavy hydrogen kind called deuterium—but nothing else. None of it. No oxygen. No iron. No uranium. None of the stuff we're made of. That comes later. At the end of 30 minutes, there should have been 75 percent hydrogen and 25 percent helium, roughly. We go out and look at the universe today, and we can see what its composition is. We can look at stars. We can look at interstellar gas clouds. We can look at gas now that's between the galaxies, very rarified, but we can look at it. We can look at it spectroscopically. We can study the spectral lines. We can determine what elements are present and in what abundances.

And, lo and behold, for the universe overall—not for this hunk of rock we call Earth, but for the universe overall—75 percent hydrogen, 25 percent helium, and a smidgen of other things. That's a prediction of the theory, Standard Model applied to conditions in the Big Bang, that is confirmed by the state of the present universe. About half a million years out, atoms formed for the first time. Before that, although we had nuclei, it was too hot to have atoms. If electrons got in orbit around nuclei, bang, another electron or an atom came along and knocked them out and we couldn't have atoms. What we had at that time, then, were nuclei, which are positively charged particles, and electrons, which are negatively charged particles, and, since charged particles interact strongly with electromagnetic waves, which consist of electric and magnetic fields, which are what exert forces on charges—electrons and protons—because of that, the universe was opaque.

As soon as an electromagnetic wave got going, it quickly got absorbed by a proton or electron that vibrated in response to its waving electric field. So the universe was opaque. But, at the instant the atoms formed—it wasn't an instant; it was a period of time about half a million years out, but, during that relatively short period of time—electrons joined into their orbits around the nuclei, making neutral atoms. Two things happened. One, as the electrons fell from higher energy states into the lowest energy states they could be in in those atoms, the quantized lowest energy states, they gave off photons. The atoms became electrically neutral, and their interaction with electromagnetic radiation, light and so on, was much less, and, therefore, the universe became, at that time, transparent, and it emitted, at that time, a burst of electromagnetic radiation.

At that time, the electromagnetic radiation consisted of photons with very high energies, but the universe was then transparent, and those photons would move throughout the universe with relatively little chance of interacting with anything. Most of them would just continue on forever. As the universe expanded, it cooled. Those photons had to do work against the gravitational attraction of all the stuff in the universe, and they, too, cooled. They lost energy. A photon can't slow down. It has to go at speed c. We know that. They lose energy, as I said in the discussion of gravitational red shift. They lose energy by going to lower frequencies because $E = hf$. So they can lose energy by lowering their frequency. They then became ultraviolet, invisible light and infrared, and they ended up in the present universe as electromagnetic waves in the microwave region of the spectrum.

That's the cosmic microwave background. It's the fossil remnant of the time in the universe when atoms first formed. At that point, the universe became transparent. And, when we detect it in our radio telescopes or you in your TV antenna (as I said, you've grabbed a photon, or one's interacted with your TV antenna), that photon has been traveling through the universe since that time 500,000 years ago, after the Big Bang 12-15 billion years ago, and this is the first time it's ever interacted with anything. Remarkable. So you're looking right back at that time. By studying that radiation, then, we have a very good picture of what conditions were like in the universe 500,000 years after the Big Bang. Again, it confirms what we seem to think we know about the Big Bang theory. Sometime around 100,000,000 years—pretty fast, in the grand scheme of things—maybe a little bit later, the galaxies begin to form, and, certainly by a few billion years, most of the galaxies have formed.

This is one of the hardest things for us to study, and, until very recently, it presented a big problem for cosmologists, people who study the history of the universe. But, with the discovery of the ripples in the cosmic microwave background, we understand the seeds of that large-scale structure that became the galaxies and clusters of galaxies and clusters of clusters of galaxies and all kinds of larger-scale structures—great walls of clusters of galaxies and then big voids in the universe. That structure was there, and we think we understand a little bit about how it happened. Now we are looking back so far in time with the Hubble Space Telescope we see these quasars that I've mentioned before that are probably black holes at the centers of primordial galaxies, and there's great amounts of energy spewing out as material falls into the black holes and heats up and radiates away a lot of its energy.

So we think we understand pretty much how galaxy formation began, and we think we're looking back far enough that we're now seeing primordial galaxies. That happened actually fairly early. By a few billion years, there are galaxies. Five billion years out, roughly 5 billion years ago—this is a 10-billion-year scenario, but, 5 to 7 billion years ago, then (it could be 12 billion years old; maybe it's 15 billion years old, I'm not sure exactly what.)—Earth forms 5 billion years ago. The solar system, the sun, Earth—all that—forms about 5 billion years ago. Earth is actually 4.5 billion years old. We can date it pretty accurately with radioactive elements in it. Earth forms, and, by about 10 billion years out, 12 billion years out, 15—it depends on exactly how old you take the universe to be—intelligence evolved, at least on Earth, and many of us believe it probably evolved lots of other places because it's probable that the standard theory of particles and forces, which contains the interaction of all the forces, somehow contains what it takes to make life and intelligence.

That may be a natural outcome of matter. It doesn't mean we believe in flying saucers and being visited by aliens. But many physicists and astrophysicists and cosmologists believe that the evolution of intelligence on Earth is not a unique event, that it's probably a natural outcome, ultimately, of the laws that govern the interaction of the quarks and leptons to make complex matter. Remarkable, how those few simple particles can evolve to such complex things. That's a very brief history of the evolution of the universe as we understand it, and it is understanding built on our theory of the biggest things in the universe, the clusters of galaxies, the galaxies, and the whole large-scale structure of the universe coupled with our understanding of a the smallest things, the elementary particles.

What's going to happen in the future? Well, general relativity gives us an overall answer. Here's a graph that has time on its horizontal axis and the "size of the universe" on its vertical axis. If you don't like the size of the universe because the universe may be infinite, but it may not, we're not sure, take this to be the typical distance between two galaxies. As time goes on, that distance increases. General relativity tells us there are basically three kinds of scenarios that could occur. There's a scenario in which the universe might expand forever. If that's what it's going to do, the universe is infinite in extent, it's called open, and it has negative curvature. Its curvature is shaped like a satellite surface. You need a two-dimensional analogy for that, the curvature of spacetime being the sum of all of the gravitational effects of all the matter in the universe.

Another possibility is that it just barely expands forever. Such a universe is also open and it's infinite, but its overall geometry is flat. It's like that of a flat sheet. There are dimples in that flatness due to the spacetime curvature associated with individual massive objects, but the overall spacetime curvature is flat, and that universe will just barely expand forever. Finally, there's the possibility that the universe will expand, reach a point where it stops expanding due to the gravitational pull of everything in it, and collapse back to another Big Bang. Such a universe is closed, it's finite, and it has positive curvature. However, I hasten to add, it has no edge and no center because it's a four-dimensional spacetime that is curved back on itself, and nobody's at the center. I don't have time to go into that, but it's a fascinating topic in itself.

Now, I have to modify this picture slightly. All these pictures show the expansion slowing down as you would expect it to do if gravity is trying to pull this expanding stuff back. But, in fact, discoveries right at the very end of the twentieth century, astronomical discoveries, one of them by my colleague Alex Filippenko (who has an astronomy course with The Teaching Company), show that there may be an acceleration of the cosmic expansion, and, if there is, there are some new ideas in physics that have to be brought in or some old ideas of Einstein's that he discarded that have to be recycled to explain what's going on there. It may be that the expansion is actually accelerating instead of slowing down. Very recent astronomical evidence suggests that that may be the case.

Which of these scenarios is favored? Well, currently there are some subtle problems involved in the Big Bang theory, and they can be explained away by assuming, as I indicated on my time line, that there was a period of extremely rapid inflation where the universe expanded many, many manyfold in a very, very tiny, short time at about 1 over 1, with 34 zeros after it, seconds after the Big Bang. What that would do is expand the universe so much that any curvature was flattened out in that overall expansion, and the universe became essentially flat. That's the current prevailing theory, and that theory accounts for several of the quandaries we have in the Big Bang theory, several subtle things that are a little bit difficult to explain otherwise. For example, why is there matter and not anti-matter? Why does the universe appear to be almost flat? Why do we not see any large-scale, overall curvature in the universe?

So there's the inflationary universe scenario. That would be a course in itself. But I just gave you a hint of what that idea is. However, a universe that is just barely going to expand forever requires more matter than we can see out there. When we look out in the universe and count up all the stars in the galaxies and we ask, "Is there enough matter to pull the universe back in to close it?" The answer is no. Is there enough to just barely halt that expansion and make us a flat universe, that middle case that seems to be favored? The answer is no. The matter that we can see is only enough to make a universe that is going to expand forever with a lot of energy left over. On the other hand, if we look at the galaxies and the matter in them, the orbits the matter in the galaxies describe suggests there exists additional dark matter in the galaxies that is causing additional gravitational effects, and we don't know what that dark matter is.

In fact, in order for the current Big Bang theory of the universe to work, and most cosmologists believe it does, we have to assume that most—it's not only a small fraction but most—of the matter that makes up the universe is in some form that we don't even know about. Some of it may be mundane stuff like dead stars but made of ordinary matter, but most of it actually has to be—and you can prove this with current cosmological theories—has to be in a form we don't even know about. It's not protons and neutrons and quarks and things. It's some totally new kind of matter, and it's a very sobering thought that we may not know about what 90 percent of the universe is made of. The universe may be made of and governed by something that we know nothing about, and the stuff that we're made of, ordinary matter, may be just a small fraction of it going along for the ride. That's a sobering thought.

What's our place in this grand scheme? Well, I want to give you two, just quick insights into that. First of all, we're very much, very much, children of these processes. Particularly, we're children of the stars. I mentioned that the Big Bang cooked up helium and a few other light elements, but it didn't make the oxygen, the carbon essential for life, the iron, the zinc, the silicon, all those other things—it didn't make those at all. They were cooked up by nuclear processes in the interiors of stars, and, when those stars exploded, they spewed that material out into interstellar space, and gradually that material would re-congeal into new stars, new solar systems, planets. Eventually life evolved. We are, in fact, living on a planet in orbit around a so-called third generation star. It contains material from two previous generations of stars that exploded and spewed this material out into interstellar space, and we are star children made, literally, of elements that were cooked up in the interior of those stars. That's our direct connection to all this.

I'd like to end this lecture with a vision by physicist Freeman Dyson, from The Institute for Advanced Study in Princeton. Dyson did a study. He said, "What happens if intelligence once evolves in an expanding universe? Could it continue to evolve forever?" He concluded that even if the universe went through eons of infinite expansion, things got further and further apart and cooler and cooler and cooler, there might come a time when an intelligent being was some kind of a crystal structure locked into some solid material or was some gas cloud or something. But, he concluded, that intelligence, once established, could continue to exist forever and could continue to communicate with itself across this vast void. So we may be near the start of a very long, infinitely long, perhaps, adventure in which intelligence continues to exist in the universe and ponders what went on before it and appreciates this vast richness that those simple laws of physics give us.

Toward a Theory of Everything
Lecture 24

We know how general relativity can lead to an understanding of the overall structure of the universe, raising the question: Is the universe closed or open? Will it expand forever, or will it eventually close? Do we, then, know everything there is to know? Can we explain everything that needs to be explained about our universe? The answer is, clearly, no, and for several reasons.

General relativity describes the properties of the universe at the large scale: the overall curvature of the universe, the curved spacetime around gravitating masses, the behavior of matter near a black hole. Quantum physics, in contrast, describes the universe at small scales. In nearly all of physics, one or the other of these two theories suffices.

There are, however, situations in which intense gravity—spacetime curvature—exists on very small scales and to describe these, we need to merge general relativity with quantum physics to make a theory of *quantum gravity*. Quantum gravity becomes important on scales of around the so-called *Planck length*, about 10^{-33} centimeters. (That's 1/1,000,000,000,000,000, 000,000,000,000,000,000 of a centimeter.) At this scale, fluctuations in the structure of spacetime required by the uncertainty principle become huge. (Recall that confining matter to a small space means a large uncertainty in velocity—and that implies a large velocity and a large energy.) Quantum gravity must be important at the very centers of black holes, where general relativity predicts that matter is crushed to infinite density in a space of zero size. Quantum gravity must have been important in the history of the universe before the *Planck time*, about 10^{-43} seconds after the Big Bang.

A brief history of physics shows a common theme: the merging of distinct fields and phenomena under ever broader, more encompassing theories. Newton's theories of motion and gravity subsumed celestial and terrestrial motion under the same set of laws. The work of Maxwell and others in the 19th century joined electricity and magnetism under the theory of electromagnetism; soon optics joined them, when Maxwell realized that light was an electromagnetic wave. Special relativity and quantum physics were successfully merged in 1948 with the theory of *quantum electrodynamics*.

Theoretical work in the 1970s, followed by experiments in the 1980s, confirmed the *electroweak theory*'s unification of electromagnetism with the weak nuclear force. Unification of the color force (described by the theory of *quantum chromodynamics*) and the electroweak force appears to be in sight; the resulting *grand unified force* would explain all physical phenomena except those involving gravity. The ultimate *theory of everything* will require a merger of general relativity with quantum physics to make a theory of *quantum gravity*.

String theory may lead the way to this *theory of everything*. String theory arose around 1970, enjoyed a brief heyday in the mid-1980s, then faded because of seemingly insurmountable problems. In 1995, string theory had a dramatic comeback and today, string theorists are hard at work exploring the theory and its implications. Some—but not all—physicists are optimistic that this work may lead to the theory of everything. However, string theory still presents as-yet-unsolved mathematical problems, and no version of string theory has yet produced a full, quantitatively correct explanation for all the forces and particles of nature.

In string theory, the fundamental entities are not particles but tiny, string-like loops, whose size is roughly the Planck length at which the incompatibility between general relativity and quantum physics arises. Different vibrations of the strings correspond to the different "elementary" particles, just as different vibrations of a violin string make different notes. In this sense, all particles are aspects of a single underlying entity—the string. String theory not only shows how the individual particles arise, but it also predicts their masses—something that the standard model of particles and fields cannot do. Because the size of the strings is roughly the Planck length and because strings are the most fundamental entities there are, it makes no sense to talk about what happens on scales smaller than the Planck length. Thus, string theory sidesteps the conflict between quantum theory and general relativity by simply avoiding the regime in which the conflict occurs. Because a string is an extended object, the interaction between two particles occurs not at a point in space and an instant in time, but is spread out over time and space.

In particular, different observers see different parts of the strings interacting at different times, so there can be no unambiguous point and time where the interaction occurs. In the mathematics of string theory, this has the effect of eliminating the infinite spacetime curvature that would occur with true point particles.

An essential requirement of string theory is that the strings exist not in the four dimensions of ordinary spacetime (three of space, one of time), but in a spacetime of as many as 11 dimensions. Unlike the dimensions we're used to, the "extra" dimensions don't extend forever, but are curled up into tiny, closed structures on scales so small even quarks can't move in the extra dimensions. But the strings are small enough to vibrate in the extra dimensions. This effect gives string theory some of its richness and its ability to explain the diversity of particles we observe in the world.

An essential requirement of string theory is that the strings exist not in the four dimensions of ordinary spacetime (three of space, one of time), but in a spacetime of as many as 11 dimensions.

An analogy can help us to understand how there can be extra dimensions we don't notice (adapted from Green, *The Elegant Universe*): Imagine a bug walking along a tightly stretched, cylindrical rope. From a distance, it looks as if the bug lives in a one-dimensional world; it can move back and forth *along* the rope, but has no other freedom of motion. But move in close and you see that the bug can move in a second dimension. This is the dimension *around* the rope. This dimension doesn't extend a great distance like the length of the rope, but wraps around in a limited space. If the bug is much bigger than the rope diameter, it won't even notice the extra dimension. The extra dimensions of string theory are like this, except that they involve shapes much more complicated than the cylinder of the rope, and there are seven of them.

If we achieve a theory of everything, will that be the end of physics? No! There are still plenty of everyday phenomena we haven't yet explained— even though we're sure their explanations follow from the known laws of physics. Remember the dark matter: We still don't have any idea what most of the universe is made of! Remember Dyson: We may have an infinity of time to explore the richness of our evolving universe. Stay tuned! ∎

Suggested Reading

Green, *The Elegant Universe: Superstrings, Hidden Dimensions, and the Quest for the Ultimate Theory.*

Kaku, *Hyperspace*, Chapters 5–9.

Questions to Consider

1. This isn't really a question, but rather a task. Look for articles in the newspaper or in magazines that you read that deal with the latest developments in science and find those that cover relativity and quantum physics. See in what way they expand on or change what you have learned in this course.

2. Do you think it is possible to have a "theory of everything?" Defend your answer.

Unification of the Fundamental Forces

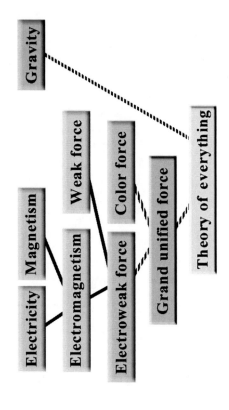

"Curled Up" Dimensions

One of the difficulties with string theories is that they require many more dimensions than the four dimensions of ordinary spacetime. Some versions of string theory require as many as 11 dimensions. So where are the extra seven dimensions?

According to string theory, these dimensions are "curled up" or "compactified" on such very small length scales that they are not noticed in our every-day lives or even in subatomic physics experiments. This diagram shows an analogy in fewer dimensions. At the top we see an ant on a rope. Viewed from a distance, it looks like the ant can move in only one dimension, namely back and forth along the rope. But if we get very close we can see that the ant can move in two mutually perpendicular directions, either along the rope or around it. The around-the-rope dimension is a "curled-up" dimension, evident only when the system is examined on very small length scales. This around-the-rope dimension is analogous to the extra dimensions in string theory.

Toward a Theory of Everything
Lecture 24—Transcript

Welcome to the final lecture, number twenty-four, aptly titled "Toward a Theory of Everything." In the last twenty-three lectures we've been through a lot of physics, and we've explained a lot about the physical world. Think back a minute. We understand a lot more about the nature of space and time. We know how electromagnetic waves work. We know how the theory of relativity, general relativity, predicts the bending of space and time by massive objects, and we understand how the motion of other objects follows as they go in the straightest possible lines in curved spacetime. We know what protons are made of. We understand that quarks bind together to make protons and neutrons and lots of other particles that were once thought to be elementary.

We've seen the whole panoply of the Standard Model of Particles and Forces and how it can explain things as small as the development of individual, so-called subatomic particles, nuclei and on upward. We also understand how it can help us explain the early times and the origin of our universe. In fact, the origin of the elements around us follow from the Standard Model of Particle and Forces. We know how general relativity can lead to an understanding of the overall structure of the universe, raising the question: Is the universe closed or open? Will it expand forever, or will it eventually close? Do we, then, know everything there is to know? Can we explain everything that needs to be explained about our universe? The answer is, clearly, no, and for several reasons.

First of all, we have no idea why there are three families of matter. We have no idea why the electron has the mass it does. We have no idea why the quark masses are distributed as they are. We don't know why there is this set of particles and not some other set of particles. You could probably envision a universe that was made with some other particles. Maybe you wouldn't be in that universe because maybe intelligent life couldn't exist, but maybe it could. Why do we have the universe we do? Why does it have the basic constituents it does? Why are there three fundamental forces, maybe reducible to one, and why do those forces have the properties they do? We don't know the answers to those questions. If we had a real theory of everything, a theory that could explain everything, it would explain, on first principles, why there are electrons, why they are the way they are, and so on. We don't have that theory yet.

What we do have, at the basis of our understanding of the universe, are two very powerful theories. One is relativity, ultimately the general theory of relativity, our theory of the large-scale structure of the universe, the curvature of spacetime as shaped by the presence of matter—our theory, ultimately, of gravity. That's one theory we have. The other we have is quantum physics, ultimately at the subatomic level as embodied in our Standard Model of Particles and Forces. A theory that tells us what particles we have, how they interact, and, ultimately, within the framework of quantum physics, how we describe, through that statistical, probabilistic wave interpretation, how these particles are going to behave and how they are, therefore, going to carry out the program of the universe, if you will.

But here's what we don't have. We don't have a theory that deals with those two together. Now, usually that isn't a problem. General relativity is important in explaining large-scale behavior of the universe. It's important in explaining the orbit of a satellite around the Earth. If we really want to do it right, we use general relativity, not Newtonian gravity. It's important in the large scale. Quantum physics is important in the small scale. We saw, when we dealt with Schrödinger's cat, that, although quantum physics may apply to large-scale objects, it probably isn't really very important there. We can get away with classical physics. We can get away, ultimately, with relativity, general relativity, our theory of gravity describing the behavior of macroscopic matter.

On the other hand, we don't need to think about general relativity. We don't need to think about spacetime curvature when we deal at the atomic and subatomic realms. It may be that there is a gravitational attraction between an electron and a nucleus of an atom. The nucleus may warp spacetime, but that is vastly overpowered by the electrical attraction, the electrical force. So gravity is completely negligible on the subatomic scale. What that says in more sophisticated terms is, we don't deal, in our everyday lives, and even in the everyday lives of physicists, with places where spacetime curvature is so extreme that the curvature of spacetime is on scales that are comparable to the size of the atom and the size of elementary particles.

Therefore, we have never had significant occasion to have to merge general relativity and quantum physics. If we did, we'd be in trouble because that's something we simply don't know how to do. When would that become important? Well, it turns out to become important on a scale called the *Planck scale*. I won't describe how we derived the Planck scale except to say that it comes from taking the constant big G that appears in the theory of gravity, the theory of general relativity, and the constant h that appears in quantum physics and combining them in a certain simple way with other constants of nature, and out comes a particular length. That length is about a 1 over 1 with 33 zeros after it, of a centimeter. It's a very tiny, tiny, tiny distance, and it's called the *Planck length*.

On scales much bigger than the Planck length, a proton is a 1 over 1 with 13 zeros after it centimeters—a 1 over 1 with 15 zeros after it meters. It's 20 orders of magnitude, 20 powers of 10, larger than the Planck scale. When we're dealing with things larger than the Planck scale, and a proton by that size is enormous compared to the Planck scale, then we don't have to worry about general relativity and quantum mechanics merging. It's only when we get down to scale sizes that are so small that we're dealing with things about 10^{-33} centimeters in size (a 1 over 1 with 33 zeros after it) that we have to begin worrying about quantum gravity, so-called quantum gravity, the merging of quantum physics and general relativity.

Why is it that that scale becomes important? I want to give you just the briefest sense of why that is. Recall the uncertainty principle. The uncertainty principle says that if we know the position of an object accurately, then we don't know anything about its velocity. If we know the velocity of an object accurately, we don't know anything about its position. What does that mean? Well, it means if we force ourselves to look at a very small scale of space and time, on a small scale of space, in particular, then anything that's in that small region, we must be very, very uncertain about its velocity. What that means is we can't know that its velocity is small because if I said I had a very tiny region here and anything that's in that tiny region is also not doing very much, that's a contradiction of the uncertainty principle because to say I have a very, very small region is to say I know very little about the velocities of many things that might be in that region, and those velocities can be very large. In fact, if I know they aren't very large, then I'm contradicting the uncertainty principle. They may very well be very large.

That kind of argument, that uncertainty principle argument, applies even to spacetime itself. What looks like to us a completely flat, smooth fabric in one of those two-dimensional analogs I've made of spacetime—maybe with some curvature over the large scale due to the presence of large masses like the Earth or the sun or a galaxy or something like that but basically smooth—if we look at very, very small levels, there are then fluctuations in the fabric of spacetime. It becomes sponge-like and like the boiling surface of boiling water.

Why? Because the uncertainty principle says if we look in too small a region we can't know how much energy is there. We can't know how much stuff is going on, and there must be lots of action going on there because, if we knew there weren't, we would be violating the uncertainty principle. So the uncertainty principle dictates that, at very small scales, this roiling and boilingness must affect spacetime itself. It turns out that the scale at which that becomes important, the scale at which the fluctuations and the structure of spacetime become large in some sense so they aren't just minor perturbations on an almost smooth background, but they become large effects—the dominant effect on the structures of spacetime is at scales about the Planck length (10^{-33} centimeters, a 1 over 1 with 33 zeros after it centimeters). That's a tiny, tiny distance, a distance we don't normally worry about, but there it is.

If we ever have to worry about distances that small, we have to have a theory of quantum gravity. We have to have merged quantum physics and general relativity, and that's what we don't know how to do. When would this be important? Well, I can give you two places or times when it might be important. One is at the very center of a black hole. Remember what happens in a black hole? In a black hole, matter crushes inexorably through that event horizon, and, after that, the crushing of the matter to an infinite density at the center of the black hole is simply inexorable, according to our current theory of black hole formation, which is the theory of general relativity. General relativity actually says that if a black hole forms, it becomes a singularity of infinite density at the center.

Now, that's probably absurd, and that prediction of general relativity is probably wrong because, as the matter shrinks towards that singularity, eventually it gets compressed into the size of the Planck length, 10 to the minus 33 centimeters, that tiny, tiny distance at which the structure of space and time is now governed by these quantum fluctuations. At that point, we don't have a theory that describes what's going on because we've been unable to merge general relativity and quantum physics. Why have we been unable to merge them? Again, because of these quantum fluctuations. General relativity is a theory about the smooth and continuous. It's not a quantum theory. It doesn't talk about discreteness. It doesn't know about things coming in little bunches and bundles. Quantum physics, on the other hand, is a theory about that discreteness, and that results in the uncertainty principle and these fluctuations in the fabric of spacetime at these very small scales.

So, what's wrong is we don't know how to make general relativity a quantum theory. We don't know how to quantize general relativity. We don't know how to describe the gravitons that would be the force carrier of particles of general relativity in any detail. We have not got a consistent theory that merges quantum physics and general relativity. By the way, you might be asking the same question about special relativity, and, just to put that question aside, by 1948 we had a good theory that merges special relativity and quantum physics. It's called *quantum electrodynamics*. Richard Feynman was one of the leaders in developing that theory, the famous physicist Richard Feynman. Quantum electrodynamics is basically a theory that re-expresses Maxwell's theory of electromagnetism, which, you will remember, led to relativity.

Maxwell's theory is completely consistent with relativity. It rephrases it in a way that's consistent with quantum physics, and that has been done, and that is the most successful, most well-tested theory of physical reality that we have, the theory of quantum electrodynamics. It assumes quantum physics, and it assumes and it subsumes special relativity. It is sometimes called *relativistic quantum mechanics* or *relativistic quantum physics*. We have that. What we don't have is a combination of general relativity, the theory of gravity, the theory of curved spacetime, with quantum physics, and, therefore, we can't explain things that are happening on space scales smaller than this incredibly tiny Planck length.

There is one situation we know beyond the black hole where that would be important to do, and that is at times earlier than the Planck time, which I mentioned in my time line in the last lecture. In that time line, one of the earliest times I put was 10^{-43} seconds (1 over 1 with 43 zeros after it) of a second after the Big Bang started. That's an infinitesimally, unimaginably tiny time, but, after that time, the universe had already expanded enough that the density was already low enough that we did not have to deal with scales as small as the Planck length, and, therefore, we could treat the overall structure with general relativity, and we could treat the subatomic details with quantum physics and the Standard Model, and nothing contradicted that. They were both approximations but very, very good approximations.

But before that time, 10^{-43} seconds (a 1 over 1 with 43 zeros after it) of a second, we have to do quantum gravity. We can't push our understanding of the universe back further without knowing how to combine quantum physics and general relativity. And, for the reasons I described, for these fluctuations in the structure of spacetime, the discontinuousness of quantum mechanics versus the continuousness of general relativity, that's a very difficult thing to do, and we don't know how to do it. If we could do it, we would achieve the final step in what is a long history of physics: Taking previously unrelated fields and finding that they're described by one underlying principle. That's particularly true of the forces of physics, these basic, fundamental forces that govern all interactions in physical reality.

I want to take a moment and take you through that history to give you a sense of just where it comes from and how far we've come and how far we still have to go. So I'm going to look at the unification of the fundamental forces through the history of physics. It's a process that's been going on, and it continues to this day. The first force I want to mention is gravity because that's certainly the first force that was experienced by humankind. I argued, a long time ago, that Newton didn't discover gravity by having an apple falling on his head. Some cave person knew about gravity. Apes swinging in trees know about gravity. Birds certainly know intuitively about gravity. So we've known about gravity for a long time. Gravity itself, though, as a formal theory emerged with Newton in the late 1600s, and Newton, for the first time, had unified terrestrial motion and celestial motion and understood them to be aspects of the same thing.

So, in putting gravity here on my picture, I've already assumed one of the important unifications. Before Newton's theory of gravity, it was not understood that the motion of the moon and the other celestial bodies was really the same thing as the motion of an apple falling here on Earth. Newton unified that with his theory of gravity. Two other forces that are important are the electric force and, associated with it, the study of electricity, electric charge, electric currents, that sort of thing, and, separately, the force of magnetism, the force of interaction between two magnets, between a compass needle and the Earth's magnetic field, the force of magnetism. Those were studied originally as two separate forces. But, as you know from the lecture on electromagnetism, the important point about the subject electromagnetism is it arose because physicists began to understand the connections between electricity and magnetism.

As those connections became more clear, as it became obvious that that was not just a coincidence but an intimate connection ultimately, it became evident that electricity and magnetism are really two aspects of the same thing. I didn't dwell on this when we did special relativity, but one of the other things that special relativity mixes, besides space and time, is electricity and magnetism. What looks like pure electricity to one observer looks like electricity and magnetism together to another observer. Electricity and magnetism are two aspects of the same thing, two sides of the same coin, and Maxwell's unification in the 1860s basically puts those together as one subject, electromagnetism. The electromagnetic force was one of the four fundamental forces that were understood to be fundamental, as far as we could tell, until about the 1970s and 1980s.

So we had electromagnetism and we've got gravity sitting here as fundamental forces at this point. The weak force was discovered in nuclear reactions. That was discovered in the early part of the twentieth century. It's a little bit obscure, but, again, it mitigates certain kinds of nuclear decays, not all of them, and it's important on the sun and making the sun shine. So it's one of the forces that are involved in nuclear interactions. It's the weak force, and it was understood as a separate, independent force of nature through much of the twentieth century. But, as I indicated earlier, in 1970, theorists succeeded in describing the weak force and electromagnetism as two aspects of the same thing, a further kind of unification. And, by the 1980s, experimental evidence accumulated to confirm that this was indeed correct. So, by the 1980s, we had the electroweak force as one of the fundamental forces.

Under different circumstances, it may manifest itself as the weak force, the electromagnetic force—and the electromagnetic force, under different circumstances, may manifest itself as the electric force and the magnetic force. By the way, one of these features of these unified forces is that, when you go to situations where the energy is very high, as it was in the early universe, what now appear to us as separate forces were all unified and just acted as one basic interaction. So things are more complex now. It's that big theme of the universe cooling and getting more complex. One result of that complexity is the fundamental force or forces then become manifest as separate, individual forces that seem distinct until our theorists and experimentalists go to work and figure out that they really are aspects of the same thing.

So we have the electroweak force. Then we have the color force, which many physicists would still call the strong force or even the nuclear force, but I like the term "color force" because one of the aspects quarks have is called "color." It has absolutely nothing to do with color, and the color force binds quarks that have different colors. They are called "red," "green," and "blue," and it means nothing in terms of regular red, green, and blue. But the force is called the "color force," and the study of quarks and how they stick together is called *quantum chromodynamics* in analogy with quantum electrodynamics, which is the relativistic electric force. So we have the color force ultimately resulting in the strong force, the nuclear force, and so on, and here's where we stand basically today.

I argued earlier, in my lecture on the Standard Model and the structure of matter, that we have three fundamental forces, electroweak force, color force, and gravity, and we don't know how to unify them further. Most physicists would argue we're quite close to understanding the electroweak force and the color force as aspects of some more fundamental underlying force. We aren't there yet, but theorists are close. We do not have the power or the energy with our accelerators to create the conditions that would show us that unification. Not yet, anyway, but theorists are working on that, and we believe it will not be very long before we understand the electroweak force and the color force as aspects of one force. That would be called *grand unification*, and we might call that force a *grand unified force*. So watch for that. That may happen in your lifetimes, if not sooner. In fact, all of this may happen in your lifetime but not sooner. The next one is a bit of a longer shot.

So there we are. We may, in a matter of years to decades, be down to two fundamental forces, a grand unified force that breaks down into color force and electroweak force, further into electric, and magnetic, and weak force, and so on. The real challenge, and that's what the topic of today's lecture is, is the further unification of gravity with the grand unified force—once we have the grand unified force—and that would give us a theory of everything because that would unite all the known forces into one. I don't know what we're going to call that force. Some people call it supersymmetry because of certain mathematical properties of the theories. Others simply call it a theory of everything because, if there's nothing left undiscovered, there's nothing new, there are no unknown forces that we haven't discovered yet, then the "theory of everything" will describe all interactions in the universe in terms of one fundamental basic force, and we'll understand how, in different circumstances, that force will appear as gravity, or as a grand unified force, or as the electroweak force, or whatever, but we will understand it as one fundamental interaction.

That's one of the Holy Grails of physics, to find the theory of everything. It may be a goal that is decades to centuries away. On the other hand, there are at least some physicists who believe we may have stumbled across a theory that may be the way to the theory of everything, and I want to spend the remainder of the course, the remaining few minutes of this lecture, describing that theory of everything. That theory is called *string theory*.

String theory is remarkably different from anything we've seen in physics before. In our Standard Model of Particles and Forces, we ultimately describe matter in terms of a few fundamental particles which, quantum mechanics aside, we think of as little, point-like objects, the quarks, the leptons, the electrons, and so on. Those are the fundamental entities. In string theory, one thinks of the fundamental entities very differently. They're considered to be strings, like a piece of string wrapped into a loop, like this loop I have here. This would be a string. It wouldn't be as big as I have it here. In fact, it would be about the Planck length long, and the Planck length is 1 over 1 with 33 zeros after it of a centimeter. It's a tiny, tiny thing.

But strings would be the Planck length in size, and there would be basically one kind of string, and the string could vibrate. It could vibrate in different ways, maybe like this, maybe like this. It could vibrate in different ways, and each of those fundamental vibrations of the string would correspond to a different elementary particle, a different quark, an electron, what have you. So all of matter would be made of one kind of thing, a string, and the things we think of as elementary particles would be the allowed vibrations of these strings. Just like a violin string has certain allowed vibrations that are the notes it can produce, the string would have certain allowed vibrations, which are the elementary particles. String theory, therefore, ought to predict something the Standard Model observer of particles and forces cannot, namely, the actual masses of the particles of the so-called elementary particles.

Now, these strings are tiny. A proton is about 10^{-13} (a 1 over 1 with 13 zeros after it) centimeters in diameter. A string is 20 orders of magnitude smaller. These are tiny, tiny entities, if they exist. Nobody has ever seen a string. Nobody's very likely to see a string in anything like the foreseeable future, but, nevertheless, we can mathematically analyze the properties of entities like this and see what they would say. In particular, we can try to analyze their properties and see if those properties correspond to the properties of the known elementary particles. Now, how is string theory going to work? How's it going to help us with the quantum gravity problem?

Let me give you two slight answers to that. The current versions of string theory seem to result in one of the modes of vibration that have characteristics like what we would expect the graviton, the quantized gravitational force carrier, to have. So that's encouraging. The other thing is, the strings are about the Planck length, and that's the smallest thing there is. We're used to thinking we can go smaller and smaller and smaller, but if the smallest unit of matter is the size of the Planck length, then it makes no sense to talk about scales smaller than that. And that, in a crude way, sidesteps the question of whether or not quantum mechanics and gravity are at odds because we never get to scales where that's an issue. So that's a simple way to get rid of the discrepancy between general relativity and quantum physics because the strings are simply the smallest entities, and they are already at a scale that's just barely big enough to avoid that conflict.

How confident are we of all of this, by the way? Well, this is a fairly new theory. The ideas of string theory were developed in the 1970s. They enjoyed a brief surge of enthusiasm in the 1980s, and then there seemed to be intractable mathematical problems. Nobody has yet produced a string theory that exactly reproduces the particles of the Standard Model. There are hints that it may be possible, but nobody's done that yet, and enthusiasm for string theory faded in the late 1980s and early 1990s. Then, in 1995, there were a number of rather stunning breakthroughs that got people excited about string theory again, and it's only been in the years from 1995 onward that people have really been enthusiastic about the possibility that string theory may give us a theory of everything.

There are still a lot of enormous complications. I have no time to go into all of them, but I want to give you just a feel for some of the aspects of string theory. I'd like to show you a slightly more sophisticated reason why, in string theory, the problem of quantum gravity goes away, why this business of the fluctuations in the structure of spacetime becomes less problematical. It's a slightly more sophisticated look at the issue. I'm going to show you a picture here that describes a simple thing, an interaction between two particles. What I'm showing in this picture is a Y-shaped structure in which the two legs of the Y are labeled A and B and they represent two particles that are moving through space and time. Time is going roughly to the right and a little bit upward in this picture.

These particles are moving along in space and time, and then they come together, and they interact, and maybe in a very simple interaction they join to form a composite particle, which I'll call C, which continues on. Now, in this description, in which particles are really true, point-like particles, that interaction between A and B occurs at a single point. It's right there, unambiguous. It occurs at a single point in space and time. It's an event. In fact, let's impose on this a particular direction of time. There it is. That arrow describes the direction of time. This is now attempting to be a three-dimensional picture of four-dimensional spacetime. I've suppressed one of the space dimensions, but I've drawn here a flat plane, a two-dimensional space, and it runs perpendicular to my time dimension, as it needs to, because the four dimensions are perpendicular in some sense.

So it's unambiguous that this interaction between particle A and particle B occurs at a fixed point in space and a fixed time in time. That's in somebody's frame of reference. In somebody else's frame of reference, space and time are mixed. That's what we find relativity does. It mixes your time dimension and my space dimension and so on when we're moving relative to each other. So somebody else's time dimension might be that way. This is sort of a mixed picture, partly in the new person's frame of reference and partly in the old one, where the A-B-C interaction still is. But the point is, again, in that fame of reference, the interaction occurs at a single instant of time and point in space, and it's that instantness, that pointness that the interaction is occurring at a really, truly point of arbitrarily small size, that is what gets general relativity and quantum physics in trouble because there we are again dealing with those tiny, tiny scales.

What happens if A and B are not true, little, tiny point particles, not even this big ball but really truly, infinitesimal points—not that but instead are string-like loops? They are moving through space like this, and there are two of them moving. Then they come on a collision course, and they collide, and they merge to make one particle. What would that look like? Well, it would look something like this: Particle A is now not a particle but a looplike of string. Particle B is another loop of string. They move along through space and time. They merge, and they go on as one string, C, representing a particle. What if we impose these time arrows and these planes representing a two-dimensional space in this situation? Well, what happens in the first case is we get a plane that's cutting through that at one particular angle. Where does the interaction take place? Well, it takes place all over that plane, not in one single point.

Now let's go to the other frame of reference where that plane that represented two spatial dimensions has a different orientation. When we go to that other situation, that other frame of reference, we find that the interaction now takes place at different places on this complex-looking spacetime structure representing the two strings coming together. There is no unique point in spacetime. There is no infinitesimal tinyness in which this interaction occurs, and the description of the interaction in string theory is necessarily extended over a volume of space and time that is large enough, given the tiny but, nevertheless, nonzero size of the strings, to alleviate those problems of the discreteness of quantum mechanics colliding with the continuousness of general relativity. Amazing.

Now, string theory requires some real weirdness to make it work. One of the weirdest things is, it no longer requires that we have 4 dimensions of 3 of space and 1 of time, but that we have 11 dimensions; 10 of space and 1 of time. How can that possibly be? Well, I'm going to give you a brief analogy to how 11-dimensional string theory can work, and I'm taking this quite liberally from an elegant book by Brian Greene called *The Elegant Universe*. It's in my bibliography, and you can read a brief paragraph about it. I urge you to read it if you'd like to know a lot more about string theory than I'm going to be able to tell you in this short time. Let's look at an analogy for how there could be 11 dimensions and we don't notice it. What would that mean?

Here's a picture that shows an ant walking along on a stretched piece of rope. We're looking at this from some distance away, and we don't notice that the rope has any particular size. If we look at that ant and say, "Well, how is that ant free to move?" The answer is to move back and forth along the rope, but that's about it. It lives in a one-dimensional world. It has only one dimension that it can exist on. On the other hand, if we zoom in on that situation with a magnifying glass and look again, we see the full size of that rope. If we look at the ant and ask what it can do, there's the ant, crawling on the rope. The ant can move back and forth along the rope like it could before, and it can move long distances back and forth along the rope, but it can also move in a perpendicular direction around the rope. The extra dimension that's there, the second dimension, is not as extended as the first dimension. It's so-called compactified. It's curled back up on itself.

That's how the extra 7 dimensions—4 of our regular extended space and time, plus 7 extra dimensions—are curled up in a compactified shape. Unfortunately, they don't describe simple shapes like spheres but some very complicated-looking shapes, described by very strange mathematics. Nevertheless, the description of string theory says there have to be 11 dimensions and 7 of them are curled up in this way, and it's the ability of the strings to vibrate, the strings are small enough—they can see these extra dimensions, like the ant can here—and they can vibrate in the extra dimensions, and that's what gives string theory its richness as a theory in describing, possibly, all of the elementary particles we see.

Well, if we get there, if we reach this theory of everything, if string theory really does explain all these things that we see in the universe around us— gravity and the elementary particles—have we come to the end of physics? Is this the last physics course once we've discovered that? Well, no, and I want to give you just a few quick reasons why it isn't. First of all, most of us who are doing physics are doing, still, old-fashioned or, a lot of us, are doing old-fashioned, classical physics. I work in the study of the sun and its outer atmosphere. We use Maxwell's equations and laws of gas behavior that have been known about for 100 years or more. We still don't know all the richness of phenomena that those laws can describe. So we have a lot to do to understand the world as we see it in terms of the theories that already exist, let alone a theory of everything.

So, if we had a theory of everything, although it would in principle tell us everything, we'd have a lot more to do to explain the phenomena around us. There may still be other levels as we push back to smaller and smaller scales or back to that time before the Big Bang. You may think, 10 to the minus 43 seconds, not much else happened before that. But there could be infinitely many different and interesting eras in which there's new physics going on there. Remember the dark matter. We don't know what most of the universe is made of. That should be a sobering thought. There's a lot more to explore about this universe. And, finally, remember my comment about Freeman Dyson's vision last time. We, the intelligent, conscious beings, may have the infinite future of a universe that is evolving toward an unimaginable richness. We may have that future ahead of us to explore and appreciate this very rich and vast universe. So stay tuned.

Timeline

1543 .... Copernicus publishes *De Revolutionibus Orbium Coelestium*, challenging the then-held view that Earth was at the center of the universe.

1686 .... Newton completes the *Principia Mathematica*, which includes his laws of motion and theory of gravity.

1801 .... Young's double-slit experiment shows that light is a wave.

1860 .... Maxwell completes the synthesis of the four equations of electromagnetism and shows that they imply electromagnetic waves that propagate at the speed of light.

1887 .... The Michelson-Morley experiment fails to detect Earth's motion through the ether.

1897 .... J. J. Thomson discovers the electron.

1900 .... Planck resolves the ultraviolet catastrophe by postulating quantization of the energy associated with hot, glowing objects.

1905 .... Einstein explains the photoelectric effect by proposing that light energy is quantized.

1905 .... Einstein publishes the special theory of relativity.

1911.. Rutherford discovers the atomic nucleus and proposes his "solar system" model for the atom.

1913.. Bohr publishes his quantum theory of the atom.

1916.. Einstein publishes the general theory of relativity.

1919.. Observations by Eddington and colleagues at a total solar eclipse confirm general relativity's predictions of the bending of light by the Sun's gravity.

1923.. DeBroglie sets forth his matter-wave hypothesis. Compton effect experiment convinces most skeptics of the reality of quanta.

1927.. Heisenberg states the uncertainty principle. Davisson and Germer show that electrons undergo interference, thus experimentally verifying DeBroglie's matter-wave hypothesis.

1929.. Hubble discovers the expansion of the universe.

1932.. Carl Anderson discovers the positron, verifying Dirac's hypothesis that antimatter should exist.

1939.. Lise Meitner identifies the process of nuclear fission.

1948.. Feynman, Tomonaga, and Schwinger produce the theory of quantum electrodynamics, successfully uniting special relativity with quantum mechanics.

1964.. Quarks proposed as fundamental constituents of matter.

1965.. Penzias and Wilson discover cosmic microwave background radiation.

1983.. Experimental verification of electroweak unification.

1994.. Existence of the top quark is experimentally verified.

1995.. Second string theory revolution increases interest in string theory as a possible "theory of everything."

1998.. Neutrinos found to have nonzero mass. Cosmic expansion of the universe found to be accelerating.

Glossary

aberration of starlight: A phenomenon whereby a telescope must be pointed in slightly different directions at different times of year, because of Earth's orbital motion. The fact of aberration shows that Earth cannot drag with it the ether in its immediate vicinity and, thus, helps dispel the notion that ether exists.

absolute motion: Motion that exists, undeniably, without reference to anything else. The relativity principle denies the possibility of absolute motion.

Big Bang: The explosive event that began the expansion of the universe.

black hole: An object so small yet so massive that escape speed exceeds the speed of light. General relativity predicts the possibility of black holes, and modern astrophysics has essentially confirmed their existence.

color force: The very strong force that acts between quarks, binding them together to form hadrons and mesons.

Compton effect: An interaction between a photon and an electron, in which the photon scatters off the electron, as in a collision between billiard balls, and comes off with less energy. The effect provides a convincing demonstration of the quantization of light energy.

Copenhagen interpretation of quantum physics: The standard view of the meaning of quantum physics, which states that it makes no sense to talk about quantities, such as the precise velocity and position of a particle, that cannot even in principle be measured simultaneously.

cosmic microwave background: "Fossil" radiation from the time 500,000 years after the Big Bang, when atoms formed and the universe became transparent.

dark matter: Matter in the cosmos that is undetectable because it doesn't glow. Dark matter, some of it in the form of as-yet-undiscovered exotic particles, is thought to comprise most of the universe.

electromagnetic wave: A structure consisting of electric and magnetic fields in which each kind of field generates the other to keep the structure propagating through empty space at the speed of light, c. Electromagnetic waves include radio and TV signals, infrared radiation, visible light, ultraviolet light, x rays, and gamma rays.

electroweak force: One of the three fundamental forces now identified, the electroweak force subsumes electromagnetism and the weak nuclear force.

elsewhere: A region of spacetime that is neither past nor future. The elsewhere of a given event consists of those other events that cannot influence or be influenced by the given event—namely, those events that are far enough away in space that not even light can travel between them and the given event.

escape speed: The speed needed to escape to infinitely great distance from a gravitating object. For Earth, escape speed from the surface is about 7 miles per second; for a black hole, escape speed exceeds the speed of light.

ether: A hypothetical substance, proposed by 19th century physicists and thought to be the medium in which electromagnetic waves were disturbances.

event horizon: A spherical surface surrounding a black hole and marking the "point of no return" from which nothing can escape.

field: A way of describing interacting objects that avoids action at a distance. In the field view, one object creates a field that pervades space; a second object responds to the field in its immediate vicinity. Examples include the electric field, the magnetic field, and the gravitational field.

frame of reference: A conceptual framework from which one can make observations. Specifying a frame of reference means specifying one's state of motion and the orientation of coordinate axes used to measure positions.

general theory of relativity: Einstein's generalization of special relativity that makes all observers, whatever their states of motion, essentially equivalent. Because of the equivalence principle, general relativity is necessarily a theory about gravity.

geodesic: The shortest path in a curved geometry, like a great circle on Earth's surface. Objects that move freely follow geodesics in the curved spacetime of general relativity.

gravitational lensing: An effect caused by the general relativistic bending of light, whereby light from a distant astrophysical object is bent by an intervening massive object to produce multiple and/or distorted images.

gravitational time dilation: The slowing of time in regions of intense gravity (large spacetime curvature).

gravitational waves: Literally, "ripples" in the fabric of spacetime. They propagate at the speed of light and result in transient distortions in space and time.

gravity: According to Newton, an attractive force that acts between all matter in the universe. According to Einstein, a geometrical property of spacetime (spacetime curvature) that results in the straightest paths not being Euclidean straight lines.

hadron: A "heavy" particle, made up of three quarks. Protons and neutrons are the most well known hadrons.

Heisenberg uncertainty principle: The statement that one cannot simultaneously measure both the position and velocity (actually, momentum) of a particle with arbitrary precision.

interference: A wave phenomenon, whereby two waves at the same place simply add together to make a composite wave. When both waves reinforce, the interference is said to be constructive and results in a stronger wave. When the waves tend to cancel each other, the interference is destructive. Interference is useful in precision optical measurements, including the Michelson-Morley experiment.

length contraction: The phenomenon whereby an object or distance is longest in a reference frame in which the object or the endpoints of the distance are at rest. Also called the *Lorentz contraction* and *Lorentz-Fitzgerald contraction*.

lepton: Collective name for the light particles electron, muon, tau, and their associated neutrinos.

mass-energy equivalence: The statement, embodied in Einstein's equation $E=mc^2$, that matter and energy are interchangeable.

Maxwell's equations: The four equations that govern all electromagnetic phenomena described by classical physics. It was Maxwell in the 1860s who completed the full set of equations and went on to show how they predict the existence of electromagnetic waves. Maxwell's equations are fully consistent with special relativity.

mechanics: The branch of physics dealing with the study of motion.

meson: A particle made up of two quarks (actually, a quark and an antiquark).

Michelson-Morley experiment: An 1880s experiment designed to detect Earth's motion through the ether. The experiment failed to detect such motion, paving the way for the abandonment of the ether concept and the advent of relativity.

neutrino: An elusive particle with very small mass that arises in weak nuclear reactions.

neutron star: An astrophysical object that arises at the end of the lifetime of certain massive stars. A typical neutron star has the mass of several Suns crammed into a ball with a diameter about that of a city.

photoelectric effect: The ejection of electrons from a metal by the influence of light incident on the metal.

photon: The quantum of electromagnetic radiation. For radiation of frequency f, the quantum of energy is $E = hf$.

Planck's constant: A fundamental constant of nature, designated h, that sets the basic scale of quantization. If h were zero, classical physics would be correct; h being nonzero is what necessitates quantum physics.

Principle of Complementarity: Bohr's statement that wave and particle aspects of nature are complementary and can never both be true simultaneously.

Principle of Equivalence: The statement that the effects of gravity and acceleration are indistinguishable in a sufficiently small reference frame. The principle of equivalence is at the heart of general relativity's identification of gravity with the geometry of spacetime.

Principle of Galilean Relativity: The statement that the laws of motion are the same in all uniformly moving frames of reference; equivalently, such statements as "I am moving" or "I am at rest" are meaningless unless "moving" and "rest" are relative to some other object or reference frame.

quanta: Discrete, indivisible "chunks" of a physical quantity, such as energy.

quark: A fundamental particle, building block of protons and neutrons, as well as all other hadrons and mesons. There are six different quarks, two in each of the three families of matter.

relativistic invariant: A quantity that has a value that is the same in all frames of reference. The spacetime interval is one example of a relativistic invariant.

relativity principle: A statement that only relative motion is significant. The principle of Galilean relativity is a special case, applicable only to the laws of motion. Einstein's principle of special relativity covers all of physics but is limited to the case of uniform motion.

spacetime: The four-dimensional continuum in which the events of the universe take place. According to relativity, spacetime breaks down into space and time in different ways for different observers.

spacetime curvature: The geometrical property of spacetime that causes its geometry to differ from ordinary Euclidean geometry. The curvature is caused by the presence of massive objects, and other objects naturally follow the straightest possible paths in curved spacetime. This is the essence of general relativity's description of gravity.

spacetime interval: A four-dimensional "distance" in spacetime. Unlike intervals of time or distance, which are different for observers in relative motion, the spacetime interval between two events has the same value for all observers.

special theory of relativity: Einstein's statement that the laws of physics are the same for all observers in uniform motion.

string theory: A description of physical reality in which the fundamental entities are not particles but tiny string-like loops. Different oscillations of the loops correspond to what we now consider different "elementary" particles. String theory is a leading candidate for a "theory of everything."

time dilation: In special relativity, the phenomenon whereby the time measured by a uniformly moving clock present at two events is shorter than that measured by separate clocks located at the two events. In general relativity, the phenomenon of time running slower in a region of stronger gravity (greater spacetime curvature).

ultraviolet catastrophe: The absurd prediction of classical physics that a hot, glowing object should emit an infinite amount of energy in the short-wavelength region of the electromagnetic spectrum.

universal gravitation: The concept, originated by Newton, that every piece of matter in the universe attracts every other piece.

wave packet: A construction made from waves of different frequencies that results in a localized wave disturbance.

white dwarf: A collapsed star with approximately the mass of the Sun crammed into the size of the Earth.

wormhole: A hypothetical "tunnel" linking otherwise distant regions of spacetime.

Bibliography

To the Student/Reader: It is difficult to find readings that exactly parallel the structure of the lectures. The readings I've chosen are designed to extend and complement the lectures, rather than to repeat the lecture material. Most of the materials chosen, even those in the "Suggested Reading" category, are aimed at lay audiences. Motivated readers will find more in-depth and mathematically oriented coverage of these topics in science textbooks, especially at the college level. The books listed here include a mix of older and more contemporary works. Many good books on the development of modern physics are now quite dated and out of print, but they still provide excellent introductions to the subject. Books listed under "Essential Reading" are in print as of 1999; some of the others are out of print but should be available in most public libraries. New books on the more contemporary subjects covered in these lectures are published each year.

Barrow, John D., and Silk, Joseph, *The Left Hand of Creation: The Origin and Evolution of the Expanding Universe* (Oxford University Press, 1993). This overview of cosmology, originally published in 1983 and updated 10 years later, is particularly good on the relation between particle physics and cosmology.

Brennan, Richard, *Heisenberg Probably Slept Here: The Lives, Times, and Ideas of the Great Physicists of the 20th Century* (John Wiley & Sons, 1997). Part biography, part science, this book details the lives and contributions of Newton (even though he's not from the 20th century), Einstein, Planck, Rutherford, Bohr, Heisenberg, Feynman, and Gell-Mann. Notably absent are the great female physicists of the century, most of whom made their contributions in nuclear physics: Nobel laureates Marie Curie (two Nobel Prizes) and her daughter Irène; Maria Goeppert Mayer, Nobel laureate for her contributions to nuclear theory; and Lise Meitner, who first identified the process of nuclear fission.

Casper, Barry, and Noer, Richard, *Revolutions in Physics* (New York, W. W. Norton, 1972). Used in college science courses for nonscience students, this well-written book describes a number of important revolutions in ideas about the physical world, including but not limited to those of modern physics.

Calaprice, Alice, ed., *The Quotable Einstein* (Princeton University Press, 1996). Need a quote by Einstein on your favorite subject? You'll find it all here—from religion to science to marriage to music to vegetarianism to abortion to capitalism to sailing, and much more!

Daintith, John, Mitchell, Sarah, Tootill, Elizabeth, and Gjertsen, Derek, *Biographical Encyclopedia of Scientists* (Institute of Physics Publishing, 1994). A handy source of brief biographical sketches of major scientists, both historical and contemporary.

Davies, Paul, *About Time: Einstein's Unfinished Revolution* (New York: Simon and Schuster, 1995). Davies, winner of the Templeton Prize for Religion, is a prolific popularizer of science and an active researcher in gravitational physics. This book takes the reader on an exploration of the nature of time, well beyond what's covered in this lecture series.

Davies, P. C. W., and Brown, J. R., eds., *The Ghost in the Atom* (Cambridge University Press, 1986). This book introduces the strange concepts of quantum physics and provides transcripts of question-and-answer sessions with some of the leading physicists concerned with interpretations of quantum physics.

Einstein, Albert, and Infeld, Leopold, *The Evolution of Physics*, (Simon & Schuster, 1967 and earlier editions). This old classic presents the conceptual background behind the development of relativity. Although not as lively as some contemporary works, it's good to hear about relativity in Einstein's own words (and those of his colleague, Infeld).

French, A. P., ed., *Einstein: A Centenary Volume* (Harvard University Press, 1979). This work was produced by the International Commission on Physics Education in celebration of the hundredth anniversary of Einstein's birth. The interested reader will find here a wealth of historical, biographical, and scientific perspective on Einstein's life and work, including translations of some original Einstein writings.

Fritzsch, Harald, *An Equation that Changed the World* (University of Chicago Press, 1994). There's more emphasis here than I would like on $E=mc^2$, but this book still provides a good introduction to relativity.

Greene, Brian, *The Elegant Universe: Superstrings, Hidden Dimensions, and the Quest for the Ultimate Theory* (W.W. Norton, 1999). This lively and very contemporary book is written at just the right level for a layperson interested in really understanding what string theory is about. Author Greene, himself a string theory researcher, concentrates on the revolution in string theory of the mid-1990s and communicates his enthusiasm for a theory he believes really has the makings of the "theory of everything."

Gribbin, John, *In Search of Schrödinger's Cat: Quantum Physics and Reality* (Bantam Books, 1984). A good book for nonscientists on the basics of quantum physics, which then goes on to give clear descriptions of such "quantum weirdness" as Schrödinger's cat and EPR experiments. As the paperback edition's cover says: "A fascinating and delightful introduction to the strange world of the quantum..."

Hafele, J. C., and Keating, Richard, "Around-the-World Atomic Clocks: Predicted Relativistic Time Gains" and "Around-the-World Atomic Clocks: Observed Relativistic Time Gains," *Science*, vol. 177, pp. 166–170 (July 1972). These back-to-back scientific papers describe in detail the background and experiments on relativistic time dilation using atomic clocks. Both special and general relativistic effects are covered. There's a good bit of math. *Science* is a general publication for scientists, so this is not just for relativity experts.

Han, M. Y., *The Probable Universe: An Owner's Guide to Quantum Physics* (Summit, PA: TAB Books/McGraw-Hill, 1993). This short book provides a brief introduction to the ideas of quantum physics, then goes on to its many contemporary practical applications, such as tunneling microscopy, microelectronic devices, and more. Color plates of tunneling microscope photos show just how far technology can go in manipulating individual atoms.

Hawkins, Michael, *Hunting Down the Universe: The Missing Mass, Primordial Black Holes and Other Dark Matters* (Little, Brown, 1997). This is a cosmology book with an emphasis on contemporary problems and observational evidence in cosmology.

Heisenberg, Werner, *Physics and Philosophy: The Revolution in Modern Science* (New York: Harper, 1962; also republished by Prometheus Books, Amherst, NY, 1999). In his later years Heisenberg turned increasingly to philosophical issues. This book sets forth his views, especially on the interpretation of quantum physics.

Hey, Tony, and Walters, Patrick, *Einstein's Mirror* (Cambridge University Press, 1997). Written in a breezy, popular style and prolifically illustrated, this book presents the ideas of relativity intermixed with lots of modern technology and astronomy. Not the most coherent presentation of relativity but fun to read.

—————, *The Quantum Universe* (Cambridge University Press, 1987). An exposition of quantum physics, stylistically very similar to the authors' relativity book described above.

Kaku, Michio, *Hyperspace* (New York: Anchor Books, 1994). Kaku, a prolific science writer, provides a trendy look at modern physics, emphasizing string theory and the search for a "theory of everything."

Kane, Gordon, *The Particle Garden* (Addison Wesley, 1995). In this book, Kane, a particle physicist and popular lecturer, gives a clear, nonmathematical explication of the standard model of particles and forces.

Lasota, Jean-Pierre, "Unmasking Black Holes," *Scientific American*, vol. 280, no. 5, p. 40 (May 1999). In this article, an astrophysicist presents contemporary evidence for the reality of black holes.

Lederman, Leon (with Dick Teresi), *The God Particle: If the Universe Is the Answer, What Is the Question?* (Boston: Houghton Mifflin, 1993). A Nobel laureate and former director of the Fermi National Accelerator, Lederman teams with Teresi, science writer and former editor of *Omni* magazine, in a lively look at the history and science of modern particle physics.

Lightman, Alan, *Ancient Light: Our Changing View of the Universe* (Harvard University Press, 1991). Lightman—scientist, science writer, and novelist— devotes a chapter to each of the main themes in modern cosmology. The book includes biographical sketches of leading cosmologists.

——————, *Great Ideas in Physics* (New York: McGraw-Hill, 1992). Lightman has chosen his favorite "big ideas" in physics. The sections on relativity provide good supplements to this course.

Lindley, David, *Where Does the Weirdness Go? Why Quantum Mechanics Is Strange but not as Strange as You Think* (Basic Books/HarperCollins, 1996). Lindley, formerly an astrophysicist at Cambridge and the Fermi National Accelerator Laboratory, later became an editor at the British science journal *Nature* and at *Science News*. *Where Does the Weirdness Go?* is an excellent introduction to quantum physics, with emphasis on such quandaries as Schrödinger's cat, EPR experiments, and the general philosophical interpretation of quantum physics.

Mather, John C., and Boslough, John, *The Very First Light: The True Inside Story of the Scientific Journey Back to the Dawn of the Universe* (Basic Books, 1996). Mather, Project Scientist at NASA for the Cosmic Background Explorer (COBE) satellite, teams with science writer Boslough to explore the science, politics, and personalities behind our most detailed knowledge of the cosmic microwave background radiation.

Mermin, N. David, *It's About Time: Understanding Einstein's Relativity* (Princeton University Press, 2005). A well-written book suitable for those who want a more mathematical approach that doesn't go beyond high-school algebra. Simple geometric diagrams reinforce the concepts and the math.

Mook, Delo, and Vargish, Thomas, *Inside Relativity* (Princeton University Press, 1987). A physicist and artist teamed up to ensure that this introduction to relativity would be truly comprehensible to nonscientists. And it is. It's not flashy, but it provides a good, solid introduction to the subject.

Moore, Thomas, *A Traveler's Guide to Spacetime* (New York: McGraw-Hill, 1995). Written as a textbook for a month-long relativity segment in sophomore-level college physics courses, this book is much more mathematical than others in this bibliography, but it provides a lively, clear, and very contemporary look at its subject. Readers wanting more math would do well to read Moore.

Padmanabhan, T., *After the First Three Minutes: The Story of Our Universe* (Cambridge University Press, 1998). An astrophysicist outlines the history of the universe. Includes good background material on star formation and other relevant astrophysical processes. The title is a takeoff on Weinberg's *The First Three Minutes*, listed below.

Pagels, Heinz, *The Cosmic Code: Quantum Physics as the Language of Nature* (New York: Simon and Schuster, 1982). This solid introduction to quantum physics for the nonscientist is, unfortunately, out of print.

Riordan, Michael, *The Hunting of the Quark* (Simon & Schuster, 1987). This book gives a history of particle physics, especially the quark concept, that emphasizes the personalities and interactions of the scientists and scientific teams involved in particle theories and experiments.

Sime, Ruth, *Lise Meitner: A Life in Physics* (University of California Press, 1996). A scholarly but eminently readable account of the life of Lise Meitner, the nuclear physicist who identified the process of nuclear fission.

Spielberg, Nathan, and Anderson, Byron D., *Seven Ideas that Shook the Universe* (Wiley, 1985). Among the "universe shaking" ideas covered in this book are some from quantum physics; see Chapters 7 and 8.

Taylor, Edwin, and Wheeler, John Archibald, *Spacetime Physics*, 2nd edition (New York: W. H. Freeman, 1992). John Archibald Wheeler of Princeton is one of the leading American physicists of the century. Wheeler and MIT physicist Edwin Taylor have produced a book aimed at college physics students but emphasizing the conceptual essence of relativity. Written in a lively but slightly quirky style, this book presents special relativity in a way that prepares the reader for the curved spacetime of the general theory. The authors do not avoid math, but the conceptual aspects of relativity are always in the forefront.

Thorne, Kip, *Black Holes and Time Warps: Einstein's Outrageous Legacy* (New York: W. W. Norton, 1994). Thorne, a leading researcher on black holes and general relativity, has written a lively and up-to-date book for nonscientists. A good introduction to relativity is followed by convincing arguments for why black holes must exist.

Weinberg, Steven, *The First Three Minutes* (Basic Books, 1988). A noted theorist and winner of the 1979 Nobel Prize for physics details the first few minutes of creation, when a lot of important events happened that set the stage for all that followed.

Will, Clifford, *Was Einstein Right? Putting General Relativity to the Test* (New York: Basic Books, 1986). This book explores experimental tests of general relativity, both the "classic" tests, such as the bending of starlight and early gravitational time dilation experiments, to the even more convincing results from modern astrophysics. Its 1986 publication date makes it a bit dated in this rapidly advancing field, but the essential evidence for general relativity is all there.

Wheeler, John Archibald, *A Journey into Gravity and Spacetime* (New York: Scientific American Library, 1990). Wheeler, described under Taylor and Wheeler, above, has written this "coffee table" book with plenty of color pictures and few equations. As in *Spacetime Physics*, the emphasis is on the essential conceptual basis of general relativity.

Wolf, Fred Alan, *Taking the Quantum Leap* (New York: Harper & Row, 1989). Wolf, a former physics professor from San Diego State University, has written yet another good introduction to quantum physics for the nonscientist. Wolf's book is particularly strong on the role of the observer

Bibliography

in quantum measurements. Simple illustrations and catchy section titles help make this book an inviting introduction to the subject.

Wolfson, Richard. *Simply Einstein: Relativity Demystified* (W.W. Norton, 2003). Your Teaching Company professor elaborates on relativity, with explanations and diagrams that augment the presentation in this course.

Notes